Disasters and Public Health

Planning and Response

Bruce W. Clements

AMSTERDAM • BOSTON • HEIDELBERG • LONDON
NEW YORK • OXFORD • PARIS • SAN DIEGO
SAN FRANCISCO • SINGAPORE • SYDNEY • TOKYO

Butterworth-Heinemann is an Imprint of Elsevier

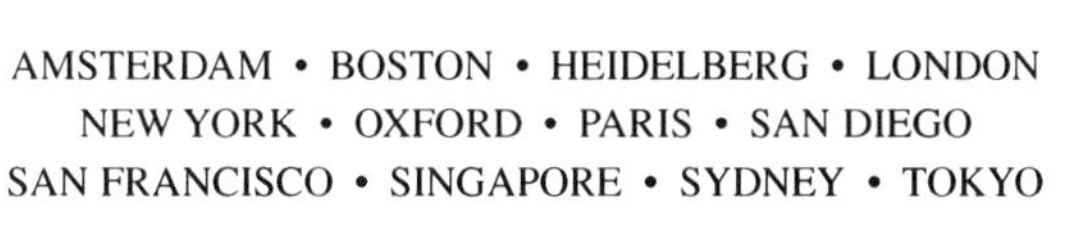

Butterworth-Heinemann is an imprint of Elsevier
30 Corporate Drive, Suite 400, Burlington, MA 01803, USA
Linacre House, Jordan Hill, Oxford OX2 8DP, UK

Library of Congress Cataloging-in-Publication Data
Clements, Bruce, MPH.
Disasters and public health: planning and response/Bruce Clements.
p. cm.
ISBN 978-1-85617-612-5
1. Disaster medicine. 2. Emergency management. I. Title.
[DNLM: 1. Disaster Planning—methods. 2. Public Health Administration—methods.
3. Civil Defense—methods. 4. Disasters—prevention & control. 5. Emergencies. WA 295 C626d 2009]
RA645.5.C527 2009
363.34'8068–dc22

2009001355

British Library Cataloguing-in-Publication Data
A catalogue record for this book is available from the British Library.

ISBN: 978-1-85617-612-5

For information on rights, translations, and bulk sales, contact Matt Pedersen, Commercial Sales Director and Rights; email m.pedersen@elsevier.com

For information on all Butterworth–Heinemann publications visit our Web site at www.elsevierdirect.com

Publisher: Amy Pedersen
Acquisitions Editor: Pam Chester
Development Editor: David Bevans
Project Manager: Andre Cuello
Designer: Eric Decicco

Typeset by diacriTech, Chennai, India

Transferred to Digital Printing in 2012

Dedication

To my wife Stacey and my sons Brandon, Brett, Brock, and Bryce

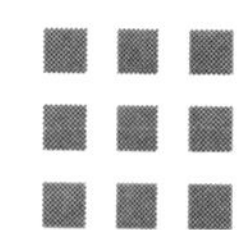

Table of Contents

Acknowledgments

Thanks to anyone who picks up this book and pursues more information on preparedness. I especially wish to thank those in health care, public health, and the first responder community, who are working diligently toward local, state, and federal preparedness. It often seems like a thankless task. Even in the face of a disaster, it is difficult to see your success in the aftermath when the focus is so often on the lessons learned from things that went wrong. It is particularly difficult to enumerate what was prevented through your efforts. It is difficult to count injuries and illnesses avoided. Just know that when you work on preparedness at any level, you are mitigating the impact of catastrophes and saving lives. You may never carry a child from a burning building but may prevent the fire. You may never rescue a family from a rooftop but may enable them to survive long enough to be rescued. You may never capture terrorists but will decrease the impact of their attacks. In that regard, those of you working toward preparedness at every level are anonymous heroes.

This book is the culmination of a variety of experiences over many years. It is also the product of encouragement from many friends and colleagues. Thanks to Alice Weis for inspiring my interest in preparedness many years ago through your storytelling skills. Thanks to Senior Master Sergeant Malcolm Jones, USAF, (Retired) for your friendship and encouragement through the years. Also thanks to Brigadier General Edith Mitchell, USAF, (Retired), Colonel Mike Hayek (deceased), and other Air Force and Air National Guard colleagues who promoted my preparedness interests and aspirations. Thanks to my colleagues at Saint Louis University, Institute for Biosecurity, especially my friend and mentor Greg Evans, I appreciate your insight, guidance, and friendship. To the members of Missouri Task Force 1, Urban Search and Rescue Team and MO-1, Missouri Disaster Medical Assistance Team, what a privilege to serve with you! Thanks to the local public health planners, epidemiologists, and others working so hard on preparedness across Missouri and to the staff of the Missouri Department of Health and Senior Services and the "DHSS Alumni," you have taught me so much. I also want to thank my friends and colleagues at Clean Earth Technologies in Earth City, MO. Especially to the Cofounders/Managing Members, Bob Morgan and Jeffrey Golden, I appreciate the opportunities you have given me to work on preparedness through private sector initiatives.

There has been an array of leaders in preparedness and related fields who have awakened me to the growing range of threats we are facing. They have been instrumental in forming my perspective on these issues. The list is too long to include everyone but the short list includes Michael Osterholm, Greg Evans, D.A. Henderson, Tara O'Toole, Linda Landesman, Eric Noji, Scott Lillibridge, Rachel Schwartz, Gil Copley, Hope Woodson, Mike Williams, Karen Webb, Victoria Fraser, Denise Murphy, Debbie Mays, Steven Lawrence, Brooke Shadel, Jeff Lowell, Ruth Carrico, Judith English, Terri Rebmann, Julie Eckstein, Nancie McAnaugh, William Jermyn, Patrick Gaffney, Rex Archer, Gary Christmann, Phil Currance, Mark Thorpe, Jim Imholte, William Patrick, Tom Hartmann, William Stanhope, and Ron Parker.

It takes a team to put a book together and I have been so fortunate to have the outstanding support of Pam Chester Dave Bevans, and Andre Cuello at Butterworth-Heinemann. I appreciate your guidance and support. It also takes friends to give you an occasional push. To Ann Rubin, KSDK News Channel 5, St. Louis, thanks for being a sounding board and helping me focus my efforts to get this book started. To Vered Kater, Jerusalem, Israel, thanks for always challenging and pushing me forward. To actress, author, and speaker Nancy Stafford, thanks for inspiring me and encouraging me to write.

Most of all, I need to thank my family for enabling me to serve in volunteer activities over the years and giving me the support I need to write. My wife, Stacey, through your dedication and support, this book was made possible. And to my boys, Brandon, Brett, Brock, and Bryce, thanks for letting me miss a few outings and ball games to work on this and other preparedness projects. Your support and love allows me to do what I do and this book is for you and all about keeping families safe.

—Bruce W. Clements St. Louis, MO

Preface

The past several decades have included the biggest industrial disasters in history with the 1984 Union Carbide accidental release of methyl isocyanate and the 1986 nuclear facility explosion at Chernobyl. The chemical, biological, radiological, and nuclear (CBRN) threats from rogue nations and terrorist organizations continue to grow. The escalation of terrorist attacks from the desire for publicity to the determination to inflict mass casualties has changed preparedness priorities and introduced many new challenges. At the same time, natural disasters are impacting more people as populations grow across high-risk areas for wildfires, earthquakes, hurricanes, and other threats. The global population is also at risk for pandemics that modern health care and public health do not have adequate tools in place to mitigate. These diverse threats are continuing to push those engaged in preparedness to do more with less. As these threats continue to grow in complexity, it is often difficult to cut through the volumes of available information and determine what is truly important. This book is a step in that direction.

Each chapter focuses on a unique disaster health threat and begins with a case study or personal account to provide some perspective on the nature of the threat. A glossary of relevant terms is included for each topic, as well as a description of how each threat emerges. The specific health threats posed by each disaster are described. Individual and community actions to reduce morbidity and mortality are then broken down into three sections, including what to do before, during, and after each disaster. Efforts were made to make each section as succinct as possible while still including the most essential information and references.

There are several extremely important issues to consider that are not addressed by this book. Most importantly are the mental health consequences of disasters. While not specifically addressed in each chapter, there are important mental health ramifications to every type of disaster described. In fact, the mental health aspect of some disasters may have a greater impact on long-term morbidity and mortality than the immediate effects on physical health. While it is an extremely important consideration, it is not the focus of this book. In addition, this book is focused on human and not animal health outcomes. There are important linkages between human and animal health that are not included in this book. In particular, the need to include care for pets in preparedness and response initiatives is not discussed. There are some vulnerable individuals who refuse to evacuate or take advantage of the services offered by responders unless their pets are included. This can place high-risk populations, such as the elderly, at unnecessary risk if local preparedness initiatives have not considered the need to accommodate pets. Finally, while special needs or vulnerable populations are discussed in many chapters, it is important to underscore the importance of this population. They are the most frequent victims of almost any disaster, and particular consideration is needed in every scenario to determine how to best identify which special needs or vulnerable populations are at risk and how to best accommodate them.

Throughout the years, I have had the privilege of being involved in a variety of disaster preparedness and response activities. As a young Air Force sergeant in the 1980s, I wrote disaster plans and provided training on chemical, biological, radiological, and nuclear (CBRN) defense through the final years of the Cold War. Although there was a resurgence of interest in CBRN preparedness during the first Gulf War in the early 1990s, people soon became apathetic again and not much attention was given to the growing threat. In 1995, we saw the Aum Shinrikyo attack the Tokyo subway system with sarin nerve agent and just weeks later, the Oklahoma City Bombing. I recall vividly the heartbreaking media coverage from the twisted remnants of the Alfred P. Murrah Federal Building. I held my newborn son and wept as the reports came in on the children lost in that tragedy. I also made the decision that day, on the evening of April 19, 1995, to contribute whatever I could to preparedness and response. I volunteered as a Hazardous Materials Technician with a new Urban Search and Rescue (USAR) Team that was forming in Missouri. That team eventually became the FEMA USAR Team, Missouri Task Force One. I went on to volunteer as a Safety Officer for a newly forming Disaster Medical Assistance Team for Missouri, MO-1 DMAT. I had the opportunity to organize a public health response team to Central America in response to Hurricane Mitch in 1998, the most deadly hurricane to hit the Western Hemisphere in the past two centuries. I also had the opportunity to work in academia as the Associate Director of the Saint Louis University, Institute for Biosecurity and as the Public Health Preparedness Director for Missouri during a record year of disaster declarations. Each experience taught me important lessons and introduced me to the hardest-working, most compassionate people I have ever met. It is my hope that the information in this book will enable that community of responders, as well as future responders, to be better prepared for the threats we will undoubtedly face in the future.

About the Author

Bruce W. Clements is a retired Air Force Public Health Officer and former Preparedness Director for the Missouri Department of Health and Senior Services. He holds undergraduate degrees in Disaster Preparedness, Bioenvironmental Engineering, and Business Administration, and a Master of Public Health Degree. His military experience includes over 23 years of service with assignments that include serving as a Nuclear, Biological, and Chemical Warfare Defense Instructor and as a Public Health Officer. He has also served as the Public Health Preparedness Director for the State of Missouri during a record year of disaster declarations (2006) and has served as an Infection Control Occupational Health Intervention Manager at BJC Healthcare. He has lectured extensively on public health preparedness topics, published peer-reviewed articles and books on preparedness, and served as a media resource for outlets such as CNN, FOX News, National Public Radio, and the Associated Press. He currently serves as a Senior Scientist and Director of Strategic Development for Clean Earth Technologies, LLC in Earth City, MO, where he is responsible for developing and managing bioterrorism and emerging infectious disease research and intervention projects. He also serves as an Adjunct Instructor for the Saint Louis University, Institute for Biosecurity, where he teaches on the public health response to terrorism and infectious disease threats.

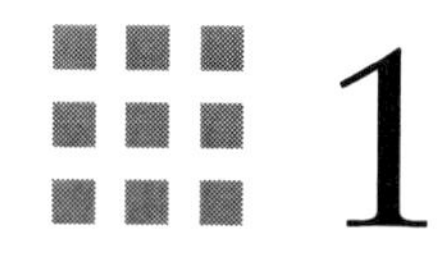

General Public Health Preparedness

Objectives of This Chapter

- Define public health emergencies.
- Explain the factors that contribute to community resilience.
- Describe the relationship between public health and emergency management in preparedness activities.
- Identify how the U.S. preparedness framework uses worst case scenarios for planning.
- List several target capabilities of U.S. preparedness.
- Explain why social capital is important.
- List ways to build social capital.
- Describe individual preparedness.
- Explain how individual preparedness is motivated.
- List the three fundamental actions that constitute individual or home preparedness.

Introduction ("Be Prepared")

"Be prepared" is the motto of the Boy Scouts of America. The organization's founder, Robert Baden Powell, selected the motto and emphasized through the principles and practices of the organization that preparing yourself requires you to think through situations in advance. The well-being of yourself and others depends on how you think about, prepare for, and react to a variety of situations (Boy Scouts of America, 1998). This creed was a central theme of Pahuk Pride 2008, a national youth leadership training initiative from the Mid-America Council of the Boy Scouts of America. This June 2008 gathering at the Little Sioux Scout Ranch in Iowa was designed to cultivate young leaders of troop programs. It required a recommendation from an applicant's scoutmaster and, by applying, a scout is acknowledging the desire and willingness to lead others.

Nearly 100 of these young men were spread out across the camping area when an alert came over a weather radio warning that dangerous weather was threatening the area. The Boy Scouts were quickly rushed into shelters and took refuge under tables and furniture in the two facilities. One of the facilities took the brunt of the storm and 4 young men died and over 40 were injured (Keen and Stone, 2008). Several of the surviving scouts were recognized for their bravery. In fact, the actions of all were laudable. Their faithfulness to their motto certainly saved the lives of many young men that day.

Disasters and Public Health: Planning and Response

They had an emergency plan and exercised it. They monitored a weather alert radio, and, when alerted, they stayed calm and took the right measures in seeking shelter. When the storm passed, they were prepared to triage and care for the injured. The actions taken by the Boy Scouts that day in Iowa stand as an excellent example of preparedness and provide insights into the concept of resiliency.

The word "resilience" is taken from the Latin word "resilire" and literally means to bounce back or rebound. It has historically been used as a social psychology term to describe those who bounce back from adversity through elasticity of spirit (Claudel, 1965). The term has migrated to other disciplines such as ecology, where it refers to the movement of an organism from one stable domain to another through the influence of turbulence (Holling, 1973). It is been used in the field of industrial safety to describe the ability of a worker to anticipate risk and make changes before harm occurs (Hollnagel et al., 2006). As it relates to preparedness, resiliency is the ability to anticipate, prepare for, and recover from the physical and psychological challenges of-man-made and natural disasters. Preparedness is the process, and resiliency is the result. How well we bounce back from the adversity of a disaster is largely dependent on how well we have prepared ourselves to face it. A resilient individual will have a more resilient home. A resilient community is comprised of resilient homes. A nation becomes resilient one person, one family, one home, and one community at a time.

Public health emergencies are multidimensional, dynamic situations that overwhelm existing healthcare and public health infrastructure resulting in adverse community health effects. Large-scale, unanticipated events can pose extraordinary challenges. This includes natural or man-made, accidental or intentional events, with chronic or acute health effects. Many public health emergencies generate a variety of these facets. The September 11, 2001, terrorist attacks on America killed thousands on the day of the attacks, but the impact on the health of those present will be felt for years to come. Chronic respiratory disease among those excessively exposed to dust and debris during clean-up and mental health issues for those exposed to the trauma of the attacks are just two examples of the lasting public health impact that 9/11 will have on New York. In that regard, the public health impact was acute and intentional on the day the planes hit the World Trade Center but also chronic and accidental for many exposed to the dust during the recovery. Sustaining the public's health through the lifecycle of a disaster such as 9/11 requires broad preparedness and profound resiliency.

Public health resilience may be achieved across communities through applying fundamental principles and services of public health in the context of emergency management. A summary listing of these basic functions was adopted by the Public Health Functions Steering Committee in 1995. This committee includes representatives from all federal public health agencies and from the major professional public health organizations such as the American Public Health Association, National Association of County and City Health Officials, Association of State and Territorial Health Officials, and others. In 1994, they adopted a vision, mission, and essential services provided by public health agencies (U.S. Public Health Service, U.S. Department of Health and Human Services, The public health workforce: an agenda for the 21st century, Appendix B, n.d., www.health.gov/-phfunctions/pubhlth.pdf).

The services described are not exclusively accomplished by federal, state, and local public health agencies. The public health infrastructure spans across public and private agencies requiring the involvement of education, healthcare, social services, mental

Vision – Healthy People in Healthy Communities

Mission – Promote Physical and Mental Health and Prevent Disease, Injury, and Disability

Ten Essential Public Health Services

- Monitor health status to identify community health problems
- Diagnose and investigate health problems and health hazards in the community
- Inform, educate, and empower people about health issues
- Mobilize community partnerships to identify and solve health problems
- Develop policies and plans that support individual and community health efforts
- Enforce laws and regulations that protect health and ensure safety
- Link people to needed personal health services and assure the provision of healthcare when otherwise unavailable
- Assure a competent public health and personal healthcare workforce
- Evaluate effectiveness, accessibility, and quality of personal and population-based health services
- Research for new insights and innovative solutions to health problems

Source: U.S. Public Health Service, Department of Health and Human Services.

health, environmental, pharmaceutical, first responders, insurers, media, and many more. There is a continual public health process of assessment, policy development, and assurance associated with the 10 essential services. Assessment includes continual evaluation of community health needs, investigation of risks, and analysis of health needs. Policy development is instituted including advocating for resources, setting priorities, and developing plans and policies to address health needs. Assurance is the management of resources, implementation and evaluation of programs, and educating the public (U.S. Institute of Medicine, 1988). Although these activities are pursued daily to maintain public health programs, in public health preparedness they occur within the context of emergency management processes. Depending on the phase of a potential or active emergency, this can introduce tremendous challenges.

In the 1970s, a U.S. Comprehensive Emergency Management Model was first established. It comprises four phases, including mitigation, preparedness, response, and recovery. Mitigation includes steps taken to anticipate and reduce the impact of a disaster. Building codes have been established in many regions likely to experience hurricanes and earthquakes. This mitigates the impact of a disaster by making structures less likely to suffer severe damage. One analogous public health mitigation measure is a vaccination program such as annual flu shots. Preparedness is advanced planning and coordination to ensure more effective response and recovery. A good example is the flurry of state and local meetings prompted by concerns over the emerging threat of pandemic influenza. This has spurred planning efforts that have enhanced overall public health preparedness. When coordinated properly, the planning accomplished for one type of emergency will advance general, all-hazards preparedness. Response includes immediate actions taken with a current or imminent disaster. The health function of the response to a major disaster may involve mass triage, mass care, mass prophylaxis, or mass fatality management. Recovery includes activities that are sustained after the initial response to begin restoring the area back to normal (Waugh, 2000). This goes far beyond clean-up and infrastructure restoration. It includes the management of the far less visible chronic health effects and mental health impact of each disaster.

Phases of Emergency Management

- Mitigation: Measures to reduce the impact of hazards
- Preparedness: Efforts to prepare for a potential hazard
- Response: Actions taken to respond to an emergency or disaster
- Recovery: Measures taken to return an area to normal following a disaster

There is a complex interchange between the functions of emergency management and public health during each phase of the emergency management process. Each essential function of public health changes according to the phase of an emergency (See Figure 1–1). For example, informing and educating the community about public health issues during mitigation and preparedness are very different messages than what is shared during a response or recovery. Developing plans that support community health efforts may become a dynamic process of instituting an emergency plan and continually evaluating and adjusting it to best meet the needs of the affected population. There is not a straightforward recipe for how these processes are adjusted and mature through the lifecycle of a disaster. Most local public health agencies also do not have the resources to sustain normal operations at the same time they are responding to the health needs of a community during an emergency. However, it is important that those serving in key emergency management and public health positions understand the dynamic nature of the public health aspects of an emergency, the processes they must work through to acquire and deploy resources, and the balancing of the rise and fall of priorities.

Preparedness Definitions

Disaster: Natural or man-made events that disrupt normal community function due to losses that exceed the ability of the affected community to manage.

Individual Preparedness: Actions taken by an individual or a family to prevent, protect against, and minimize physical and emotional damage that results from a disaster.

National Planning Scenarios: Fifteen threats that provide a framework to prioritize and develop response capabilities in the United States.

Public Health Emergency Preparedness: Activities taken by healthcare and public health organizations to ensure effective response to emergencies that impact health, especially events that have timing or scale that overwhelms normal capacity.

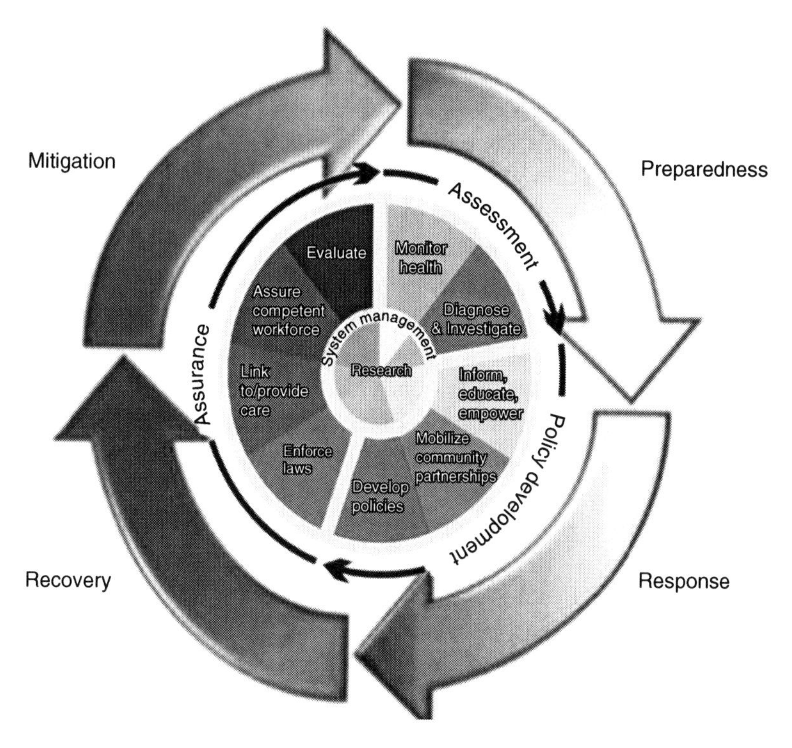

FIGURE 1–1 Essential Public Health Services and the Phases of Emergency Management.

Resiliency: The ability to anticipate, prepare for, and recover from the physical and psychological challenges of-man-made and natural disasters.

Social Capital: The resources derived from an intangible community network of relationships.

Target Capabilities: The core abilities needed to prevent, protect, respond, and recover from the 15 National Planning Scenarios.

Universal Task List (UTL): A listing of hundreds of tasks that define what is needed to prevent, protect, respond, and recover from the 15 National Planning Scenarios. It provides the basis for training and exercises in the United States.

A National Approach: U.S. Preparedness Infrastructure

To advance U.S. preparedness and resiliency, President Bush began issuing Homeland Security Presidential Directives following the terrorist attacks of 2001. These directives are intended to enhance preparedness and reduce risk by defining the federal infrastructure, including the establishment of threat conditions, development of a National Incident Management System, instituting a national domestic all-hazards preparedness goal, and a variety of related measures. At the heart of the overall U.S. preparedness initiative are four components, including the National Preparedness Vision, 15 National Planning Scenarios, UTL and the Target Capabilities List.

The vision for the National Preparedness Guidelines is **A NATION PREPARED with coordinated capabilities to prevent, protect against, respond to, and recover from all hazards in a way that balances risk with resources** (U.S. Department of Homeland Security, 2007a). As we break down this brief sentence, it quickly becomes apparent that it poses a massive set of challenges. The first major challenge is coordination. This includes coordination between the public and private sectors, horizontal coordination at each level of government, vertical coordination between federal, state, and local agencies, and even international coordination of resources during large-scale disasters. The geographic size and political fragmentation of the United States make this exceedingly difficult. There are dozens of federal agencies that must attempt to orchestrate planning with all 50 states, U.S. territories, hundreds of Native American reservations, and thousands of local governments. The states facilitate coordination with Native American and local governments adding another essential layer to the process, but between most of these agencies and layers of government, there is an innate tension. Preparedness funding formulas fuel controversy. Federal agencies are pushed by Congress to require added preparedness deliverables from state and local agencies while often providing less funding. The state and local governments, in turn, lobby Congress for changes in the funding structure and requirements. Tensions persist between metropolitan areas who bear the burden of protecting the majority of the population and their rural counterparts who struggle to acquire adequate levels of funding to protect their residents as well. Since the approval of preparedness funding rests with Congress, there is also a tendency to focus on issues rather than infrastructure. As those serving in Congress hear concerns from their constituency on specific issues, there is an inclination to address citizen concerns by redirecting existing funds to focus on "hot issues," such as bioterrorism or pandemic influenza. This often results in reductions of core preparedness funds that support the very infrastructure needed to respond to "all-hazards" while adding separate funding and

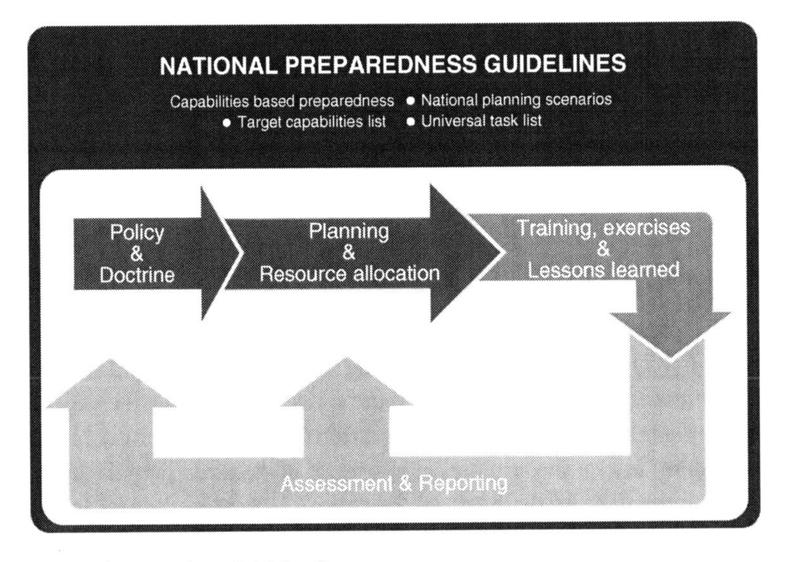

FIGURE 1–2 The National Preparedness Guideline Process.

performance requirements focused on the issue of the day. Although there are specific scenarios that require unique approaches, the majority of possible emergency situations are best met with a fundamental, consistent infrastructure. This is especially true of public health preparedness but applies across the board to all emergency response agencies.

Due to the unprecedented U.S. disasters of the first years of the twenty-first century, such as the 2001 terrorist attacks and Hurricane Katrina, the National Response Plan was modified in 2008. It has now evolved into the National Response Framework. Although the National Response Plan has always been a framework, the name had not accurately reflected it. The introduction of the National Response Framework is a natural progression for our response infrastructure. More information on the new framework can be found at http://www.fema.gov/emergency/nrf/.

To capture the depth and breadth of possible disasters, 15 National Planning Scenarios have been developed to provide a framework that encompasses the range of worst case U.S. catastrophes. The idea is that the combined challenges posed by these scenarios include sufficient diversity to achieve "all-hazards" preparedness. They include a nuclear detonation, a radiological dirty bomb attack, five biological events, four chemical events, two natural disasters, a conventional explosion, and a cyber attack. The biological scenarios include diseases that are person-to-person transmissible and nontransmissible, animal and human, food borne and environmental, intentionally released and naturally occurring and require management of mass prophylaxis, mass care, and mass fatalities. The chemical agents include both chemical warfare agents and toxic industrial chemicals with immediate and long-term health effects. As a group, these scenarios stress nearly all possible components of the U.S. response infrastructure and address the unique coordination challenges required for an effective federal, state, and local response.

The use of these dire scenarios to frame our preparedness efforts poses a variety of challenges. When realistic casualty numbers associated with these scenarios are incorporated into training and field exercises, participants may develop a pessimistic attitude both in terms of the likelihood of the event and the viability of an effective response. As a result, fundamental interest in preparedness measures may suffer as essential participants and partners conclude that it is simply too difficult to achieve preparedness in the context of these scenarios. In addition, if efforts are expended toward the worst of these large-scale events, there is the potential to overlook smaller, more likely threats. However, it is true that all the scenarios are plausible and with that realization there is a moral imperative to focus resources on each of them. Furthermore, it was observed through the subsequent development of the UTL that many of the tasks required were consistent across many or all scenarios.

U.S. Department of Homeland Security (DHS), 15 National Planning Scenarios

1. Nuclear Detonation: 10-Kiloton Improvised Nuclear Device
2. Biological Attack: Aerosol Anthrax
3. Biological Disease Outbreak: Pandemic Influenza
4. Biological Attack: Plague
5. Chemical Attack: Blister Agent
6. Chemical Attack: Toxic Industrial Chemicals
7. Chemical Attack: Nerve Agent
8. Chemical Attack: Chlorine Tank Explosion
9. Natural Disaster: Major Earthquake
10. Natural Disaster: Major Hurricane
11. Radiological Attack: Radiological Dispersal Devices
12. Explosives Attack: Bombing Using Improvised Explosive Devices
13. Biological Attack: Food Contamination
14. Biological Attack: Foreign Animal Disease (Foot and Mouth Disease)
15. Cyber Attack

The UTL is a "living" document developed from the planning scenarios. As training and exercises identify gaps, new tasks are added. The list provides consistent language and references so all public and private agencies at every level have a common understanding and definition of the tasks required to achieve preparedness outcomes. The UTLs comprised hundreds of tasks and divided into four "Mission Areas" or taxonomies: Prevent, Protect, Respond, and Recover (U.S. Department of Homeland Security, 2005). Once the detailed UTL was defined; there was a need to refine them into specific, measurable capabilities. The result was 37 Target Capabilities.

Target Capabilities

Common Capabilities

- Planning
- Communications
- Community Preparedness and Participation
- Risk Management
- Intelligence and Information Sharing and Dissemination

Prevent Mission Capabilities

- Information Gathering, Recognition of Indicators, and Warning
- Intelligence Analysis and Production
- Counter-Terror Investigation and Law Enforcement
- Chemical, Biological, Radiological, Nuclear, and Explosives (CBRNE) Detection

Protect Mission Capabilities

- Critical Infrastructure Protection
- Food and Agriculture Safety and Defense
- Epidemiological Surveillance and Investigation
- Laboratory Testing

Respond Mission Capabilities

- On-Site Incident Management
- Emergency Operations Center Management
- Critical Resource Logistics and Distribution
- Volunteer Management and Donations
- Responder Safety and Health
- Emergency Public Safety and Security
- Animal Disease Emergency Support
- Environmental Health
- Explosive Device Response Operations
- Fire Incident Response Support
- WMD and Hazardous Materials Response and Decontamination
- Citizen Evacuation and Shelter in Place
- Isolation and Quarantine
- Search and Rescue (Land-Based)
- Emergency Public Information and Warning
- Emergency Triage and Pre-Hospital Treatment
- Medical Surge
- Medical Supplies Management and Distribution
- Mass Prophylaxis
- Mass Care (Sheltering, Feeding, and Related Services)
- Fatality Management

Recover Mission Capabilities

- Structural Damage Assessment
- Restoration of Lifelines
- Economic and Community Recover

These broad capabilities are assessed by breaking down specific critical tasks associated with each one and establishing performance measures and matrices. For example, to display the "Medical Surge" target capability, there is the need for a written plan. The metric to measure that specific critical task is simply a yes or no, either a written plan exists or it does not. Other measures have time requirements. For example, a State Medical Coordinating System must be active within 2 hours of event notification (U.S. Department of Homeland Security, 2007b). The complete assessment of any one of these target capabilities is a difficult and time consuming process but is an essential activity to validate progress made from the billions of dollars invested in U.S. preparedness in the first decade of the twenty-first century.

Every nation has a unique approach to their preparedness activities. For example, the United States instituted a Homeland Security Advisory System in 2002 that uses five threat conditions or levels: Low = Green, Guarded = Blue, Elevated = Yellow, High = Orange,

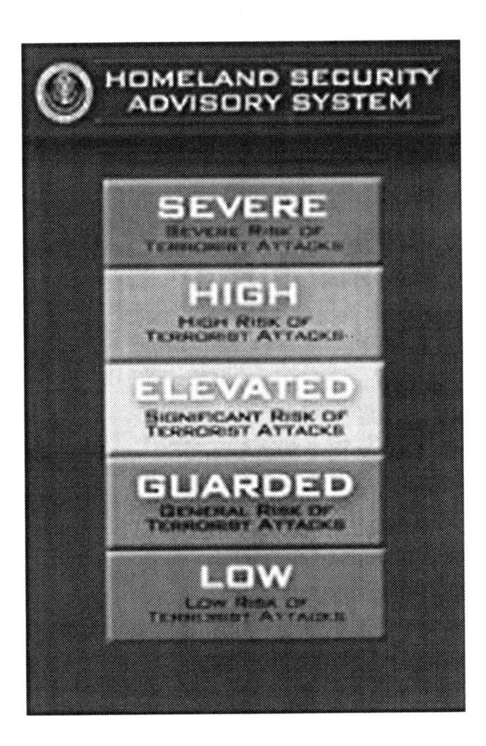

FIGURE 1–3 U.S. Department of Homeland Security Advisory System.

and Severe = Red. There are specific actions recommended for each. Based on intelligence reports, the status can be changed for the entire nation, specific regions, or particular sectors such as chemical facilities or airports (The White House, 2002).

The recommended actions for general citizens grow with each level. At the "Low" (Green) level, it is recommended that everyone develop and practice a family emergency plan, create a household "Emergency Supply Kit," know how to shelter-in-place and how to turn off utilities, and seek out additional preparedness training and volunteer activities. Each time the level is raised, citizens are encouraged to update their home emergency supplies and plans and become increasingly alert of what is going on around them (U.S. Department of Homeland Security, Citizen guidance on the homeland security advisory system, n.d., www.dhs.gov/xlibrary/assets/CitizenGuidanceHSAS2.pdf).

Other nations use similar alert approaches. Australia established a three-tier terrorism alert system in 1978 (Low, Medium, and High) and moved it to a four-level system in 2003. The four-level system added one additional level (Low, Medium, High, and Extreme). The "Extreme" level indicates an attack has occurred or is imminent (Australian Government Attorney-General's Department, 2008). Similar to the United States, the British now have a five-tier alert system (Low, Moderate, Substantial, Severe, and Critical). It was reduced from a seven-tier to a five-tier system in 2006 (BBC News, 2006).

Fostering Community Resilience

Since the terrorist attacks of 2001 and Hurricane Katrina in 2005, tremendous resources have been distributed among local governments across the United States to ensure consistent disaster planning, training, and exercises among the first response communities. Although substantial progress has been made, much more remains to be completed. As progress is made using the national framework described, there are several important elements still lacking in many communities. One of the most essential community preparedness assets is social capital. It is also one of our greatest gaps. It is an intangible bond that holds a community together but is tough to define and even more difficult to measure. Perhaps, the most appropriate definition as it relates to preparedness is "the sum of the actual and potential resources embedded within, available through, and derived from the network of relationships possessed by an individual or social unit. Social capital thus comprises both the network and the assets that may be mobilized through that network (Nahapiet and Ghoshal, 1998)." In the context of public health preparedness, this concept suggests that the network of individuals and their collective resources represent a more robust and resilient capacity than the sum of individuals alone. In other words, a community has much more resiliency if neighbors know and are willing to assist each other. There are several factors contributing to a recent decline in U.S. social capital. The mobility and suburbanization of our society means that people spend more time commuting to and from their jobs, taking time away from community involvement. Individuals are also more likely to relocate when they retire or change jobs. As a result, many citizens are not rooted in their community. There are also dramatic changes in how we spend our leisure time. Some who at one time may have joined a softball or bowling league, volunteered with an organization such as the Red Cross, or been active in community or faith-based activities, now choose to spend more time on the Internet or watching television. For many, even their sources for news are moving away from their communities. Readership of local newspapers continues to decline as people get their news from 24-hour cable news networks and from the Internet. These and other factors contribute to a decline in the number and quality of social connections people make in their communities and the resulting social capital. As social capital erodes, people are less likely to know and look out for their neighbors or trust their local government officials (Putnam, 1995). These are essential elements of a disaster resilient community. If people know their neighbors, they are more likely to assist

FIGURE 1–4 Findlay, Ohio, September 15, 2007—Students from the University of Findlay help to repair "Java Station" a local coffee house that was damaged by flooding in the downtown area of the town. Many volunteers and organizations have helped home owners and business owners recover from recent damaging floods in the community. Photo by John Ficara/FEMA.

Building Social Capital in Support of Preparedness

1. Engage social organizations such as the Lion's Clubs and Jaycees
2. Use the Internet to network with those who are homebound (by necessity or choice), including e-mail, instant messaging, chat rooms, and threaded discussions
3. Reach out to faith-based organizations and invite participation
4. Contact local little league and adult sports clubs and offer preparedness information
5. Send information to students at universities and colleges
6. Distribute information to children at elementary and secondary schools
7. Provide materials through adult day care and child care programs
8. Work with senior and disability outreach organizations
9. Invite participation of local businesses (both large and small)
10. Partner with local chapters of various professional organizations

each other during a crisis. They are also more likely to be aware of vulnerable neighbors in special need of aid following a disaster.

Volunteer activities are vital to community preparedness. There are nearly 1.2 million firefighters in the United States and over 820,000 of them are volunteers (Karter and Stein, 2007). With nearly 305 million people in the United States, that is one firefighter for every 254 people. There are about 841,000 trained and licensed Emergency Medical Service (EMS) personnel in the United States, and from this pool, over 60% of ambulance services rely on volunteer EMS professionals for a portion of their workforce. There is one EMS

FIGURE 1–5 New Orleans, LA, January 21, 2006—Volunteers from AmeriCorps carry a floor joist at a Habitat for Humanity home site in New Orleans. The organizations have teamed up to provide housing for Hurricane Katrina victims. Volunteer agencies are important FEMA partners who provide needed service to victims. Photo by Marvin Nauman/FEMA.

professional for every 363 people (National Association of Emergency Medical Technicians, EMS fast facts, www.naemt.org/aboutEMSAndCareers/ems_statistics.htm). Recent statistics also indicate that there are about 1.1 million sworn state and local law enforcement authorities (U.S. Department of Justice, 2008). That is one sworn officer for every 277 U.S. residents. There are about 738,000 physicians and 2.3 million nurses. That is one physician for every 414 people and one nurse for every 133 people (U.S. Census Bureau, Statistical abstract of the United States: 2004–2005, www.census.gov/Press-Release/www/releases/archives/health_table150.pdf). Finally, consider the number of staffed hospital beds in the United States in light of the National Planning Scenarios. There are about 947,000 staffed beds in the entire nation (American Hospital Association, 2007). That is one bed for every 322 people. Even if all public and private assets were available for every major disaster, which they are not, and even if these resources could be coordinated and managed seamlessly, which they cannot, it is clear that all aspects of preparedness and response will fall short without volunteer support.

U.S. Citizen to Critical Resource Ratios

1. One nurse for every 133 people
2. One firefighter for every 254 people
3. One law enforcement official for every 277 people
4. One staffed hospital bed for every 322 people
5. One EMS professional for every 363 people
6. One physician for every 414 people

A community wide disaster will affect all residents of an area, including first responders and their families. Some first responders will not be available. Although an overwhelmed community can count on assistance eventually coming from state and federal agencies, as well as nongovernmental organizations (NGOs), there is a small window of opportunity where life saving interventions are most effective in the aftermath of most disasters. To uphold a community facing a disaster, there must be a core group of volunteers at the local level. There are a variety of avenues into local volunteer opportunities. Many faith-based organizations are among the first to respond and are often the last to leave following a disaster. There are also federal initiatives such as the Community Emergency Response Team Program. It provides fundamental training to citizens in how to help themselves and their community during a disaster. Members are trained in team organization and management, fire safety, search and rescue, and first aid. The goal is to provide some community level response capacity when professional first responders are not yet available. The Fire Corps is designed to provide nonemergency responder manpower to fire and EMS personnel during a disaster. Medical Reserve Corps volunteers include practicing and retired medical and public health professionals who can be deployed to support an area where the healthcare system has been overwhelmed. The Volunteers in Police Service train volunteers to assist local law enforcement agencies and the USAonWatch Program is the national organization of Neighborhood Watch Programs. These programs serve as the central focus of national volunteer initiatives but many other opportunities exist as well, including NGOs such as

FIGURE 1–6 Windsor, Colorado, May 26, 2008—The message written on the side of this home is a result of volunteers who came to Windsor to help residents recover from the recent tornado. Photo by Michael Rieger/FEMA.

the Red Cross, Salvation Army, United Way, and countless others. Effective management of a public health emergency is impossible without volunteer organizations such as these.

Motivating Individual Preparedness

The requisite steps for becoming a prepared person include the acknowledgement of risk and responsibility, staying informed, making a plan, and getting involved. The risks vary according to where you live, work, or travel so it is important to consider the full range of potential disasters that could affect you, your family, your friends, and neighbors. In the United States, most major disaster risks are well understood in each geographical region. The Atlantic and Gulf Coasts areas are susceptible to hurricanes. The West Coast and Midwest have major fault lines where earthquakes may occur. Extreme weather events affect the entire nation, but tornados are more likely in the Midwest, and blizzards are more likely in the North. The threats of emerging infectious diseases and terrorism have no predictable geographic distribution and are considered threats everywhere. In spite of the clear threats posed across the nation, most people maintain awareness of risk and yet fail to adequately prepare.

> *If we have not gotten our message across, then we ought to assume that the fault is not with our receivers.*
>
> Baruch Fischhoff, Department of Engineering and Public Policy, Carnegie-Mellon University, 1985

In the efforts to prepare the public to minimize the health impact of disasters, the first question is "What should I be prepared for?" The 15 U.S. planning scenarios are worst case situations, but several of them have never occurred in the United States. Many citizens may have a difficult time comprehending the threat of a nuclear detonation or an anthrax release. If preparedness is framed with these scenarios, some dismiss the threats and take no action toward preparedness. This attitude prevails even among many first responders. However, if we look over the past 100 years of U.S. disasters, there are several important lessons to be learned.

Although disasters consistently occur, there are periodic extreme disasters that impact the health and safety of hundreds or thousands of citizens. The problem is not convincing residents of the hazards that are common to their geographic region. The challenge is convincing them that they may be directly affected, that the disaster may be far from typical, and that they can do some relatively simple things to prepare. They also need to understand that there are unlikely but potential disasters for their region. This includes blizzards in the South and hurricanes in New England. The impact of these disasters can be greatly diminished by individual preparedness.

Major U.S. Disasters of the Past Century

- 1900 Galveston Hurricane: On September 8, 1900, one of the worst hurricanes in history slammed into Galveston, Texas with winds of about 120 miles per hour and a storm surge over 15 feet high that covered the entire island. It has been estimated that the number killed in Galveston was about 6000 and about that many more were lost across the inland areas as well (Ramos, 1998–1999).
- 1906 San Francisco Earthquake: On April 16, 1906, one of the most devastating earthquakes in history hit San Francisco. It initiated an enormous fire that burned across the city. Over 700 people died from the earthquake and fires, and 225,000 were left homeless out of a population of about 400,000 (U.S. Geological Survey, 2008).
- 1910 Idaho and Montana Wildfire: In late August 1910, a wildfire swept across northern Idaho and western Montana destroying over three million acres and taking 86 lives (Devlin, 2000).
- 1918 Influenza Pandemic: As World War I was coming to an end, an outbreak of a particularly virulent strain of influenza swept around the world and infected nearly a third of the global population. Hundreds of thousands died in the United States and about 50 million died globally (Taubenberger and Morens, 2006).
- 1925 Tri-State Tornado: On March 18, 1925, the most destructive tornado on record touched down in Missouri and stayed on the ground for over 200 miles (320 km) across Illinois and Indiana. Over 700 died and 3000 were injured. Over 7000 people were directly affected (Burgess, 2006).
- 1938 New England Hurricane: In September 1938, a rare New England Hurricane struck New York, Connecticut, and Rhode Island. More than 500 lives were lost and over 57,000 homes destroyed (boston.com, 2005).
- 1952 Polio Outbreak: Just before Jonas Salk's polio vaccine work was completed, the United States was hit with a severe polio outbreak. There were an estimated 57,000 cases with about 3000 fatalities and over 20,000 paralyzed (Fischman, 2005).

- 1964 Alaskan "Good Friday Earthquake" and Tsunami: On March 27, 1964, the second most powerful earthquake ever recorded, magnitude 9.2, hit Alaska. The majority of the 132 deaths associated with this disaster are attributed to the tsunami that was triggered by the quake that killed people as far south as California. The jolt from the earthquake was so powerful that it capsized boats in the Gulf Coast (U.S. Geological Survey, 2004).
- 1974 Tornado Outbreak: On April 3 and 4, 1974, the worst outbreak of tornados in U.S. history occurred. It included 148 tornados across 13 states in a 16-hour period. Over 300 people were killed and over 5,000 injured (National Oceanic and Atmospheric Administration [NOAA], 25th anniversary of the 1974 tornado outbreak, n.d., www.publicaffairs.noaa.gov/storms/).
- 1980 Eruption of Mt. St. Helens: May 18, 1980, Mount St. Helens volcano in the state of Washington violently erupted killing 57 and causing widespread environmental damage. It was the worst volcanic disaster in the history of the United States (Tilling et al., 1990).
- 1993 Storm of the Century: In March 1993, an enormous storm hit the Eastern United States. It has been called the White Hurricane, Blizzard of 1993, and simply, "The Big One." It was a massive low pressure weather system that reached from Cuba to Canada causing storm surges, tornados, and some of the southernmost U.S. blizzard conditions ever recorded. The death toll was 270 and 48 were missing at sea (Lott, 1993).
- 2005 Hurricane Katrina: On August 29, 2005, Hurricane Katrina made landfall on the Gulf Coast leaving a path of destruction beyond anything observed in recent U.S. history. It killed over 1300 and is considered the most destructive natural disaster in American history (The White House, 2006).

The barriers to individual preparedness are still not well understood but emerging research is offering some insight into why some take actions to prepare while others do not. The type of disaster plays an important role in the perception of risk and subsequent actions toward preparedness. Previous individual experience with disasters and gender plays significant roles as well (Ho et al., 2008). In addition to risk perception, social influences and access to resources influence the actions of individuals facing a potential disaster. Overcoming these barriers to stimulate personal preparedness and appropriate public response requires an adapted and comprehensive communication initiative (Riad et al., 1999). Effective public motivation requires comprehensive risk communication. There are several cardinal rules for effectiveness:

- Accept and involve the public as a legitimate partner.
- Plan carefully and evaluate your efforts.
- Listen to the public's specific concerns.
- Be honest, frank, and open.
- Coordinate and collaborate with other credible sources.
- Meet the needs of the media.
- Speak clearly and with compassion.

Source: Seven Cardinal Rules of Risk Communication. Pamphlet drafted by Vincent T. Covello and Frederick H. Allen., U.S. Environmental Protection Agency, Washington, DC, April 1988, OPA-87-020.

Although these rules were developed primarily for use during response and recovery, they are not only for emergencies. These same principles may be adapted to better communicate preparedness before emergencies occur. For example, the public is often not included in preparedness activities to the extent they can and should be. As federal dollars flow down to state and local agencies accompanied by mandates to carry out training and exercises, these actions often focus on several key public agencies and do not involve private sector partners to the level they will be involved during a real incident. Engaging the public is difficult but essential if the exercises are going to successfully identify gaps and shortfalls that must be addressed before an incident.

Rhetoric changes first, behavior changes next, and attitudes change last.

Peter Sandmann (Holing, 1996)

Engaging the media prior to disasters is vital. Those who are responsible for disaster planning and response must understand that disasters attract media attention and the actions of officials will be judged not only based on outcomes but on the perceptions conveyed by the media. The First Amendment of the Constitution guarantees freedom of the press and they are not going away. They are also familiar and trusted by the community, making them one of your most important partners before, during, and after disasters. The worst mistakes to make with the media are to ignore them or attempt to control them. They will get a story with or without your input and if you are in a position of responsibility and they get their story without your input, it is not likely to be in anyone's best interest. It is best to engage them prior to a disaster. Invite them to participate in exercises, even though they likely will not. Always be prepared to provide them with current information during a crisis, including a single-page summary of important facts. However, you should never tell them what to say or how to say it. Any perceived attempt to control them will not be received well. If they are doing their job well, they will usually want more access and more information than you are able to provide. Do not let that be discouraging. Simply give them a story in an honest, timely manner. Most media representatives want to support community preparedness and contribute whatever helpful information they can during a crisis. Do not judge good journalists according to experience with bad ones. Instead, judge bad journalists against your experience with good ones. They are one of your most important allies to deliver the message of individual preparedness, as well as specific preparedness messages for seasonal threats such as heat waves and hurricanes.

Defining Individual Preparedness

Personal and home preparedness is the bedrock of a resilient community. U.S. public interest in preparedness surged in the aftermath of the terrorist attacks of 2001 and following Hurricane Katrina. However, confusion persists over what constitutes preparedness. While some recommend a 72-hour supply of essential items, others insist on a week, a month, or even a year. In establishing a home emergency kit, more is better but 72 hours is the minimum. Although it may take more or less time than that for governmental and NGOs to deploy resources and begin offering assistance, it typically takes about 3 days. The size and severity of a situation will determine if it will take more or less than 3 days. Pulling together several days of supplies can make a situation manageable that may otherwise be devastating. The problem is that based on financial benefit, most major businesses have established a just-in-time delivery system. If major transportation routes are impacted by any type of emergency, the resources in an affected region can quickly be diminished. Those who have not made preparations will engage in a last-minute rush for supplies and many will not get what they need.

> *I believe that disasters turn into catastrophes when there is a failure in pre-planning and mitigation. Mitigation is best summed up by "An ounce of prevention is worth a pound of cure." While disaster response is sexy and gets the limelight, it is before the disaster in chambers like these with elected officials like you that in the end will make the real difference. Success in the next catastrophe that hits Seattle, or any other jurisdiction will be predicated on two things: What investment has been made in mitigation? How prepared is your citizenry to be on their own for a minimum of three days, and in reality for a major event, seven days?*
>
> Eric Holdeman, Director, King County Office of Emergency Management (Holdeman, 2007)

Individual and home preparedness begins with a written plan. The U.S. Department of Homeland Security has developed a basic template with associated pocket cards (Figure 1–7A and B). The plan needs to consider a variety of scenarios. Consider what you would need to do to sustain yourself at home for several days without utilities or communication. If your home is damaged, you need to have a neighborhood meeting location identified nearby where everyone in the family can meet. If you are at work, school, or someplace else, you need to identify a regional location where you will meet. You should also identify an out of town evacuation location and an out of town contact. During a disaster, it is often difficult to place a call within the impacted area but still possible to place an out of town call. A common out of town contact should be shared among family members with instructions to call them and let them know how and where you are. Other important information such as insurance, medical information, and social security numbers should be documented. Quick reference cards should be filled out and carried with each family member. These simple measures can have tremendous benefits in the aftermath of a disaster by facilitating coordination and communication. Not only will it give significant others peace of mind, it also can help identify who may be missing and in need of emergency assistance. The U.S. DHS has provided a simple template to assist with this planning.

In Missouri, the Department of Health and Senior Services has developed a program called "Ready in 3." This program encourages individuals to take three simple steps to become better prepared: create a plan, prepare a kit, and listen for information. This parallels the http://www.ready.gov structure. Recommendations for an emergency kit include support for every family member and pets for at least 3 days. This is an excellent baseline level of preparedness and is kept simple and inexpensive to allow those short on time or resources to make a step in the right direction.

The U.S. DHS makes more suggestions. These include duct tape and plastic sheeting to seal the home to shelter-in-place following an airborne release of a dangerous substance. They also suggest having household bleach available to chlorinate water for drinking. These supplies require a larger investment but address a broader range of disaster scenarios. Using some of these materials also requires more information and

Missouri DHSS Ready in 3 Recommendations

- Bottled water (at least one gallon per person, per day, for at least 3 days)
- Canned or dried food (at least a 3-day supply of nonperishable food items for each person)
- Manual can opener
- Battery-powered radio
- Flashlight
- Extra batteries for flashlight and radio
- First-aid kit
- Extra supply of prescription medicine
- Clean clothes and sturdy shoes
- Extra credit card and cash
- Sturdy trash bags
- Formula/baby food for infants in the home
- Food and water for pets in the home

Source: Missouri Department of Health and Senior Services, http://www.dhss.mo.gov/Ready_in_3/Kit.html

U.S. DHS: Additional Recommendations

Their kit recommendations include all of the items in Missouri's Ready in 3 Program plus:

- Whistle to signal for help
- Dust mask, plastic sheeting, and duct tape
- Moist towelettes
- Tools to shut off utilities
- Local maps
- Important family documents
- Emergency references such as a first-aid book
- Sleeping bags or warm blankets
- Household chlorine bleach
- Fire extinguisher
- Matches in a waterproof container
- Mess kits or disposable utensils
- Paper and pencil
- Books, games, and other activities for children

Source: U.S. Department of Homeland Security, http://www.ready.gov/america/_downloads/checklist.pdf

training. Although carrying out actions such as shutting off utilities sounds very simple, many residents have never done it before and will not know where to begin. This goes beyond the most basic level of preparedness suggested by the Missouri Ready in 3 program but is also a prudent set of recommendations for the general public.

Family Emergency Plan

Make sure your family has a plan in case of an emergency. Before an emergency happens, sit down together and decide how you will get in contact with each other, where you will go and what you will do in an emergency. Keep a copy of this plan in your emergency supply kit or another safe place where you can access it in the event of a disaster.

Out-of-Town Contact Name: Telephone Number:

Email:

Neighborhood Meeting Place: Telephone Number:

Regional Meeting Place: Telephone Number:

Evacuation Location: Telephone Number:

Fill out the following information for each family member and keep it up to date.

Name: Social Security Number:
Date of Birth: Important Medical Information:

Name: Social Security Number:
Date of Birth: Important Medical Information:

Name: Social Security Number:
Date of Birth: Important Medical Information:

Name: Social Security Number:
Date of Birth: Important Medical Information:

Name: Social Security Number:
Date of Birth: Important Medical Information:

Name: Social Security Number:
Date of Birth: Important Medical Information:

Write down where your family spends the most time: work, school and other places you frequent. Schools, daycare providers, workplaces and apartment buildings should all have site-specific emergency plans that you and your family need to know about.

Work Location One
Address:
Phone Number:
Evacuation Location:

School Location One
Address:
Phone Number:
Evacuation Location:

Work Location Two
Address:
Phone Number:
Evacuation Location:

School Location Two
Address:
Phone Number:
Evacuation Location:

Work Location Three
Address:
Phone Number:
Evacuation Location:

School Location Three
Address:
Phone Number:
Evacuation Location:

Other place you frequent
Address:
Phone Number:
Evacuation Location:

Other place you frequent
Address:
Phone Number:
Evacuation Location:

Important Information	Name	Telephone Number	Policy Number
Doctor(s):			
Other:			
Pharmacist:			
Medical Insurance:			
Homeowners/Rental Insurance:			
Veterinarian/Kennel (for pets):			

Dial 911 for Emergencies

FIGURE 1–7A U.S. Department of Homeland Security Family Emergency Plan template.

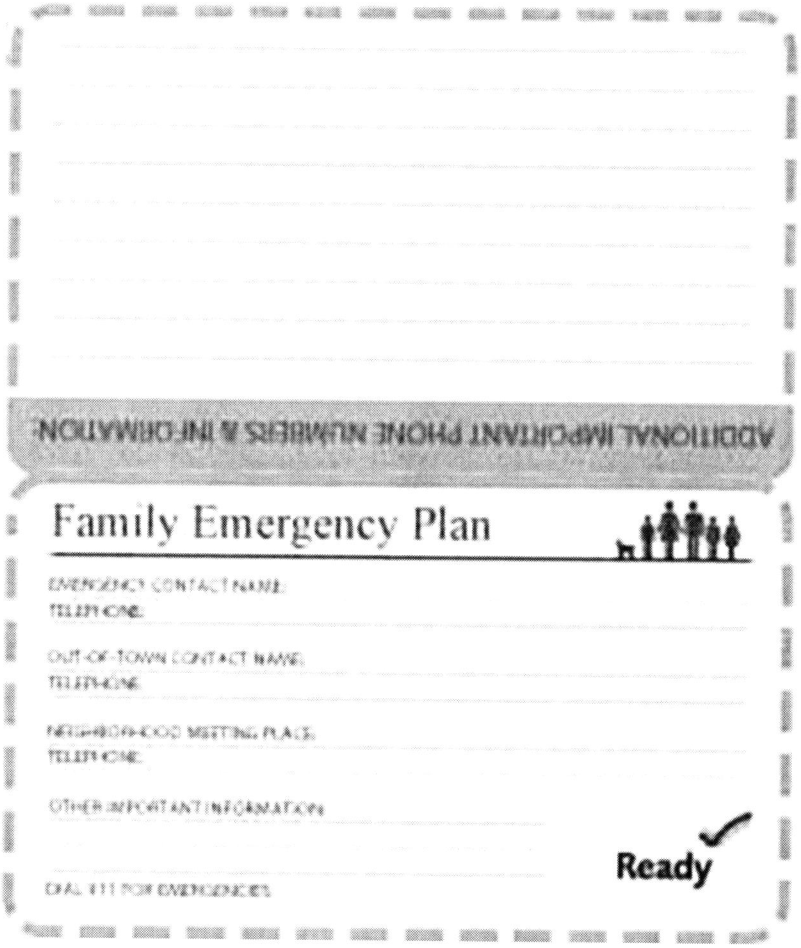

FIGURE 1–7B Pocket Card.
Source: U.S. Department of Homeland Security, http://www.ready.gov/america/_downloads/familyemergencyplan.pdf.

There are some who choose to exceed state and federal agency recommendations. They are motivated by a variety of factors. Some are traditional survivalists who adopt preparedness as a central part of their lifestyle. Some have truly experienced crisis or hunger, and as a result, they stockpile food and supplies for future eventualities. There are also those who prepare as a religious duty so they can sustain themselves and help others during a crisis. For example, this is a tenet of Mormonism where church leaders have established guidelines for storing a supply of food, clothing, and when possible, even fuel for one year to prepare for war, famine, and other catastrophes (Isackson, D., Are we obedient people? Our lives may depend on it! *Meridian Magazine*, n.d., www.meridianmagazine.com/lineuponline/050902people.html). As a result, they have developed several excellent resources, including online food calculators where anyone can go online and enter their family demographics and get the shopping list for a 1-year supply of food. Brace yourself for some substantial estimates if you calculate a 1-year stockpile with one of these references or Websites. For example, a family with two adults and two small children would need to stockpile about 1 ton of supplies. Although this is an excellent preparedness posture, it is out of reach for many who do not have the resources to invest in it or the means to store and maintain it.

One-Year Food Supply Estimates for a Family of Two Adults and Two Children

- 896 pounds of grains (wheat, flour, rice, etc.)
- 224 pounds of dry and evaporated milk
- 178 pounds of sugars (honey, sugar, molasses, etc.)
- 168 pounds of legumes (beans and lentils)
- 40 pounds of fats and oils (shortening, mayonnaise, peanut butter, etc.)
- 28 pounds of cooking essentials such as baking soda, salt, and so on
- 56 gallons of water
- 4 gallons of bleach

Source: Latter Day Saints, Food Storage Calculator. http://lds.about.com/library/bl/faq/blcalculator.htm

To reduce the health impact of many major disasters, a 72-hour supply of food, water, and other essential items is an essential first step. A week or more is better. These measures will provide "all-hazards" individual preparedness. Although nobody can be prepared for every scenario, a home preparedness kit can be what fills the essential gap between the impact of a disaster and the arrival of assistance regardless of whether it is a tornado or terrorist attack. When combined with a basic plan and following instructions from local officials, the health impact of many incidents will be greatly reduced if the general public can be motivated to take these steps.

Summary

This introductory chapter of out text introduces the relationship between public health and emergency management in the development of the U.S. preparedness infrastructure. The U.S. National Preparedness Guidelines are based on federal, state, and local preparedness capabilities that are generated through the use of 15 National Planning Scenarios. The scenarios are used for training and exercises to identify and assess Universal Tasks and Target Capabilities at each level of government. This is the structure by which preparedness is developed and evaluated in the United States. However, public officials do not have the capacity to meet the needs arising from major disasters without the assistance of volunteer organizations and the cooperation of the public. A key aspect of public and volunteer engagement is social capital. This interconnectedness of community networks has waned in recent years and to effectively manage a major incident, it needs to be reignited. In addition, citizens must be motivated to take simple measures such as completing a plan, preparing a kit, and following the advice of officials. These steps will mitigate the severity of major disasters and sustain the physical and mental health of those affected by catastrophe.

Websites

All-Hands.net, Emergency Management Community Portal: www.all-hands.net/.

American Red Cross: www.redcross.org/.

Association of Schools of Public Health, Centers for Public Health Preparedness: http://preparedness.asph.org/.

Association of State and Territorial Health Officials, Public Health Preparedness Program: www.astho.org/?template=preparedness.html.

Australia Natural Hazards and Emergency Management: www.ga.gov.au/hazards/management/-preparedness.jsp.

CDC/Red Cross: Preparedness Today: What You Need to Know: www.-redcross.org/preparedness/cdc_-english/CDC.asp.

Centers for Disease Control and Prevention (CDC), Emergency Preparedness and Response: www.bt.cdc.gov/.

Interaction, American Council for Voluntary International Action: www.interaction.org/.

International Association of Emergency Managers: www.iaem.com/.

International Committee of the Red Cross: www.icrc.org/.

National Association of County and City Health Officials, Public Health Preparedness Program: www.naccho.org/topics/emergency/.

National Association of Emergency Managers: www.nemaweb.org/.

National Environmental Health Association, Terrorism and All-Hazards Preparedness Program: www.neha.org/research/terrorism/index.html.

National Voluntary Organizations Active in Disaster: www.nvoad.org/.

ReliefWeb (United Nations Office for Coordination of Humanitarian Affairs): www.reliefweb.int/.

Salvation Army: www.salvationarmy.org/.

U.S. Citizen Corps Program: www.citizencorps.gov/.

U.S. Department of Homeland Security Business Preparedness Information: www.ready.gov/business/.

U.S. Department of Homeland Security Children's Preparedness Information: www.ready.gov/kids/index.html.

U.S. Department of Homeland Security Individual Preparedness Information: www.ready.gov/.

U.S. Federal Emergency Management Agency, Preparedness Guidance: www.fema.gov/areyouready/.

United Kingdom Individual Preparedness Information: www.preparing-foremergencies.gov.uk/.

United Kingdom News and Information for Emergency Practitioners: www.ukresilience.gov.uk/.

References

American Hospital Association. (2007). Fast facts on U.S. hospitals. Washington, DC. October 22, 2007. Retrieved August 11, 2008, www.aha.org/aha/content/2007/pdf/fastfacts 2007.pdf.

Australian Government Attorney-General's Department. (2008). Australia National Counter-Terrorism Alert Level. Retrieved July 2, 2008, www.ag.gov.au/agd/www/nationalsecurity.nsf/AllDocs/F2ED4B7E7B4C028ACA256FBF00816AE9?OpenDocument.

BBC News. (2006). New terror alert system launched. August 1, 2006. Retrieved July 3, 2008, http://news.bbc.co.uk/2/hi/uk_news/5233562.stm.

boston.com. (2005). The great hurricane of 1938. The Boston Globe. n.d. Retrieved April 27, 2008, www.boston.com/news/globe/magazine/galleries/2005/0724/hurricane1938/.

Boy Scouts of America. (1998). *Boy scout handbook*. 11th ed. Pineville, NC: Boy Scouts of America; p. 54.

Burgess, D. W. (2006). The tri-state tornado of 18 March 1925, Part I: Re-examination of the damage path. 23rd-Conference on Severe Local Storms. American Meteorological Society. Retrieved June 2, 2008, http://ams.confex.

Claudel, P. (1965). The American elasticity. In: *Works in prose*. La Pleiade. Paris, France: Gallimard; pp.1204–1208; in French.

Devlin, S. (2000). Taming "the dragon": the big burn of 1910. Missoulian. Retrieved July 2, 2008, www.missoulian.com/specials/1910/tame.html.

Fischman, J. (2005). Pushing back polio: 50 years ago this week, daring scientists beat a germ that was crippling the nation. *U.S. News and World Report*. April 10, 2005. Retrieved May 24, 2008, http://health.usnews.com/usnews/health/articles/050418/18polio.htm.

Holdeman, E. (2007). Testimony Seattle city council. Retrieved April 22, 2008, www.metrokc.gov/prepare/docs/Eric_Corner/07-07-03_Holdeman_SeattleCC_Testimony.pdf.

Holing, D. (1996). It's the outrage, stupid. *Tomorrow* 6(2).

Holling, C. S. (1973). Resilience and stability of ecological systems. *Annu Rev Ecol and Syst* 4:1–23.

Hollnagel, E., Woods, D., & Leveson, N., eds. (2006). *Resilience engineering: concepts and precepts*. Aldershot, UK: Ashgate.

Ho, M. C., Shaw, D., Lin, S., & Chiu, Y. C. (2008). How do disaster characteristics influence risk perception? *Risk Anal* 28(3):635–643.

Karter, M. J., & Stein, G. P. (2007). U.S. Fire Department profile through 2006. Quincy, MA: National Fire Protection Association.

Keen, J., & Stone, A. (2008). Boy Scouts recount survival stories of deadly Iowa tornado. *USA Today*. June 13, 2008. Retrieved August 7, 2008, www.usatoday.com/weather/storms/tornadoes/2008-06-12-boyscouts_N.htm.

Lott, N. (1993). The big one! a review of the March 12–14, 1993 "Storm of the Century," Technical Report 93–01. National Climatic Data Center, Research Customer Service Group.

Nahapiet, J., & Ghoshal, S. (1998). Social capital, intellectual capital, and the organizational advantage. *Acad Manage Rev* 23(2):242–266.

Putnam, R. D. (1995). Bowling alone: America's declining social capital. *J Democracy* 6(1):64–78.

Ramos, M. (1998–1999). *After the great storm: Galveston's response to the hurricane of Sept. 8, 1900*. Denton, TX: Texas State Historical Association. Texas Almanac. Retrieved July 1, 2008, www.texasalmanac.com/history/-highlights/storm/.

Riad, J. K., Norris, F. H., & Ruback, R. B. (1999). Predicting evacuation in two major disasters: risk perception, social influence, and access to resources. *J Appl Soc Psych* 29(5):918–934.

Taubenberger, J. K., & Morens, D. M. (2006). 1918 influenza: the mother of all pandemics. *Emerg Infect Dis* 12(1):15–22. Retrieved May 11, 2008, www.cdc.gov/ncidod/EID/vol12no01/05-0979.htm.

The White House. (2002). Remarks by Governor Ridge at Announcement of Homeland Security Advisory System. Washington, DC. March 12, 2002. Retrieved July 3, 2008, www.whitehouse.gov/news/releases/2002/03/20020312-11.html.

The White House. (2006). The federal response to Hurricane Katrina: lessons learned. Retrieved June 11, 2008, www.whitehouse.gov/reports/katrina-lessons-learned.pdf.

Tilling, R. I., Topinka, L., & Swanson, D. A. (1990). Eruptions of Mount St. Helens: past, present, and future. USGS. Retrieved May 2, 2008, http://vulcan.wr.usgs.gov/Volcanoes/MSH/Publications/MSHPPF/MSH_past_present_future.html.

U.S. Department of Homeland Security. (2007a). *National preparedness guidelines*. Washington, DC: U.S. Department of Homeland Security.

U.S. Department of Homeland Security. (2007b). *Target capabilities list: a companion to the national preparedness guidelines*. Washington, DC: U.S. Department of Homeland Security.

U.S. Department of Homeland Security, Office of State and Local Government Coordination and Preparedness. (2005). *Universal Task List: Version2.1*. Washington, DC: U.S. Department of Homeland Security.

U.S. Department of Justice, Office of Justice Programs. (2008). State and local law enforcement statistics. *Bureau of justice statistics*. September 17, 2008. Retrieved August 15, 2008, www.ojp.usdoj.gov/bjs/sandlle.htm.

U.S. Geological Survey. (2004). 40th anniversary of "Good Friday" earthquake offers new opportunities for public and building safety partnerships. USGS. March 26, 2004. Retrieved May 27, 2008, www.usgs.gov/newsroom/article.asp?ID=106.

U.S. Geological Survey. (2008). The great 1906 San Francisco earthquake. USGS. Retrieved July 2, 2008, www.earthquake.usgs.gov/-regional/nca/1906/18april/index.php.

U.S. Institute of Medicine, Committee for the Study of the Future of Public Health. (1988). *The future of public health*. Washington, DC: National Academy Press.

Waugh, W. L. (2000). *Living with hazards, dealing with disasters: an introduction to emergency management*. Armonk, NY: M.E. Sharpe.

Bioterrorism

Objectives of This Chapter

- Differentiate bioterrorism from emerging infectious disease outbreaks.
- Discuss some of the controversies surrounding bioterrorism preparedness.
- List the highest priority biological threat agents.
- Identify the epidemiological clues indicating a possible bioterrorism attack.
- Describe the common characteristics of Category A biological agents.
- Explain the criteria used to classify bioterrorism agent threats.
- Identify diagnosis and treatment recommendations for critical agents.
- Describe various types of bioterrorism surveillance approaches.
- Explain how environmental surveillance is used to identify the presence of suspicious organisms.
- Recognize the infection control precautions needed for responding to bioterrorism.

They were two men going about their daily lives. Both had reputations as pillars of the community in the Washington, DC, area. Joseph Curseen, Jr. (Figure 2–1b), was 47 and had followed in his father's footsteps. He enjoyed his job with the postal service and had not used a day of sick leave in 15 years (Becker and Toner, 2001). His coworker and friend, Thomas Morris, Jr. (Figure 2–1a), was 55. Mr. Morris was the first to experience mysterious symptoms after suspicious letters with anthrax threats were processed at the Brentwood postal facility where both men worked. After several days of a "flu-like" illness and a trip to his doctor's office, Mr. Morris had been told by his physician that his illness was likely due to a virus. On October 21, 2001, he placed a 911 phone call when his health took another turn for the worst. He had begun vomiting while at work and called for help. He told the 911 operator that he thought he may have been exposed to anthrax. Unfortunately, it was too late to intervene (CNN.com, 2001; Morris, 2001). Mr. Morris died that day and Mr. Curseen's death followed the next day. These were two of the five people who died following exposure to anthrax-laced letters sent to U.S. Senators and media outlets in the weeks following the terrorist attacks of September 11, 2001 (Figure 2–2). Another 17 people were sickened by associated anthrax exposures. The Brentwood postal facility was decontaminated and reopened in 2003 under the name: The Joseph Curseen, Jr., and Thomas Morris, Jr., Processing and Distribution Center. A dedication plaque on the building says: "We are poorer for their loss but richer for having been touched by these dedicated, hard-working heroes. We will never forget."

The "anthrax letters" of 2001 set public health preparedness initiatives in motion that continue today. Billions of dollars have been spent to develop new drugs and vaccines, place biothreat agent detection monitors in major cities and at important events,

FIGURE 2–1 Thomas Morris, Jr. (left) and Joseph Curseen, Jr. (right).
Source: U.S. Postal Service. Used with permission.

enhance laboratory capabilities, expand epidemiological monitoring, increase hospital decontamination and surge capacity, and accomplish planning, training, and exercises across a wide range of related professions. Details have recently emerged suggesting that Dr. Bruce Ivins, an employee of the U.S. Army Medical Research Institute of Infectious Diseases (USAMRIID) at Fort Detrick, Maryland, was responsible for the letters. Although there is strong circumstantial and forensic evidence pointing to Dr. Ivins, his recent suicide will keep the details cloaked in mystery (Lichtblau and Wade, 2008).

Bioterrorism Definitions

Bacteremia: The presence of bacteria in the blood. Fever, chills, tachycardia, and tachypnea are common manifestations of bacteremia.
Bubonic: Relating to an inflamed, enlarged lymph gland.
Encephalitis: Inflammation of the brain due to infection, toxins, and other conditions.
Fever: An abnormal elevation of body temperature (>100 °F), usually as a result of an infection.
Hemorrhagic: Related to bleeding or hemorrhage.
Hypotension: Abnormally low blood pressure; seen in shock.
Maculopapular: A rash of the skin consisting of both spots (macules) and elevations of the skin (papules).
Papular: Characterized by the presence of small, circumscribed, solid elevations of the skin.
Petechiae: Purplish or brownish red discoloration, visible through the epidermis, caused by bleeding into the tissues.
Pneumonic: Related to an inflammation of the lungs.
Prodrome: An early symptom of a disease.

FIGURE 2–2 One of the anthrax-tainted letters that passed through the Brentwood postal facility.
Source: Federal Bureau of Investigation. http://www.fbi.gov/page2/august08/anthrax_gallery1.html.

Sepsis: The presence of microorganisms or their toxins in tissues or in the blood. Systemic disease caused by the spread of the microorganisms or their by-products via the circulating blood is commonly called septicemia.

Septicemic: Related to a systemic disease associated with the presence of their by-products in the blood.

Septic shock: Shock due to circulatory inadequacy caused by gram-negative bacteria (bacteremia). It is less often the result of the presence of other microorganisms (fungus or virus) in the blood (fungemia; viremia).

Shock: An imbalanced condition of the hemodynamic equilibrium, usually manifested by failure to oxygenate vital organs.

Basic Facts about Bioterrorism Threats

Bioterrorism is the use or threatened use of biological agents as weapons of terror. Current U.S. laws make the threat alone, even in the absence of dangerous biological material possession or use, a severe crime. The biological material used in an act of bioterrorism may be lethal or nonlethal, a common bacteria or virus, the toxic by-product of a pathogen, a rare organism, or even a specially engineered organism, never before diagnosed or treated. Although most public concerns focus on the threat to humans, bioterrorists may also attack crops or livestock. Even a small attack on these resources may result in serious economic consequences for any nation targeted. Regardless of the focus of an attack or the agent used, the ultimate goal of bioterrorism is the same as conventional terrorism. These acts are intended to instill fear in the targeted population in support of terrorist goals. It continues to be one of our most complex preparedness challenges. It is distinct from the other disasters described in this book. Most disasters have a focal point. Resources are clearly needed in a relatively well-defined area following an earthquake, hurricane, or other natural disaster. The hazardous chemical exposures of greatest concern are typically acute and those affected are quickly seen among those exposed. Naturally occurring diseases often have a relatively slow moving timeframe where an epidemic unfolds over weeks or months. Conversely, bioterrorism can cause a large, delayed surge of severely ill people spread out across multiple communities or even multiple continents. Organisms or other biological materials can be released in the air, or placed in food or water sources. With incubation periods of days or weeks for many agents, it is difficult to find all who may have been exposed to an intentional release and successfully treat them to avert full blown disease. As a result, it is likely that the response will include an enormous number of potentially exposed individuals in multiple regions. Unlike many other disasters, a window of opportunity exists for a bioterrorism response. Most conditions can be successfully averted if they are caught early and prophylactically treated. As treatment is delayed, morbidity and mortality increase daily as the end of the incubation period is reached. The challenge of responding increases with some high-threat organisms that are spread from person-to-person. These materials can also be engineered to optimize release effectiveness or even make them impossible to treat with standard therapies. A unique aspect of the public health and healthcare response to these scenarios is that their organizations are placed in the vanguard of a response. In most other disaster scenarios, the public health and healthcare roles are in support of emergency management and first responders. In the case of bioterrorism, the majority of response will be identifying and treating those at risk.

There is a long history associated with biological weapons. (See Table 2–1) Thousands of years ago, before any understanding of pathogenic organisms, there was an initial observation that various biological materials could cause illness. A variety of plants were known to be poisonous; but in addition there was a growing understanding that feces, blood, tissue from dead bodies, or entire dead bodies of sick humans or animals, were able to cause illness and could be used as weapons. This crude use of filth and carcasses was the beginning of biological warfare and bioterrorism. The twentieth century saw a rapid acceleration in life sciences but behind the scenes, known to relatively few, was a growing group of scientific professionals, including microbiologists, physicians, and many other disciplines working covertly on the development of biological weapons. Although most nations acknowledge that these weapons should be banned and signed the Biological Weapons Convention in 1972 stating they would no longer pursue them for offensive purposes, the life sciences advances and technology continue to trickle down. As these amazing advances move us forward with new pharmaceuticals and powerful tools to manipulate microorganisms, it also places a growing body of potentially dangerous information within reach of those who may intend to use it for harm. A new arms race is on. Rather than competing super powers, it is criminals, terrorists, and rogue nations against the world. Rather than detecting an incoming threat on radar screens, as it was during the Cold War, this time it may be detected through the lens of a microscope.

U.S. Bioterrorism Preparedness Controversies

Bioterrorism became a well recognized part of American vernacular after five anthrax-tainted letters circulated on the East Coast of the United States in September and October of 2001. The letters were responsible for 22 cases of anthrax disease and five deaths (Mina et al., 2002). These events led to serious consideration of other possible biological threats and led to strong debates on issues, such as who should receive smallpox vaccinations and how available should antibiotics be to the public for preparedness purposes. Federal, state, and local public health preparedness efforts accelerated to unprecedented levels in the years immediately following 2001. The growing apprehension of Americans over these emerging threats resulted in bioterrorism preparedness becoming a moral imperative. In recent years, indifference has crept in and bioterrorism preparedness has become a much lower priority.

Table 2–1 Select Historical Events Involving Biological Weapons and Bioterrorism

<1000 c. BC	Scythian archers tipped arrows with blood, manure, and tissue from dead bodies.
5th c. BC	Assyrians poisoned enemy wells with rye ergot (*Claviceps purpurea*), a fungus containing mycotoxins.
590 BC	Athenians poisoned enemy water supplies with hellebore, an herb purgative, during the Siege of Krissa.
3rd c. BC	Persian, Greek, and Roman literature describes the use of dead animals being used to contaminate enemy water supplies.
184 BC	Carthaginian General Hannibal ordered his sailors to hurl clay pots filled with poisonous snakes onto the decks of enemy ships during a naval battle. Hannibal won the battle.
1155	Emperor Barbarossa poisons wells with decomposing human bodies.
1346	Tartur army catapulted deceased bodies of plague victims over city walls during the siege of Caffa.
1495	Spanish sell wine mixed with the blood of lepers to their enemies.
1763	British distribute variola virus contaminated blankets to Native Americans resulting in a smallpox outbreak.
1797	Napoleon floods fields around Mantua to promote malaria.
1915–18	Germans attempt to infect Allied horses with anthrax and glanders.
1932–45	Japanese operate Unit 731 in Manchuria conducting experiments that included infecting prisoners with a variety of lethal pathogens.
1942	British test anthrax bombs on Gruinard Island off the coast of Scotland.
1950–69	US and USSR grow offensive biological weapons programs.
1969	US President Nixon ends the US offensive biological weapons program.
1972	US and USSR sign the Biological Weapons Convention agreeing to an end to offensive programs.
1978	Assassination of Bulgarian exile Georgi Markov in London with an injected ricin pellet.
1979	Accidental anthrax release from a secret Soviet facility in Sverdlovsk kills 66.
1984	In the Dalles, Oregon, the Rajneeshee cult contaminated local salad bars with salmonella sickening more than 750 people.
1990	Japanese Aum Shinrikyo cult unsuccessfully attempts botulinum toxin releases in Tokyo.
1991	US troops receive anthrax vaccinations.
1991	After the first Gulf War, UN inspectors begin inspections of biological weapons capability in Iraq. Iraqi government officials confirm they had researched the use of anthrax and botulism.
1993	Aum Shinrikyo cult unsuccessfully attempts a second botulinum toxin attack on the wedding of the Crown Prince. Later the same month they unsuccessfully attempted to release anthrax from a Tokyo high rise.

(Continued)

Table 2–1 *(Continued)*

2001	Anthrax contaminated letters mailed to US Senate offices and media outlets sickening 22 and killing five.
2004	Ricin sent to US Senate Majority Leader Bill Frist's office.

Adapted from Frischknecht F. (2008). The History of Biological Warfare. In Richardt A, Blum M (eds.) Decontamination of Warfare Agents: Enzymatic Methods for Removal of B/C Weapons, Wiley-VCH and Eitzen, E. and Takafuji, E. (1997). Historical Overview of Biological Warfare. In Office of the Surgeon General, Department of the Army (ed.) Textbook of Military Medicine: Medical Aspects of Chemical and Biological Warfare.

This is most apparent in the declining public health preparedness budgets that provide the needed infrastructure to quickly identify a bioterrorism attack and respond effectively to it.

The past several years of U.S. bioterrorism preparedness initiatives have suffered increasing scrutiny as well. The growing bio-preparedness research infrastructure is considered by some to be unnecessary and possibly even increase the risks of future incidents. This is difficult to dispute in light of the discovery that the source of the 2001 anthrax letters was a worker at the USAMRIID research facility (Figure 2–3) (Regaldo et al., 2002). Other safety and security problems have surfaced at the U.S. Army Fort Detrick labs, as well as other research labs across the country (U.S. General Accounting Office, 2008; Weiss and Snyder, 2002; Williamson, 2003). This raises questions concerning the potential for increasing bioterrorism and biocrime risks from new facilities being established to research and develop measures to defend against these threats. However, there is adequate precedent for safe handling and management of extremely hazardous materials. In a wide range of academic and industrial endeavors, safe-handling of dangerous products occurs daily with few problems. Should an accident or security breach occur, it is generally felt

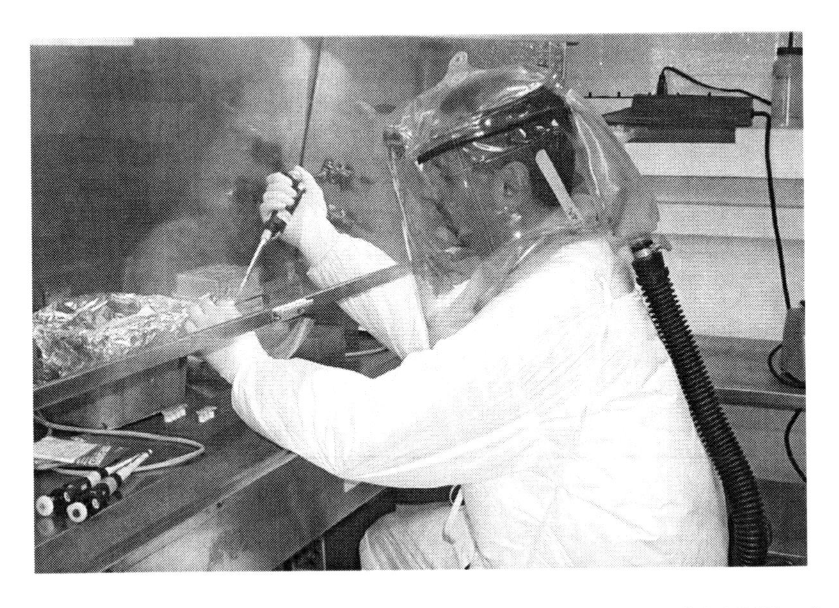

FIGURE 2–3 Researcher at the U.S. Army Medical Research Institute of Infectious Diseases (USAMRIID) at Fort Detrick, Maryland.

Source: Department of Defense, USAMRIID. http://www3.niaid.nih.gov/NR/rdonlyres/E68D5347-2172-4983-81AC-CD5739E8B8B0/0/BD_BSL_1.JPG.

that the benefits outweigh the risks. As with other risky activities, research of dangerous pathogens is essential and has important benefits that must be weighed against the risks. Sufficient measures must be in place and enforced to ensure adequate screening and security in the life sciences industry, as there are in other industries or initiatives. We cannot make preparedness decisions based upon suspicion of ourselves and cease important work that could eventually diminish the impact of even the worst case bioterrorism scenarios.

There is growing criticism of the U.S. investment in public health preparedness. The large infusion of bioterrorism preparedness funding had mixed results in the years immediately following the 2001 terrorist attacks. Many public health programs are historically underfunded. Although many of these public health programs such as public health laboratories, communicable disease, and environmental health benefited from preparedness resources, accusations of supplanting and misappropriation of resources persist. Supplanting is the use of federal preparedness funding to replace state or local resources. The problem is if the preparedness dollars are focused appropriately to cultivate the public health infrastructure that identifies and responds to bioterrorism, it turns out to be the same infrastructure that detects and responds to usual outbreaks such as food borne and seasonal illnesses. Although some supplanting violations may be obvious, the distinction can be tricky. When preparedness spending is done wisely, there is typically a dual use of resources. This is a hallmark of an effective preparedness program. If the preparedness function of public health is stove piped and not integrated into daily public health activities, it will not be effective during times of crisis.

Bioterrorism preparedness activities include the development and practice of a mass emergency distribution of pharmaceuticals, risk communication training, and Incident Command System training for public health and healthcare workers. All these preparedness activities came into play during Rhode Island's December 2006 response to a deadly outbreak of *Mycoplasma pneumoniae*. This infection causes a serious neurological illness. A cluster of five children at the same school were identified with the infection and one died. Although this was not the result of an act of bioterrorism, the local and state public health authorities used the same process developed for mass dispensing of antibiotics for an anthrax or plague response to identify other children at risk, communicate the risk to the public, and dispense prophylactic medications to over 1000 people over a holiday weekend. They reached 100% of those at risk (Association of State and Territorial Health Officials, 2008).

> *Overall, the effort was a great example of an effective partnership and collaboration between the state, CDC, EMA, Department of Education, and the towns. If it weren't for the emergency preparedness funds that CDC has provided us and training all our staff have received in ICS (which we utilized extensively during this episode), I don't believe we could have accomplished everything we did over a holiday weekend....*
>
> David Gifford, MD, MPH, Director,
> Rhode Island Department of Health

Chain of Infection

The human routes of entry for bioterrorism agents depend upon the characteristics of the agent and the method of release. Some agents can only cause infection when entering a specific portal of the body, whereas others cause different kinds of infections when entering through different portals. For example, *Bacillus anthracis* spores can be inhaled, causing inhalational anthrax disease, or can enter through a cut on the skin causing cutaneous anthrax disease. Though the agent is the same, the method of exposure results in very different disease conditions. There are also preferred portals of exit that influence transmissibility of pathogenic organisms. Respiratory diseases usually exit an infectious person through the nose and mouth, whereas gastrointestinal infections usually exit in the feces. These organisms can be passed to other people through direct contact with symptomatic or asymptomatic human reservoirs of disease, or can sometimes be spread through living (nonhuman) or inanimate reservoirs (See Figure 2–4).

The goal of any bioterrorism infection control intervention is the same as many other infectious disease challenges. The chain of transmission must be broken. There are opportunities at every step of the infection process to break the chain and intervention may be required at multiple steps in the process. This includes reducing host susceptibility by boosting immunity through vaccination or drug therapy, blocking the portal of entry using personal protective equipment, or even destroying reservoirs that can continue to sustain the infection process such as mosquitoes or rodents.

At each link in this chain, there is the potential for tremendous variation. For example, across any vulnerable population, there will be a range of physiological responses among those exposed to a biothreat agent. This is due to a variety of complex environmental and immunological variables. A group of individuals receiving the same exposure to an organism will often display a broad assortment of responses. Some will not be infected at all, while others are infected and yet display no symptoms. Others in the same cohort may display moderate symptoms, severe symptoms, or even die from their infection. These varying responses are due to differences in the exposures and immune function of those exposed. Many factors influence these differences including age, physical condition, stress level, higher breathing rates that increase the dose, and many other variables.

Categorization of Threats

Since the U.S. anthrax attacks of 2001, bioterrorism preparedness planning, training, and exercises have primarily focused on two worst case scenarios of an aerosolized anthrax release and a smallpox outbreak. These are just two of the many options a potential bioterrorist may choose. In 1999, a group of military and civilian experts were convened to narrow the list of possible biological threats and prioritize them based on the poten-

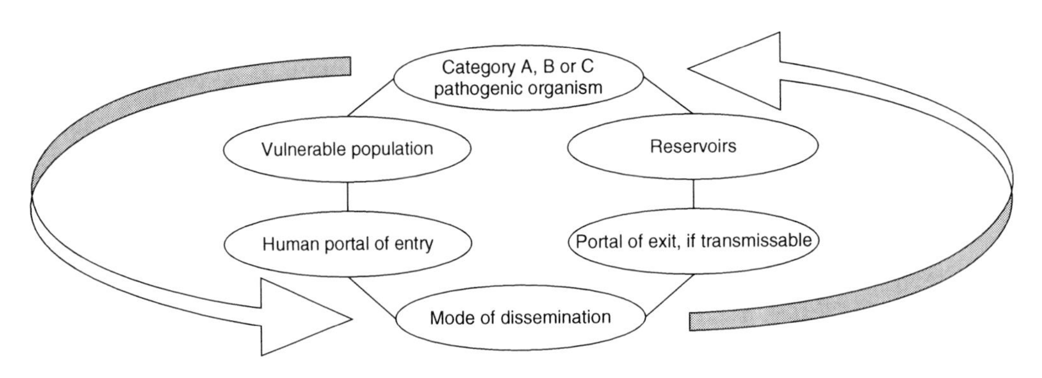

FIGURE 2–4 The bioterrorism chain of infection.

Table 2–2 Critical Biological Agents for Public Health Preparedness

Disease	Biological Agent(s)
Category A	
Anthrax	*Bacillus anthracis*
Botulism	*Clostridium botulinum* (botulinum toxins)
Plague	*Yersinia pestis*
Smallpox	*Variola major*
Tularemia	*Francisella tularensis*
Viral hemorrhagic fevers	Filoviruses and Arenaviruses (e.g. *Ebola, Lassa Fever*)
Category B	
Brucellosis	*Brucella spp.*
Food safety threats	*Salmonella spp., Escherichia coli* O157:H7
Glanders	*Burkholderia mallei*
Melioidosis	*Burkholderia pseudomallei*
Psittacosis	*Chlamydia psittaci*
Q fever	*Coxiella burnetii*
Ricin toxin	*Ricinus communis* (Castor Beans)
Staphylococcal enterotoxin B	Exotoxin from *Staphylococcus aureus* bacterium
Typhus fever	*Rickettsia prowazekii*
Viral encephalitis	*Alphaviruses* (*VEE, EEE, WEE)
Water safety threats	*Vibrio cholerae*, Cryptosporidium parvum
Category C	
Nipah virus encephalitis	Paramyxoviridae family
Hendra virus encephalitis	Paramyxoviridae family
Hantavirus pulmonary syndrome	Hantavirus

*Venezuelan equine (VEE), eastern equine (EEE), and western equine encephalomyelitis (WEE) viruses.
Adapted from Centers for Disease Control and Prevention, Bioterrorism Agents/Diseases, By Category. http://www.bt.cdc.gov/agent/agentlist-category.asp.

tial impact they could have when used as a weapon. All available open source literature as well as classified information was reviewed. The list that emerged from their efforts included three categories of pathogens that were designated as Category A, B, and C (Rotz et al., 2002) (Table 2–2). Category A pathogens are high priority organisms and toxins posing the greatest threat to public health. This category includes agents that cause the highest morbidity and mortality with a likelihood of subsequent public panic. They are capable of being spread over a large area and are sometimes also transmitted easily from person-to-person. The scenarios associated with Category A agents necessitate extraordinary actions on the part of public health and healthcare organizations making them the focus of many public health preparedness activities. Throughout the early cold-war era of the 1950s and 60s, most of these organisms were successfully weaponized in the offensive biological weapons programs of the United States and the former Soviet Union. Category B agents are fairly easy to disperse but have lower morbidity and mortality than the Category A agents and can be successfully addressed through enhancing laboratory capabilities and epidemiological monitoring. Category C agents are emerging infectious organisms that could become easily available at some point in the future and used as a weapon. These agents may be used as weapons and could have a major public health impact (Centers for Disease Control and Prevention, Bioterrorism agents/diseases by category, http://emergency.cdc.gov/agent/agentlistcategory.asp).

Health Threats: Category A Organisms

Anthrax (*Bacillus anthracis*)

Anthrax disease is caused by the gram-positive, spore-forming bacterium *Bacillus anthracis* (See Figure 2–7). The disease occurs naturally among mammals such as cattle, sheep, goats, and camels. However, it can also occur in humans when they are exposed to infected animals or to anthrax spores used as a weapon. *B. anthracis* is considered, by many biological warfare experts, to be the perfect biological weapon because of its ability to change into a spore when it is subjected to adverse environmental conditions that destroy other pathogenic organisms. Once in the spore form, it can survive for decades under extreme hot or cold conditions. Historically, most cases of anthrax result from occupational exposures among leather and wool industry workers. The organism and disease get their name from the Greek word "anthrakitis," which is a kind of coal. The name originated from the appearance of cutaneous anthrax that causes a black lesion on the skin.

There are three forms of anthrax disease. These distinct disease presentations depend upon the route of exposure. Cutaneous anthrax is the most common form of anthrax disease and is fatal in about 5–20% of those who are not treated. It occurs following the exposure of compromised skin to anthrax spores. Inhalational anthrax disease is the most lethal form of disease with more than 80% mortality for those who are not treated. It occurs following the inhalation of anthrax spores and is the most likely form of disease to be seen following an act of bioterrorism. Gastrointestinal anthrax disease is highly lethal but very rare. It usually occurs following the ingestion of live (nonspore) *B. anthracis* in

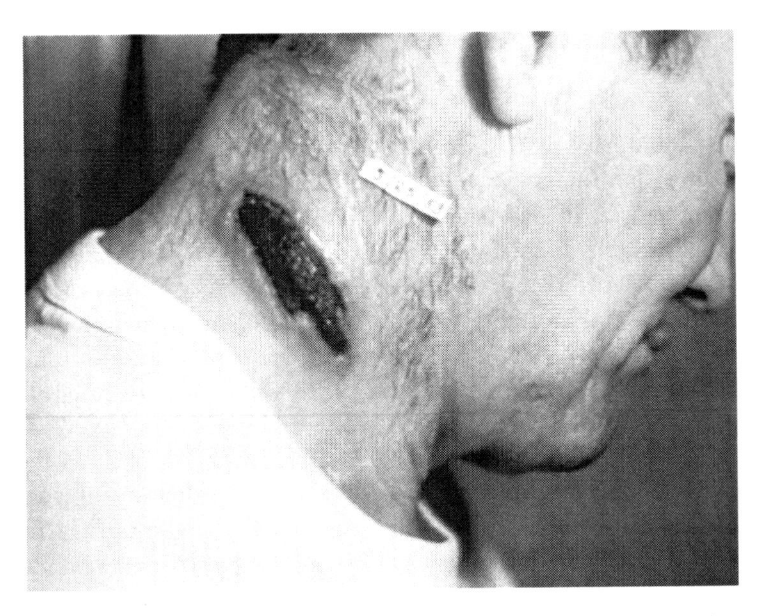

FIGURE 2–5 Cutaneous anthrax lesion on the neck. Photo Courtesy of CDC/Public Health Image Library PHIL ID#1934.

contaminated, undercooked meat. Ingestion of the spore form of *B. anthracis* is unlikely to result in a fatal disease because of the transport time in the gastrointestinal system. The spores are likely to pass through the GI system before they have the opportunity to germinate. However, vegetative, nonspore *B. anthracis* organisms can quickly begin to release toxins damaging the GI tract and can quickly cause illness when ingested.

Cutaneous anthrax is the most common form of anthrax disease. It is a localized skin infection that results from compromised skin exposure to anthrax spores. It is typically seen on commonly exposed skin such as the head, neck, hands, and arms. Areas of compromised skin, such as cuts, abrasions, or chronic dermatological conditions, are more prone to infection than the intact skin. Once the spores become embedded in the skin, they germinate and begin to release a toxin. The toxin causes an edema. The edema progresses into an ulcer over the course of several days and the ulcer develops into a depressed, painless lesion called an eschar (See Figure 2–5). Though it is sometimes mistaken for a spider bite, the anthrax eschar remains painless whereas a spider bite causes severe pain. The eschar develops a black leathery appearance for a week or two and then loosens and falls off. Cutaneous anthrax is the most common form of occupational anthrax disease and may also result from an intentional release of spores. Cutaneous anthrax comprised 11 of the 22 cases of disease resulting from the 2001 anthrax letters and none were fatal (Inglesby et al., 2002). If it is untreated, it may lead to a systemic disease with up to 20% mortality. If treated, this form of anthrax disease has very low (<1%) mortality (Lew, 1995).

Inhalational anthrax has the highest mortality of any form of anthrax disease. It results from inhaling a sufficient number of anthrax spores to initiate the disease process. This form of anthrax disease is very rare and even a single case requires notification of both public health and law enforcement authorities. In the 20 years preceding the 2001 anthrax attacks, there had not been a single case of inhalation anthrax in the United States (Inglesby et al., 2002). When anthrax spores are inhaled, they are transported by pulmonary macrophages, part of the lungs' defense system, to the mediastinal lymph nodes in the center of the chest. Over a period of several days, the spores germinate into vegetative bacilli and begin to release a toxin. This produces a two-stage illness that includes prodromal and fulminant phases. The prodromal phase lasts from a few hours to several days. It includes nonspecific, flu-like symptoms including fever, malaise, shortness of breath, nonproductive cough, and nausea. This stage of the disease is easily confused with influenza or other common infections. However, inhalational anthrax does not normally include runny nose or sore throat. Many flu-like illnesses have those symptoms (Centers for Disease Control and Prevention, 2001). In addition, influenza does not normally include shortness of breath, vomiting, and mediastinal pain. These are common symptoms of inhalational anthrax. This prodromal period is sometimes followed by a brief improvement before progressing to the next phase of illness. The second phase is a high-grade bacteremia called the fulminant stage. It includes symptoms such as fever, respiratory distress, profuse sweating, cyanosis, and shock. Patients displaying these symptoms usually progress to death within days. If an outbreak is known or suspected, initial diagnosis of inhalational anthrax is made using the described signs and symptoms. A widened mediastinum on a chest X-ray, as displayed in Figure 2–6, is considered highly suspect for inhalational anthrax. However, this clue may or may not be present at the initial patient evaluation or can result from other respiratory conditions. Confirmation testing may be accomplished using blood cultures and polymerase chain reaction (PCR) of blood or pleural fluid.

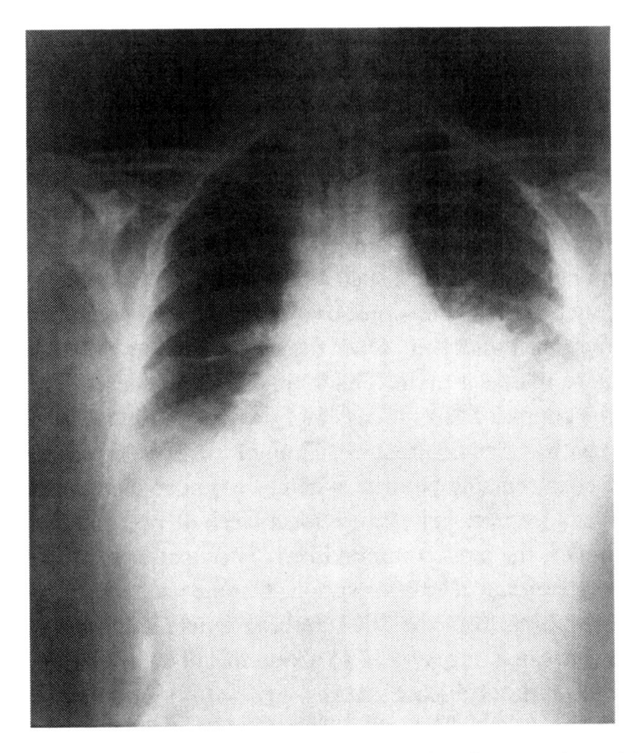

FIGURE 2–6 Chest radiograph 22 hours prior to death showing widened mediastinum due to inhalation anthrax. Photo Courtesy of CDC/Public Health Image Library PHIL ID#1118.

The gastrointestinal form of the anthrax disease is caused by the ingestion of vegetative *B. anthracis*. This may occur from consumption of undercooked contaminated meat from an infected animal and can be very difficult to diagnose. Initial presentation may include nausea and vomiting quickly progressing to bloody diarrhea and sepsis. It can lead to the same sort of fatal sepsis as untreated cutaneous and inhalational anthrax. Although gastrointestinal anthrax may result from an intentional release of anthrax, the spore form of anthrax is far less likely to generate gastrointestinal cases (Inglesby et al., 2002).

Treatment recommendations for inhalational anthrax are based on animal testing and a very small number of human cases. Antibiotic therapy should be initiated as soon as possible when exposure is suspected. The 2001 anthrax cases included 11 inhalational anthrax cases. Six survived due to the rapid initiation of multiple antibiotic therapies and aggressive supportive care (Jernigan et al., 2001). If an epidemiological or criminal investigation identifies populations at risk, postexposure prophylaxis should be provided to all potentially exposed persons. If the right antibiotics are administered quickly enough, especially before symptoms begin, the disease can be prevented. This is the greatest challenge in managing a mass exposure to *B. anthracis*. Getting the right antibiotics into the exposed population quickly enough to halt the progression from exposure to disease is very difficult. Although patient contacts (e.g., family, friends, healthcare providers) and others that are not originally exposed to a *B. anthracis* release do not require prophylaxis, there are likely to be difficulties in concisely identifying all at risk, particularly if it is an aerosolized urban release scenario. Between the need to expand the potential exposure footprint to include everyone at risk and the demands of the "worried well" that are not necessarily at

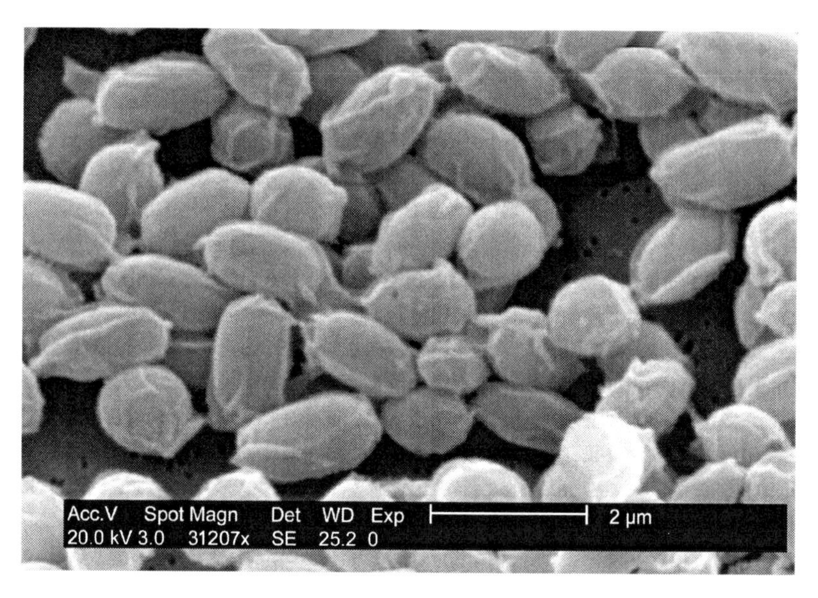

FIGURE 2–7 Under a very high magnification of 31,207X, this scanning electron micrograph (SEM) depicted spores from *Bacillus anthracis* Sterne strain bacteria. Photo Courtesy of CDC/Public Health Image Library PHIL ID#2266.

risk but insist upon prophylaxis anyway, the list of those receiving prophylactic antibiotics may easily be in the hundreds of thousands within a 2- to 3-day timeframe.

The greatest short-term challenge in managing a response to a large scale anthrax release is the identification and prophylaxis of the entire population at risk. Although this preparedness scenario has been exercised by most major U.S. cities, these exercises have shown that there is insufficient capacity to provide prophylaxis to the entire population, many known gaps, and many other things that are simply unknown. This kind of mass prophylaxis has never been done before on this scale. The long-term challenges may even prove to be greater. It will require the safe and complete remediation of a contaminated urban area. Although the inside remediation of several anthrax contaminated buildings has been accomplished, a wide area urban restoration project has never been done before.

There is a licensed, cell-free anthrax vaccine called Biothrax (Anthrax Vaccine Adsorbed) manufactured by Emergent Biosolutions (formerly BioPort) in Lansing, Michigan, the only current U.S. manufacturer of anthrax vaccine. Use of this vaccine has been restricted to military personnel and to groups with occupational exposure risks, including lab personnel required to handle suspicious powders for testing, and veterinarians or others exposed to potentially contaminated animals or animal products (Emergent Biosolutions, 2002). However, it is considered as an investigational new drug (IND) for mass prophylaxis during public health emergencies. The Advisory Committee on Immunization Practices (ACIP) recommends using anthrax vaccine in combination with a prophylactic antimicrobial regimen to reduce the recommended time of antibiotic therapy from 60 days to 30 days. This includes a three-dose vaccine regimen (0, 2, and 4 weeks) in combination with 30 days of antimicrobial prophylaxis (Centers for Disease Control and Prevention, 2002). Following the mailed anthrax incidents of 2001, more than 10,000 people were placed on a 60-day regimen of antibiotic therapy. However, adherence to the full 60-day antibiotic regimen was low. Less than half (42%) of the individuals placed on this prophylactic therapy were able to complete it (Shepard et al., 2002). Combining anthrax vaccinations with antimicrobial prophylaxis reduces the course of prophylactic therapy and may facilitate complete prophylaxis for those at risk by shortening the length of time antibiotics are taken to a tolerable timeframe.

Although there is a new anthrax vaccine in development, the existing vaccine formulation was developed in the 1960s and has not changed much since it was licensed by the FDA in 1970. One of the problems with the vaccine is the dosing schedule. The initial regimen includes six injections over the first 18 months with an annual booster thereafter. As a result, when large cohorts of people are vaccinated over an extended period of time, nearly any health issue arising in the group during that time may be erroneously associated with the vaccine. It has been an ongoing issue with this vaccine. Many illnesses and side effects that are biologically implausible from any vaccine have been attributed to Biothrax (Anthrax Vaccine Adsorbed). These debates have been fueled by the added controversies of the vaccine production facility transition from being a state-owned operation to being privately held, poor facility inspection results, and questions about vaccine efficacy and safety. Recent efforts of the Department of Health and Human Services to speed the production of the next generation anthrax vaccine led to the failed contract effort with a small California company called VaxGen. According to a recent Government Accounting Office report, the contract was premature and had impracticable expectations (U.S. General Accounting Office, 2007).

Those at greatest risk of infection are individuals exposed to the initial release of anthrax spores. Depending upon the weather and other environmental conditions, the spores will usually settle within several hours. They will then pose ongoing risks of secondary aerosolization from passing foot traffic, vehicles, and other disruptions. If a *B. anthracis* release occurs in an urban area, the environmental decontamination of surfaces will be a complicated and extremely expensive process. The clean-up will last for years and require many millions of dollars in direct cost; this is in addition to the high costs of business disruption and the personal impact of possible loss of life and required abandonment of property for years during clean-up. The social impact will be far reaching.

Exposed individual and patient decontamination for *B. anthracis* is only warranted in the immediate aftermath of a known release. Human decontamination simply consists of clothing removal and soap and water shower. Bleach and other harsh chemicals are not necessary. It is improbable that a release would be detected for at least 24 hours. During that time, those exposed will likely change clothes and shower. The chances of mass *B. anthracis* decontamination are remote. If a covert release is successfully accomplished, patients seen days or weeks later will also not pose a risk to providers and do not require decontamination. The one positive aspect of managing the aftermath of a *B. anthracis* attack is that the disease does not pose a risk of person-to-person transmission.

Botulism (*Clostridium botulinum* Toxin)

Botulism is caused by a group of neurotoxins produced by the anaerobic, spore forming, gram-positive bacillus *Clostridium botulinum*. This family of botulinum toxins cause paralytic disease and are regarded as the most potent poisons in the world (Gill, 1982).

There are three forms of botulism disease including foodborne, wound, and infant (also called intestinal) botulism. Foodborne botulism results from ingesting the toxin in contaminated food. It causes descending paralysis that can eventually lead to respiratory failure and death. Wound botulism occurs when a wound is contaminated with *Clostridium botulinum*. This form of botulism disease is most commonly associated with injected drug use or traumatic injury. Infant (intestinal) botulism occurs when susceptible infants ingest *Clostridium botulinum* spores in food that germinate in their intestines. This ingestion results in difficulty feeding, constipation, weakness, and muscle hypotonia, also called "floppy baby syndrome" (see Figure 2–8). There are eight separate *Clostridium botulinum* toxin types, including A, B, C alpha, C beta, D, E, F, and G. All these types except two (C alpha and D) cause botulism in humans. Botulinum toxin could be used as a weapon of terror to contaminate water or food supplies or could be delivered as an aerosol. If an aerosol release is used, symptoms are likely to be similar to food botulism with a delayed onset. Although there have been no documented human inhalational botulinum cases, animal studies have shown that aerosolized botulinum toxin can be absorbed by the lung (Shapiro et al., 1998). *Clostridium botulinum* is the only toxin listed as a Category A biological agent by the CDC. The other five Category A agents are replicating bacterial or viral agents. This toxin earned a place near the top of the threat list because of widespread availability and tremendous potency as a biological poison.

There are cases of botulism that regularly occur in the United States because of improper canning and food handling. Across the entire nation, fewer than 200 cases typically occur each year (Centers for Disease Control and Prevention, 1998). Foodborne

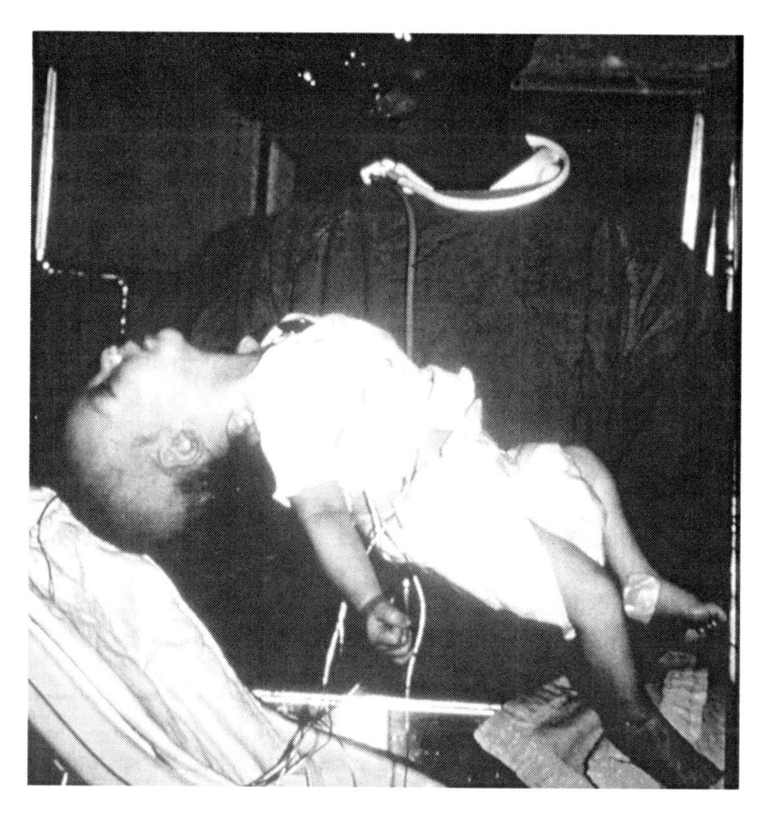

FIGURE 2–8 Six-week old infant with botulism that is evident as a marked loss of muscle tone, especially in the region of the head and neck. Photo Courtesy of CDC/Public Health Image Library PHIL ID#1935.

botulism is the most common form of botulism disease and is caused by eating foods that are contaminated with *Clostridium botulinum* that has released some of the toxin. If intentional food contamination becomes the delivery method selected by a terrorist or criminal, it would likely be limited to a small-scale attack. The toxin is easily destroyed with normal cooking and is not stable enough to survive for long-time periods in bulk food items. Regardless of the size or source of a botulism outbreak, any cases are considered to be a public health emergency that must be reported and investigated immediately. The source of the case or cases may still be accessible to the public and able to cause additional cases.

Intestinal botulism occurs when *Clostridium botulinum* spores are ingested resulting in intestinal colonization. The germinating spores release the illness-causing toxin. This form of the disease typically occurs in infants and is often attributed to feeding them honey. Although adults have immune systems that can tolerate and manage the minuscule number of botulinum spores that often exist in honey, infants do not. However, in children or adults, when an infected wound is not apparent and a specific food source cannot be identified, it is typically attributed to intestinal colonization. If an infected wound is observed, it may be wound botulism. This form of disease occurs when *Clostridium botulinum* infects a wound, multiplies, and releases the toxin causing systemic illness. Like foodborne, intestinal, and wound botulism occur naturally. However, the fourth type of disease, inhalational botulism, does not occur naturally and is the most likely form expected to result from an aerosolized botulinum toxin release. It has never been seen in humans before, and there is limited animal data to provide clues on what to expect if it occurs.

The onset of botulism symptoms depends upon the exposure dose. Incubation for foodborne botulism ranges from several hours to just over a week (St. Louis et al., 1998). The lack of human data on inhalational botulism disease progression makes it impossible to predict exactly how these cases may progress. Initial botulism symptoms typically include blurred vision and dry mouth. For small exposures, these symptoms may be all that the symptoms a patient experiences before recovery. Significant exposures may include slurred speech, difficulty swallowing, and descending peripheral muscle weakness. Severe disease involves the respiratory muscles, leading to respiratory failure and death. Presumptive diagnosis is based on the descending paralytic symptoms described. This diagnosis may be confirmed through testing available at reference labs but takes several days to complete. Botulism is sometimes mistaken for Guillain-Barré syndrome, myasthenia gravis, or other diseases of the central nervous system (CNS) but differs from other CNS conditions in the descending symmetrical paralysis and absence of sensory nerve damage (Arnon et al., 2001).

The greatest challenge in managing a response to a large scale botulism outbreak will likely be a shortage of ventilators. As the descending paralysis moves down from the top of the head to the neck and chest of those exposed, it will eventually cause respiratory failure and require mechanical assistance to maintain respiration.

Botulism treatment requires supportive care with careful respiratory support and rapid administration of antitoxin. Fortunately, it is not transmitted from person to person, so patient care does not include a negative pressure patient room. In a mass exposure, these therapies pose tremendous challenges to an effective public health response. The need for ventilators may quickly exceed available capacity in an affected community. In addition, the antitoxin efficacy is dependent upon how early it is started in the course of treatment (Tackett et al., 1984). This highlights the need for well-trained healthcare providers who maintain a high index of suspicion. It also shows how important it is to maintain a logistical system that can rapidly deliver antitoxin anywhere it is needed. The Centers for Disease Control and Prevention maintain the antitoxin supply and will release it for use through state health departments following a quick consultation. The botulism antitoxin cannot be used prophylactically for exposed individuals. It can only be used when initial symptoms appear. Although there is no botulism vaccine available, there are some promising advances. A new Human Botulism Immune Globulin Intravenous (BIG-IV) has shown great promise in reducing the length of treatment for infant botulism (Arnon et al., 2006). This could become the basis of new therapies for other forms of botulism disease as well.

Plague (*Yersinia pestis*)

Plague is a disease caused by *Yersinia pestis*, an anaerobic, gram-negative bacterium. The natural host for this organism is a rat and the disease is usually transmitted to humans through a flea bite from a flea that has fed on an infected rat and then on a human. There are about 10–15 human plague cases that naturally occur this way in the Western United States. Human cases also naturally occur in portions of South America, Africa, and Asia. Between 1000 and 3000 cases occur globally each year (Centers for Disease Control and Prevention (CDC), Division of Vector Borne Disease, Plague Home Page, www.cdc.gov/ncidod/dvbid/plague/index.htm). There are three types of plague disease including bubonic, pneumonic, and septicemic. The bubonic form of plague is the most likely form of the disease to be seen from naturally occurring infections. However, a terrorist attack could potentially involve the release of infected fleas resulting in bubonic cases. Pneumonic plague is more likely to be associated with an aerosolized release of *Yersinia pestis* during a bioterrorism attack. Sustained person-to-person transmission is possible through airborne droplets from infected individuals. This potential secondary transmission complicates the management of a large bioterrorism related outbreak and poses significant infection control challenges. Untreated pneumonic plague has a mortality rate of nearly 100%. It spreads through the lymph system and may involve multiple organ systems. All forms of plague can progress to the septicemic or pneumonic forms of the disease, but the worst case bioterrorism scenario is an aerosolized release of *Yersinia pestis* that causes a large number of primary cases that continue to spread the infection to others before it is brought under control through public health interventions.

Yersinia pestis was isolated and characterized by Alexandre Yersin who traveled to Hong Kong in 1894 to study a plague outbreak that had taken tens of thousands of Chinese lives. The disease was eventually named after him. He seldom used his first name, referring to himself simply as Yersin and he refused to attend most medical and scientific meetings of his day. He was an explorer and traveler trained in Europe who was fascinated by "Indo-China." After discovering *Yersinia Pestis*, he went on to become the founder and director of the Medical School of Hanoi (Bibel and Chen, 1976).

The word "plague" is often used to describe lethal outbreaks from a variety of infectious organisms. However, it is a specific disease that results from an infection by the gram-negative bacillus, *Yersinia pestis*. Plague has an infamous history killing millions of people in pandemic outbreaks occurring around the world across the centuries. Since plague still occurs naturally in many parts of the world, it is important to understand the baseline occurrence of naturally occurring plague so that you can recognize a suspicious case. If human cases are identified in a nonendemic area, in persons without risk factors, or in the absence of confirmed rodent cases, a terrorist release of *Yersinia pestis* should be considered as a possibility. There are three forms of plague disease: pneumonic, bubonic, and septicemic. Early symptoms for all forms of plague disease include classic flu-like symptoms such as fever, chills, myalgia, weakness, and headaches. Each form of plague poses different challenges and unique symptoms.

Pneumonic plague is the least common and most severe form of the plague disease. This form of disease results from the inhalation of infectious particles either from an intentional aerosolized release or from a cough or sneeze of an infectious person. Although it sometimes occurs secondarily to bubonic plague, it will also cause primary pneumonic disease if it is released in an act of bioterrorism. This is the most likely form of plague disease resulting from aerosolized *Yersinia pestis* in an act of bioterrorism.

Yersinia pestis can only remain viable for about an hour as an aerosol and will quickly die when exposed to sunlight and heat outside of a living host. After exposure to aerosolized *Yersinia pestis*, individuals will display symptoms in 1–6 days (Inglesby et al., 2000). Based on limited epidemiological data, the mortality rate for pneumonic plague in the United States is about 57% (Centers for Disease Control and Prevention, 1997). Patients will initially present with dizziness, chest discomfort, and a productive cough with blood-tinged sputum. By the second day, the symptoms may rapidly worsen and include chest pain, coughing, sputum increase and may include dyspnea, hemoptysis, cardiopulmonary complications, and circulatory collapse.

Bubonic plague is the most common form of plague disease. It is spread through the bite of an infected flea or by handling contaminated infected animals. Although this form of plague is less likely to be the result of bioterrorism, it is possible to intentionally spread bubonic plague through an intentional release of infected fleas. The mortality rate for this form of the disease is approximately 13% (CDC, 1997). Patients with bubonic plague will present with swollen, tender, and painful buboes (enlarged lymph nodes). These symptoms typically occur within 24 hours of the onset of early flu-like symptoms and are generally located near the area of inoculation. Most frequently, they occur in the inguinal or femoral lymph node region of the groin (See Figure 2–9). Buboes (swollen lymph glands) may erupt on their own or require incision and drainage.

Septicemic plague is the least common form of plague disease and it can result from either pneumonic or bubonic disease. It is a systemic infection that results from introduction of *Yersinia pestis* into the bloodstream. The mortality rate for this form of the disease

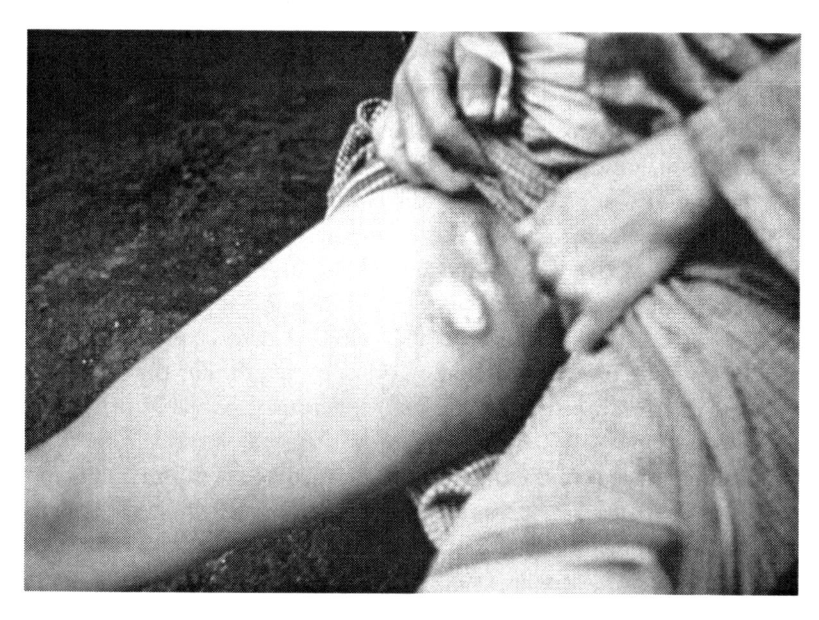

FIGURE 2–9 This plague patient is displaying a swollen, ruptured inguinal lymph node, or bubo. Photo Courtesy of CDC/Public Health Image Library PHIL ID#2047.

is about 22% (CDC, 1997). Patients with septicemic plague will initially present with nausea, vomiting, and diarrhea and may develop skin lesions, gangrene, and necrosis later in the infection process.

Rapid diagnosis and treatment are essential to reduce plague morbidity and mortality. Without quick antibiotic treatment, plague can cause death in several days. Initial diagnosis can be accomplished through microscopic examination of clinical samples of lymph node aspirates, sputum, or spinal fluid. Definitive diagnosis is accomplished using sputum or bubo cultures or IgM and IgG antibody testing available through reference labs. In addition, it is important to identify antibiotic resistance patterns to select the most effective treatment. Rapid diagnosis and antibiotic therapy are the keys to successfully managing a plague outbreak. Once known or suspected cases of pneumonic plague occur, assume all patients presenting with flu-like symptoms are positive and treat them immediately. The rapid progression of the disease does not allow much time for definitive diagnosis during a possible mass exposure or outbreak. The antibiotic selection is based on the results of susceptibility testing and the age and condition of the patient. The aim of treatment includes both the treatment of active disease and the prophylactic treatment of suspected exposures. There was a killed whole-cell vaccine for use against bubonic plague but that vaccine was not very effective and has been discontinued. There are several research initiatives currently under way to develop a new plague vaccine.

Any identified cases of plague must be reported to the local public health agency. If an intentional release is suspected, local law enforcement and the Federal Bureau of Investigation (FBI) must also be notified. As the response is initiated, infection control for bubonic and septicemic forms of plague disease will consist of standard precautions. If pneumonic plague is identified, it requires patient isolation and droplet precautions until the patient has been on antibiotic therapy for at least 48 hours and shows signs of improvement.

Smallpox (Variola Major, Variola Minor)

Smallpox disease is the most destructive infectious disease in human history. It has claimed more human lives than any other pathogen. It is caused by variola viruses, which are members of the orthopoxvirus family and it can only survive in humans. Smallpox was eradicated in the 1970s through a World Health Organization vaccination campaign. Since being declared eradicated over three decades ago, it now only exists in the government research labs of Moscow, Russia, and at the CDC in Atlanta, Georgia. Unfortunately, there are suspicions that not all nations destroyed their lab supplies of the organisms. If a rogue nation still has some of the virus and supplies it to a terrorist organization, it could cause a devastating blow to international public health. For that reason, any single confirmed case today would be considered an international public health emergency.

There are two forms of this virus including variola major and variola minor. Variola major is the most likely form of the organism to be used as a weapon and is the causative agent for classic smallpox disease. It was the most common form of disease seen in historical epidemics and has a mortality of about 30%. Variola minor is a less common organism with a very low mortality rate of less than 1% (Fenner et al., 1988). The viruses are transmitted person to person through inhalation or direct contact with viral particles that are spread by infected individuals through their respiratory secretions or from pustules or scabs on their skin. Once an individual is exposed, an asymptomatic incubation period follows. It lasts for about 2 weeks. This incubation period is followed by a flu-like, prodromal illness for 2–5 days. Near the end of this prodromal period, the infected individual begins to develop pustules in the throat and mouth. They are then infectious, before any pustules appear on the skin. Once the rash begins on the skin, it has a characteristic, centrifugal pattern that appears mostly on the face, arms, and legs. It quickly becomes papular, pustular, and then scabs over. The scabs separate in 2–3 weeks. For those who do not survive, death often comes in the second week of this disease stage. If the patient has survived, the loss of the scabs is the end of their infectiousness. Survivors have a variety of complications, including severe scarring, encephalitis, secondary bacterial infections, conjunctivitis, and blindness.

During a confirmed smallpox outbreak, diagnosis should be made based on the clinical presentation of the patient alone. The greatest challenge will be the diagnosis of an index case. Most healthcare workers have never seen a smallpox case and many may miss the initial diagnosis. The most similar patient presentation is chickenpox. Smallpox lesions begin in the mouth and throat and then spread to the face and extremities. The lesions are concentrated centrifugally on the face, arms, and legs (See Figure 2–10). In addition, the lesions appear on the palms of the hands and the soles of the feet. Across the body, the lesions are uniform and in the same stage of development (vesicles, pustules, or scabs). In contrast, chickenpox concentrates centrally on the trunk, not centrifugally. The chickenpox lesions are not seen on the palms of the hands or the soles of the feet and are not uniform in appearance. Chickenpox lesions are usually in various stages of development with pustules erupting next to scabs.

Smallpox is contagious person to person and usually spreads through respiratory droplets or direct contact with infected persons or contaminated materials. Historically, each primary case will infect three or four other people. The exception to this is if infected individuals have a severe cough, they may infect dozens of other susceptible contacts (Wehrle et al., 1970). Those at highest risk of secondary infection include household contacts and healthcare providers. In a healthcare setting, airborne and contact precautions must be instituted. The patient should be in a negative pressure room and personnel and/or visitors need to wear proper respiratory protection, gown, and gloves.

FIGURE 2–10 A young boy in Bangladesh with the classic maculopapular rash evident on his torso, arms, and face. 1974. Photo Courtesy of CDC/Public Health Image Library PHIL ID#10660.

All suspected cases should be immediately isolated and observed. Confirmed cases should remain in hospital isolation until all the scabs separate. Home isolation may be the only option during a large outbreak. Over a century ago, there were separate hospitals designated for smallpox patients to keep them out of the other facilities. That approach may need to be adopted again in the event of a new epidemic. Anyone identified as a contact of a smallpox case should be vaccinated immediately and then monitored twice daily for 17 days for a fever. If a fever greater than 38 °C (>101 °F) develops, it could be the prodromal stage of smallpox and the patient should be isolated (Fenner et al., 1988).

Smallpox vaccine does not contain Variola virus. It contains an attenuated-strain of vaccinia virus, a close cousin of Variola. It is genetically similar enough to smallpox that a small vaccinia infection on the skin resulting from the vaccination will provide protection against smallpox as well. The original vaccine was made from calf lymph and is no longer produced. The decades old stockpile of this vaccine was being used by DOD and for homeland security initiatives for the past several years but in 2007, the Food and Drug Administration approved the next generation of smallpox vaccine (Food and Drug Administration, FDA approves new smallpox vaccine, www.fda.gov/consumer/updates/smallpox090407.html). It is very similar to the old vaccine in that it is still delivered with a

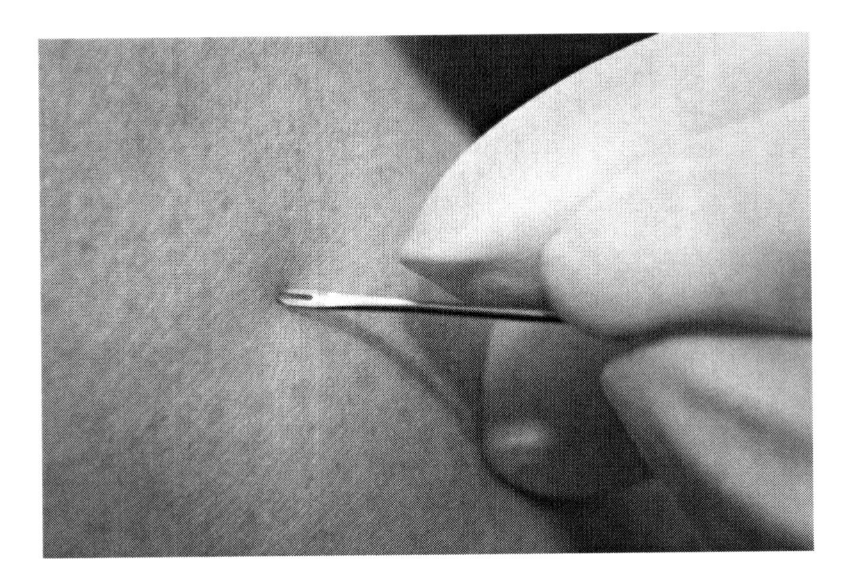

FIGURE 2–11 CDC clinician demonstrates the use of a bifurcated needle during the 2002 Smallpox Vaccinator Workshop. Photo by James Gathany. Courtesy of CDC/Public Health Image Library PHIL ID#2815.

bifurcated needle to initiate a vaccinia skin infection (Figure 2–11). It is called ACAM2000 and the primary difference is in the production process. It is made using cell cultures. Although it will have fewer side effects than the original vaccine, there are serious risks of complications with the vaccine and it should not be offered to the general public unless there is an imminent threat of a smallpox attack or outbreak.

Vaccination may also be used prophylactically to reduce morbidity and mortality. It has been proven effective if it is provided within 4 days of exposure (Dixon, 1962). An exposure is defined as suspected inhalation of viral particles from an initial bioterrorism release, face-to-face contact with a confirmed smallpox case, or direct contact with potentially contaminated materials or lab specimens. Other antiviral drugs, such as Cidofovir, may also improve patient outcomes if it is given within 1–2 days of an exposure (Huggins et al., 2001). It is impossible to say which prophylactic options may be most effective for humans in the absence of existing smallpox cases and exposures.

There are three smallpox disease variants. Each disease type is identified by the presentation of the rash. Classical smallpox is the most common and historically accounts for >90% of cases (Henderson et al., 1999). The other types of smallpox disease are malignant and hemorrhagic. They appear much different than typical smallpox and are difficult to diagnose. Malignant smallpox cases have a velvety, non-pustule rash. It was associated with high mortality. Hemorrhagic smallpox looks like a diffuse erythematous rash leading to petechiae and hemorrhages. It was often misdiagnosed as meningococcemia and is also usually fatal. One of the planning challenges we face in preparing for the possibility of a smallpox attack is the question of how the initial patients may present symptoms. As our assumptions about *variola major* infections are based on the natural presentation of smallpox disease, it may be easy to misdiagnose individuals infected from an aerosolized inhalation exposure that results in a large number of individuals with atypical presentations.

Fortunately, the variola viruses are not stable or persistent in the environment. If they were released as an aerosol, they would naturally be inactivated within about 24 hours. Because the organism dies out rapidly in the external environment, the buildings and areas in the path of an intentional release do not need decontamination and no long-term environmental recovery issues following a release (Henderson et al., 1999).

Tularemia (*Francisella tularensis*)

Tularemia is caused by *Francisella tularensis*, a very hardy and highly infectious aerobic organism. Exposure to as little as 10 organisms can result in the disease, also known as "rabbit fever" or "deer fly fever" (Saslaw et al., 1961). Naturally occurring disease is almost always rural and although it has occurred in every state in the U.S. except for Hawaii, it primarily occurs in the Midwest. It usually occurs as individual cases or as small, isolated outbreaks and has never been reported as a large epidemic. However, the infectiousness and availability of the organism makes it a potential weapon for terrorists. Any cases in urban or nonendemic areas, or among populations without clear risk factors, should be investigated as a possible bioterrorism incident.

Humans become infected by *Francisella tularensis* in a several ways. As with other Category A organisms, the route of exposure plays an important role in determining the form of tularemia disease. It is a very complex disease that presents as pneumonic, typhoidal, ulceroglandular, glandular, oculoglandular, or oropharyngeal. Natural infections usually occur from handling infected small mammals such as rabbits or rodents or through the bites of ticks, deerflies, or mosquitoes that are carrying the infection from feeding on infected animals. Most of these natural infections are ulceroglandular or glandular tularemia. If an infected animal is eaten without being thoroughly cooked, or if contaminated food or water is ingested, it may cause oropharyngeal tularemia. It is also possible to get the organism on the hands and inoculate the eye resulting in oculoglandular tularemia. The types of tularemia disease likely to result from an intentional aerosol release of *Francisella tularensis* are pneumonic and typhoidal tularemia. Morbidity and mortality vary greatly depending upon the type of tularemia disease. The most common forms of the disease, such as ulceroglandular tularemia (50–85% of all cases), have a low mortality of less than 2% (See Figure 2–12) (Dennis, 1998). However, primary pneumonic cases have a much higher mortality of 30–60% if untreated (Evans et al., 1985).

Francisella tularensis inhalation will usually cause flu-like symptoms after a 2–5 day incubation period. However, these symptoms can occur as soon as 1 day or as long as 14 days after exposure (Dennis et al., 2001). The flu-like clinical presentation may include the rapid onset of fever, chills, myalgia, weakness, and headaches. These symptoms sometimes worsen to include a nonproductive cough, sore throat, and inflammation of the bronchi, lungs, and lymph nodes. Diagnosis is complicated by the fact that there are no rapid diagnostics available for tularemia. Although there are promising initiatives under way to develop this capability, the initial identification of an intentional tularemia outbreak will depend upon an astute physician using a presumptive diagnosis based on the clinical presentation. This is particularly difficult for clinicians that have never seen a case. If a large number of otherwise healthy individuals present with these symptoms, particularly in nonrural regions with low incidence of naturally occurring tularemia, the index of suspicion should be high that it could be tularemia from an intentional release. If inhalational tularemia is suspected, respiratory secretions and blood samples may be used for confirmatory testing. When a case of pneumonic tularemia is suspected or confirmed, providers should assume all patients presenting with flu-like symptoms are positive and treat appropriately with antibiotics. The antimicrobial selection is based on the results of

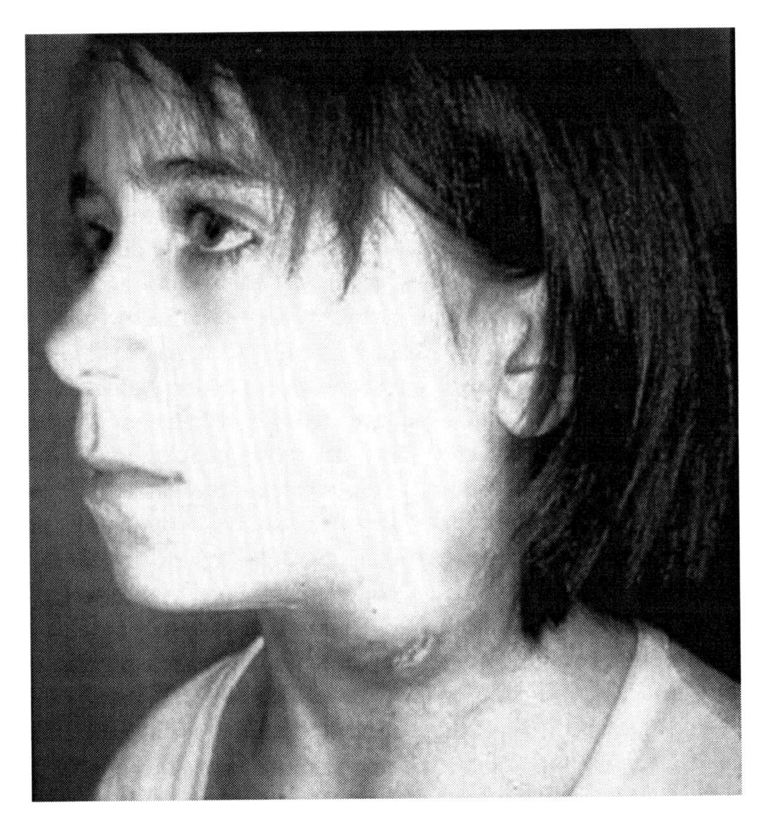

FIGURE 2–12 Girl with ulcerating lymphadenitis colli due to tularemia, Kosovo, April 2000.
Source: Centers for Disease Control and Prevention, http://www.cdc.gov/ncidod/eid/vol8no1/01-0131G1.htm.

susceptibility testing to be sure that there is no drug resistance, the age and condition of the patient, and the goal of the therapy. The goal may be treatment of suspected disease or prophylactic antimicrobial treatment for suspected exposures. A vaccine also exists for tularemia that uses an attenuated live vaccine strain (LVS) (Oyston and Quarry, 2005); however, it continues to be under review by the Food and Drug Administration. The strain used and the process of attenuating it have raised safety and efficacy questions that have slowed approval and it is not yet available to the general public.

Francisella tularensis can survive for long periods of time in a cool, moist environment. However, it is unlikely that it could survive for a long-period of time in an outdoor environment following an aerosol release. It is also not likely to be reaerosolized. Therefore, environmental decontamination following aerosol release of *Francisella tularensis* is not necessary. Individuals suspected of being contaminated by an aerosol should launder clothing and shower with soap and water. Even though *Francisella tularensis* is highly infectious for individuals directly exposed to the organisms, it is not spread from person to person.

Viral Hemorrhagic Fevers (Filoviruses, Arenaviruses, Bunyaviruses, and Flaviviruses)

The term "viral hemorrhagic fevers" (VHFs) describes a severe multisystem syndrome caused by one of four different viral families (Arenaviruses, Filoviruses, Bunyaviruses, and Flaviviruses). These are all RNA viruses that naturally reside in animal hosts and arthropod vectors. These hosts usually include rats and mice but for some VHF pathogens, such as Ebola and Marburg, it is still not certain where they naturally reside in nature. Transmission to humans results from infected hosts or vectors exposing humans through their body fluids and excretions. Some VHF viruses can spread from person to person through close contact with infected people or exposure to their body fluids. VHF illness begins as flu-like symptoms including fever, fatigue, dizziness, muscle aches, and weakness. The symptom become worse as vascular damage takes place. As the disease progresses, severe symptoms occur including hypotension, facial edema, pulmonary edema, and mucosal hemorrhages. Some patients will bleed from body orifices, internal organs, and even under the skin. Although the bleeding is sometimes severe, that is rarely the cause of death. VHF death is usually the result of multiorgan failure and cardiovascular collapse. The case fatality rates for VHFs have a wide range that differs by organism and by outbreak. For example, it was only 0.5% for an Omsk hemorrhagic fever but as high as 90% for a Zaire Ebola outbreak (Cunha, 2000; Muyembe-Tamfum et al., 1999). None of these organisms are airborne transmitted in natural outbreaks, but the concern is that they may be delivered as an aerosol during a bioterrorism attack. If a particularly infectious and lethal VHF agent is used for an aerosol attack, it could be devastating.

The RNA viruses that cause VHFs are dependent upon animal and insect hosts for their survival, and the naturally occurring forms of these diseases are limited to the area immediately surrounding their primary hosts. Humans are not normally part of their lifecycle but are sometimes inadvertently included by encountering an animal reservoir or infectious vector. Outbreaks are erratic and unpredictable but usually isolated.

VHF pathogens were placed on the Category A list of threats because of their potential lethality and infectiousness at low doses when delivered as an aerosol. Although there has never been an aerosolized VHF attack, animal studies have demonstrated the severe risk of infection posed by aerosolization of these pathogens (Johnson et al., 1995; Stephenson et al., 1984). If this scenario occurs, there will be an incubation period ranging from a couple days to 3 weeks depending on the VHF pathogen used, the individual dose received, and other variables. A prodromal illness of flu-like symptoms occurs for about 1 week. Severe infections may also display bleeding from orifices, under the skin, and in internal organs, eventually leading to multiorgan failure. Initial diagnosis is based on the clinical presentation and requires a high clinical index of suspicion (Borio et al., 2002). Treatment is limited mostly to supportive care. There are no antivirals specifically approved for VHFs and it is not known if any will be effective.

Those at greatest risk during historical outbreaks have been healthcare workers. To ensure their safety, infection control precautions must be strictly enforced with suspect VHF patients. This includes patient isolation in a negative pressure room, strict hand hygiene, and use of double gloves, impermeable gown, face shield, eye protection, and leg/shoe coverings (See Figure 2–13). In addition, HEPA respiratory protection should be worn by anyone entering the patients room (Borio et al., 2002). The risk to prehospital contacts may be somewhat lower because the patients will likely be in the earlier stages of disease and less infectious.

Prevention and Detection

Prevention of bioterrorism threats begins with solid regulations to prevent acquisition of pathogens through strict control of potential biothreat agents, rapid detection of

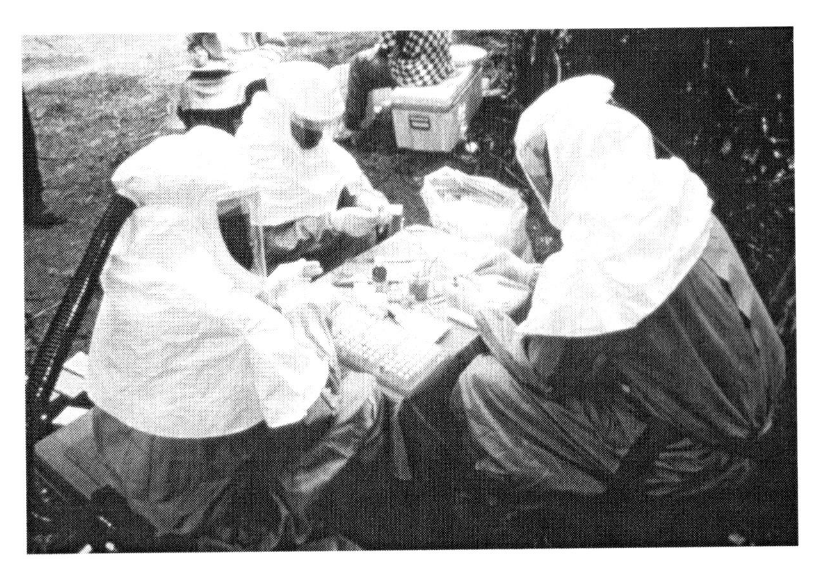

FIGURE 2–13 Scientists from CDC and Zaire prepare samples collected from animals near Kikwit, Zaire, during a 1995 Ebola outbreak. Photo Courtesy of CDC/Public Health Image Library PHIL ID#6136.

pathogenic releases, and protection of the susceptible populations through a strong preparedness infrastructure. In an effort to shore up these capabilities, the U.S. government passed a variety of preparedness promoting legislation beginning with the Public Health Security and Bioterrorism Preparedness and Response act of 2002 (Public Health Security and Bioterrorism Preparedness and Response Act of 2002, www.fda.gov/oc/bioterrorism/Bioact.html). This act was divided into five sections and was designed to address the most pressing bioterrorism threats identified following the 2001 anthrax letters.

Public Health Security and Bioterrorism Preparedness and Response Act of 2002, Five Sections

Title I: National Preparedness for Bioterrorism and Other Public Health Emergencies
Title II: Enhancing Controls on Dangerous Biological Agents and Toxins
Title III: Protecting Safety and Security of Food and Drug Supply
Title IV: Drinking Water Security and Safety
Title V: Additional Provisions

This legislation began a funding stream that strengthened overall public health preparedness for bioterrorism and a variety of other threats. Title one provided much needed resources to both the Department of Health and Human Services and the Centers for Disease Control and Prevention. It also accelerated work on the Strategic National Stockpile and provided funding to state and local health departments and local hospitals for enhanced preparedness initiatives. Title two created much needed controls over pathogenic agents and toxins, making them more difficult to acquire. This included overlapping regulatory authority by USDA and DHHS to control organisms that pose threats to plants, animals, and humans. This included a requirement for registration by anyone possessing hazardous biological material and it established strict controls for safety and security. Title three was established to protect food and drugs. It included enhanced inspection of imported foods, food safety and security, and upgrading of animal and plant health inspections. Title four increased security requirements for drinking water, and title five included several additional provisions primarily directed toward prescription drugs.

The laws instituted through the Public Health Security and Bioterrorism Preparedness and Response Act of 2002 to control pathogenic organisms were not only a result of the 2001 anthrax letters but in part a response to the exploits of microbiologist Larry Wayne Harris. He mail ordered and received pathogenic organisms. It was discovered later that he had ties to white supremacist groups and allegedly threatened to release anthrax in Las Vegas. Although the strain he purchased was a harmless strain of *Bacillus anthracis* spores used in vaccine development, it highlighted a vulnerability that needed to be addressed.

In 2005, another major piece of public health legislation was passed. The Pandemic Preparedness and Response Act was a response to growing public concerns over pandemic influenza. Although the aims of the legislation were good, at least part of the implementation was not. Although requiring state and local public health organizations to meet planning and preparedness goals for pandemics is an important step, and public health funding is always welcome, the pandemic dollars provided through this legislation came after a large cut to the basic public health preparedness budget previously established for bioterrorism. Even though the infrastructure required for pandemic response and bioterrorism are parallel, the cuts in basic preparedness funding meant losing trained staff who could meet the requirements of the pandemic preparedness funding. Unfortunately, the way the dollars were originally disbursed, some key staff needed to meet the pandemic preparedness objectives had already lost their jobs from the preparedness budget cuts. In other words, funding was given with one hand by federal health agencies to state and local public health agencies and taken with the other. To sustain adequate public health preparedness infrastructure, core funding needs to be sustained and not cut and then supplemented by additional funding for perceived emerging threats.

Cultivating the Perceptive Provider

Limiting the impact of a bioterrorism attack requires healthcare providers with sufficient training and support to remain diligent. Over an extended period of time, and in the absence of recent attacks, this diligence becomes more difficult to maintain. There are nonbioterrorism threats that are encountered regularly that draw the focus away from preparedness for unlikely events. Though this is understandable, it will not be excusable if a sentinel case is missed. The diversity of terrorist threats has resulted in substantial investments in public health and healthcare preparedness. The costs are well into the billions and are likely to continue for many years. Much of this spending takes the form of high-tech solutions including new epidemiological surveillance software or new detection and monitoring devices, such as the BioWatch Program. Although this technology may compliment other preparedness initiatives, there is not a technological solution that can replace an astute, perceptive healthcare provider. Throughout history, astute physicians and other well-informed providers have been the sentinels of our nation's public health. Typically, they identify emerging problems for public health, rather than public health identifying the problems in advance for providers (Thacker, 1994). This notion was confirmed in 2001 when clinicians and laboratory professionals who had recently received preparedness training, diagnosed the index case of inhalational anthrax (Bush et al., 2001). The preparedness education and training of healthcare providers will continue to be one of the most important measures any nation can invest in to mitigate the impact of a bioterrorism attack through rapid detection. We must consistently challenge and sharpen the intuition and skill of healthcare providers so that they are able to recognize a bioterrorism attack early, and take the appropriate actions.

Disease Surveillance

Disease surveillance is an essential and familiar function of local and state public health agencies. Regular disease surveillance includes mandatory disease reporting by local healthcare providers, data entry and analysis by local or regional public health agencies, and additional analysis, reporting, and allocation of needed resources by state and federal public health agencies. However, the bioterrorism agents of concern have short incubation periods that require rapid identification to facilitate early treatment to reduce morbidity and mortality. Most local disease surveillance activities do not process and analyze information quickly enough to ensure a prompt identification of a suspicious trend. They often lack the sensitivity needed to detect small changes in population health, and are not specific enough to distinguish normal fluctuations in illness from unusual or suspicious cases. There is no single solution to this gap in surveillance capabilities. An ideal system will include a network of systems that is able to simultaneously analyze multiple variables, effectively compare changes and trends, and control confounders and background noise (Pavlin et al., 2002). The realization of a complete, integrated, effective, uniform system throughout the United States or any other nation is many years away. In the United States, the largest system in development is called BioSense. It has a growing list of data sources including state and regional epidemiological surveillance systems, healthcare facilities, national labs, VA Hospitals, Department of Defense, and others. The plans are to eventually incorporate data from poison control center calls and-over-the counter drug sales (Centers for Disease Control and Prevention (CDC), BioSense program fact sheet, www.cdc.gov/biosense/files/fact_sheet.pdf). It is the hope of CDC officials overseeing the construction of this surveillance network that trends in a sufficient number of healthcare and public health data sets will be capable of detecting growing epidemics before current reporting systems are able to detect subtle clues. Until it is fully developed and implemented across the nation, we will continue to depend upon a patchwork of surveillance and detection systems.

Active surveillance for special events is currently the most exhaustive and effective epidemiological surveillance available. This comprehensive, symptomatic surveillance is accomplished at the Olympic Games, Superbowl, political conventions, and other high visibility events. This is a labor intensive and expensive endeavor requiring additional staff placement in key points of healthcare delivery to gather information. The information collected focuses on respiratory, neurological, and dermatological clues that may quickly indicate an intentional biological release has occurred. Although this may appear to be an ideal method to gather accurate information in a timely manner, it is far too expensive to sustain on a daily basis. Other active surveillance approaches include the monitoring of available information in public health and healthcare databases. This data may include ambulance runs, chief complaints of 911 calls, hospital admissions, school or work absenteeism, or trends in pharmaceutical or laboratory demands. The challenge of surveillance is twofold. First, the correct baseline is often difficult to determine. The success of surveillance depends upon the accurate determination of a trigger or threshold signaling an unusual event or trend. This is only successful if the numerous confounders in the data are controlled. For example, how can 911 calls or absenteeism trends distinguish between a suspicious event and the onset of "cold and flu season" or how can sales trends in certain pharmaceuticals distinguish between an increase in flu-like symptoms and price discounts involving those products? These are challenging problems but progress continues and important strides are being made to advance the surveillance process.

The trigger for activating a bioterrorism response is difficult to define. Bioterrorism encompasses such a wide array of pathogens and pathogenic by-products that it is impossible to define a "typical" bioterrorism event and establish appropriate triggers for response. In 1984, the Dalles, Oregon experienced a large outbreak of *Salmonella* gastroenteritis. More than 700 people became ill after eating at 10 local salad bars. The epidemiological investigation could not identify a single food as the source of the illnesses. Months later, the criminal investigation identified the origin of the outbreak as intentional contamination of area salad bars by members of a religious cult (Torok et al., 1997). In this case, the agent used was not lethal and initially appeared as a large, but accidental foodborne outbreak. A bioterrorism scenario will unfold in a very different way if a lethal agent is used. If the agent is transmissible from person to person or if it is delivered by spraying an aerosol, the "footprint" or focused area of an outbreak could expand across large regions. Clues that should raise the suspicion of providers include anything that is out of the ordinary. Every community has trends in seasonal illnesses with some variation from year to year. There may be obvious differences emerging, such as out of season "flu-like" outbreaks or more severe cases among groups that normally do not experience serious illness. An illness may be uncharacteristic for a particular region such as plague in New Jersey or the strain may be unusual or resistant to treatment. Clues that should raise suspicion may range from more subtle trends to the obvious clues that may be provided through claims made by aggressors or evidence identified by law enforcement authorities. Maintaining a high index of suspicion increases the likelihood of discovering an event sooner and the swift discovery of an attack translates into more lives saved. Examples of the epidemiological clues that may suggest an act of bioterrorism are listed in Table 2–3.

Table 2–3 Bioterrorism Epidemiological Clues

1) A large epidemic, with greater case loads than expected, especially in a discrete population.
2) More severe disease than expected for a given pathogen, as well as unusual routes of exposure.
3) Unusual disease for a given geographical area, found outside the normal transmission season, or impossible to transmit naturally in the absence of the normal vector for transmission.
4) Multiple, simultaneous epidemics of different diseases.
5) A zoonotic disease outbreak as well as a human outbreak, as many of the potential threat agents are pathogenic to animals.
6) Unusual strains or variants, antimicrobial resistance patterns disparate from those circulating.
7) Higher attack rates in those exposed in certain areas, such as inside a building if the agent was released indoors, or lower rates in those inside a sealed building if an aerosol was released outdoors.
8) Intelligence that an adversary has access to a particular agent or agents.
9) Claims by a terrorist of the release of a biological agent.
10) Direct evidence of the release of an agent, with findings of equipment, munitions, or tampering.

Source: Julie A. Pavlin, Walter Reed Army Institute of Research, Washington, D.C. (Pavlin, 1999).

Environmental Monitoring

Other forms of surveillance include environmental monitoring and standoff detection. Detector research and development continues to rapidly move forward. Biological monitoring includes a range of technologies from basic bioaerosol collectors to advanced laser particle analyzers. There are a variety of point source detection technologies in development, including nucleic acid, antibody, whole cell, and mass spectrometry analysis technologies. These detectors may enhance preparedness for specific high-risk events or buildings but they will always have limitations. The greatest limitation in the foreseeable future is cost. This emerging technology is expensive and is not affordable for most state and local agencies. These detectors are not sensitive enough to detect the full range of biological threats at a low level or specific enough to distinguish between normal background biological material in the air and the pathogens of greatest concern. These devices also must be prepositioned in areas most likely to experience an attack. This requires comprehensive and accurate risk assessments of each community. Once a risk assessment is accomplished, it must be continually monitored and updated according to community activities, events, and changes. At their best, detectors can only cover a limited area. This includes portable detectors that may be used for special events and more stationary detectors such as the BioWatch system. BioWatch is a surveillance program in dozens of major cities across the United States, where discretely placed air samplers continually draw air through a filter that is collected daily and tested for DNA associated with biothreat agents. Although this is an important tool, it cannot be successful without and will never replace well-trained physicians or epidemiological surveillance systems.

Immediate Actions

Once a biological threat has been detected, the biothreat agent and population at risk must be quickly identified. Because most of the severe illnesses resulting from an act of bioterrorism will have an incubation period, there is a window of opportunity for prophylactic treatment. This is true for even the most lethal agents. Decisions must be made rapidly and the response needs to begin immediately. Communication must be quickly established with the population at risk. If there are specific measures that may be taken to reduce their risk or the risk of others, it must be shared. For example, if pneumonic plague is the disease of concern, those at risk may be advised to isolate themselves from family members to avoid the risk of spreading the illness. At the same time, hospitals will be preparing for patient surges, whereas public health officials may be setting up Points of Dispensing or coordinating other strategies for mass prophylaxis. However, each response will be unique according to the agent, method of release, population at risk, environmental conditions, etc. There are far too many variables to concisely summarize the immediate actions to take. However, the initial message will consist of the preparedness mantra promoted by the Red Cross and most public preparedness organizations. People should have a communication plan in case difficulties arise in the local communication infrastructure due to the crisis. They should maintain a home preparedness kit in case an elective or mandatory quarantine is put in place to limit the spread of disease. Most importantly, they should stay informed and follow the instructions provided.

Just as the public is being informed of the appropriate measures to take, those working in healthcare, public health, and the first response community will be provided with detailed instructions on how to respond. For example, infection control recommendations and requirements will be shared (See Table 2-4). If standard precautions are suggested, that will include isolating blood and all other body fluids, secretions, nonintact skin, and mucous membranes of those who may be infected. Use gloves and gowns to prevent exposure to blood and other potentially infectious fluids. Use mask and eye protection or a face shield if there are activities that are likely to generate splashes or sprays of blood, body fluids, secretions, or excretions. Appropriate hand washing is always necessary. Other precautions may include:

- Droplet Precautions: Provide private patient rooms or cohort patients with the same infectious agent. Use a mask if within three feet of the patient.
- Contact Precautions: Provide private patient rooms or cohort patients with the same infectious agent. Use gloves when entering the room and a gown if clothing is likely to have contact with an infectious patient, or contaminated surfaces or equipment.
- Airborne Precautions: Requires a negative pressure isolation room and appropriate respiratory protection such as the N95 respirator which must be fit-tested.

Recovery

Following any bioterrorism event, there are likely to be strong debates over the need for decontamination and clean-up. With the notable exception of anthrax, clean-up will not be difficult for most pathogenic organisms. To ensure that an effective cleaning agent is used, the Environmental Protection Agency has established a listing of "antimicrobial products" under the Federal Insecticide, Fungicide, and Rodenticide Act (FIFRA). This is a much different antimicrobial than the antibiotic treatment of humans that often uses the same nomenclature. In the context of an environmental clean-up,

Table 2–4 Sample Patient Management Guidelines for Bioterrorism-Related Illnesses

IMPORTANT PHONE NUMBERS: **Infectious Diseases** __________ **Infection Control** __________ **ER** __________ **USAMRIID 301-619-2833,** **CDC Emergency Response Office 770-488-7100**	BACTERIAL AGENTS	Anthrax	Brucellosis	Cholera	Glanders	Bubonic Plague	Pneumonic Plague	Tularemia	Q Fever	VIRUSES	Smallpox	Encephalitis	Viral Encephalitis	Viral Hemorrhagic Fever	BIOLOGICAL TOXINS	Botulism	Ricin	T-2 Mycotoxins	Staph. Enterotoxin B
Isolation Precautions																			
Standard Precautions for all aspects of patient care		X	X	X	X	X	X	X	X		X	X	X	X		X	X	X	X
Contact Precautions (gown & gloves; wash hands after each pt encounter)				X***	X*	X*					X			X				X*	
Airborne Precautions (negative pressure room & N95 masks for all individuals entering the room)											X			X**					
Droplet Precautions (surgical mask)							X							X**					
Patient Placement																			
No restrictions		X	X	X	X			X	X			X	X			X	X	X	X
Cohort 'like' patients when private room unavailable				X***	X*	X	X				X			X				X*	
Private Room				X***	X*	X*	X				X			X				X*	
Negative Pressure											X			X**					
Door closed at all times											X			X**					
Patient Transport																			
No restrictions		X	X	X	X	X		X	X			X	X			X	X	X	X
Limit movement to essential medical purposes only				X***	X*	X*	X				X			X				X*	
Place mask on patient to minimize dispersal of droplets							X				X			X**					

Cleaning and Disinfection																
Routine cleaning of room with hospital approved disinfectant	X	X	X	X	X	X	X	X	X	X	X		X	X	X	X
Disinfect surfaces with 10% bleach solution or phenolic disinfectant												X				
Dedicated equipment (disinfect prior to leaving room)			X***	X*	X*				X			X			X*	
Linen management as with all other patients	X	X	X	X	X	X	X	X		X	X	X	X	X	X	X
Linens autoclaved before laundering or wash in hot water with bleach added									X							
Postmortem Care																
Follow principles of Standard Precautions	X	X	X	X	X	X	X	X	X	X	X	X	X	X	X	X
Droplet Precautions (surgical mask)						X										
Contact Precautions (gown & gloves)				X*	X*				X			X			X*	
Avoid autopsy or use Airborne Precautions & HEPA filter						X			X			X**				
Routine terminal cleaning of room with hospital approved disinfectant	X	X	X	X	X	X	X	X	X	X	X		X	X	X	X
Disinfect surfaces with 10% bleach solution or phenolic disinfectant												X				
Minimal handling of body; seal body in leaf-proof material												X				
Cremate body whenever possible									X							
Discontinuation of Isolation																
48 hrs of appropriate antibiotic and clinical improvement						X										
Until all scabs separate									X							
Until skin decontamination completed (1 hr contact time)															X	
Duration of illness			X***	X*	X*							X				

STANDARD PRECAUTIONS—Standard Precautions prevent direct contact with all body fluids (including blood), secretions, excretions, non-intact skin (including rashes), and mucous membranes. Standard Precautions routinely practiced by healthcare providers include: splash/spray and gowns to protect skin and clothing during procedures.

*Contact Precautions needed only if the patient has skin involvement (bubonic plague: draining bubo) or until decontamination of skin is complete (T-2 Mycotoxins).

**A surgical mask and eye protection should be worn if you come within 3 feet of pt. Airborne Precautions are needed if patient has cough, vomiting, diarrhea, or hemorrhage.

*** Contact Precautions needed only if the patient is diapered or incontinent.

Source: Saint Louis University, Center for the Study of Bioterrorism and Emerging Infections, October 2001.

Adapted with permission from an original table designed by LTC Suzanne E. Johnson, RN, MSN, CIC, Walter Reed Army Medical Center.

antimicrobials are substances with chemical or physical properties that destroy or repress the growth of pathogenic organisms. Thousands of products are registered with the U.S. Environmental Protection Agency (EPA) for various applications. Over half of the products registered are specifically for controlling harmful organisms in healthcare settings. These antimicrobial products fall into two broad categories including those that disinfect, sanitize, or limit the growth of organisms and those that protect various objects and surfaces from contamination. There is a regulatory distinction between products that are used directly on food, those that are used on or in a living human body, and those products that are used to control microbes in the environment. The Food and Drug Administration regulate those used directly on food or people. The EPA regulates the substances used for microbial control in the environment. They will have the final word on which antimicrobials will be used under different circumstances on different surfaces.

Summary

Bioterrorism is a particularly challenging public health preparedness issue. It requires a robust public health infrastructure to detect and manage an incident, a healthcare system with astute providers and well managed surge capacity, and a variety of disciplines prepared to follow and enforce public health recommendations. Bioterrorism preparedness requires hard work and a solid infrastructure consisting of a system of systems. These preparedness systems will all have weaknesses and the only way to identify and solve those flaws is to conduct thorough training and conduct regular multiagency exercises of local plans. The list of potential pathogens that could be used by terrorists is long. Although there will be unique aspects to any scenario, the establishment of a solid infrastructure that is well practiced in some of the more challenging Category A scenarios will provide the foundation for what is needed.

Websites

American Society for Microbiology: www.asm.org/.

Association for Professionals in Infection Control and Epidemiology: www.apic.org/.

Association of Public Health Laboratories, Emerg-ency Preparedness and Response: www.aphl.org/aphlprograms/ep/Pages/default.aspx.

Center for Biosecurity, University of Pittsburgh Medical Center: www.upmc-biosecurity.org/.

Center for Infectious Disease Research and Policy, University of Minnesota: www.cidrap.umn.edu/cidrap/.

Center for the Study of Bioterrorism, Saint Louis University: bioterrorism.slu.edu/.

Centers for Disease Control and Prevention, Emergency Preparedness and Response Program, Bioterrorism: www.bt.cdc.gov/bioterrorism/.

The Food and Drug Administration — Drug Preparedness and Response to Bioterrorism: www.fda.gov/cder/drugprepare/.

Infectious Diseases Society of America, Bioterrorism Resource Page: www.idsociety.org/BT/ToC.htm.

U.S. Army Medical Research Institute of Infectious Diseases: www.usamriid.army.mil/.

References

Arnon, S. S., Schechter, R., Inglesby, T. V., et al. (2001). Botulinum toxin as a biological weapon: medical and public health management. *JAMA* 285:1059–1070.

Arnon, S. S., Schechter, R., Maslanka, S. E., Jewell, N. P., & Hatheway, C. L. (2006). Human botulism immune globulin for the treatment of infant botulism. *N Engl J Med* 354:462–471.

Association of State and Territorial Health Officials. (2008). *States of preparedness: health agency progress*. 2nd ed. Washington, DC: Association of State and Territorial Health Officials. www.astho.org/pubs/StatesofPreparedness2008fin.pdf.

Becker, E., & Toner, R. (2001). A nation challenged: the victims; Postal workers' illness set off no alarms. *The New York Times*. October 21, 2001. http://query.nytimes.com/gst/fullpage.html?res=9C0DEED71431F937A15753C1A9679C8B63.

Bibel, D. J., & Chen, T. H. (1976). Diagnosis of plague: an analysis of the Yersin-Kitasato controversy. *Bacteriol Rev* 40:633–651.

Borio, L., Inglesby, T., Peters, C. J., et al. (2002). Hemorrhagic fever viruses as biological weapons: Medical and public health management. *JAMA* 287:2391–2405.

Bush, L. M., Abrams, B. H., Beall, A., et al. (2001). Index case of fatal inhalational anthrax due to bioterrorism in the United States. *N Engl J Med* 345:1607–1610.

Centers for Disease Control and Prevention (CDC). (1997). Fatal human plague—Arizona and Colorado, 1996. *Morb Mortal Wkly Rep* 46:617–620. www.cdc.gov/mmwr/preview/mmwrhtml/00048352.htm.

Centers for Disease Control and Prevention (CDC). (1998). *Botulism in the United States 1899–1996: handbook for epidemiologists, clinicians, and laboratory workers*. Atlanta, GA: CDC. www.cdc.gov/ncidod/dbmd/diseaseinfo/botulism.pdf.

Centers for Disease Control and Prevention. (2001). Considerations for distinguishing influenza-like illness from -inhalational anthrax. *MMWR* 50(44):985–987. www.cdc.gov/mmwR/PDF/wk/mm5044.pdf.

Centers for Disease Control and Prevention. (2002). Notice to readers: use of anthrax vaccine in response to terrorism: supplemental recommendations of the Advisory Committee on Immunization Practices. *MMWR* 51(45):1024–1026. www.cdc.gov/mmwr/preview/mmwrhtml/mm5145a4.htm.

CNN.com. (2001). Postal worker suspected he had anthrax. November 8, 2001. http://archives.cnn.com/2001/HEALTH/conditions/11/07/911.anthrax/index.html.

Cunha, B. A. (2000). *Tickborne infectious diseases: diagnosis and management*. New York, NY: Marcel Dekker.

Dennis, D. T. (1998). Tularemia. In: Wallace, R. B., ed. *Maxcy-Rosenau last public health and preventive medicine*. 14th ed. Stamford, Conn: Appleton & Lange; pp. 354–357.

Dennis, D. T., Inglesby, T. V., Henderson, D. A., et al. (2001). Tularemia as a biological weapon: medical and public health management. *JAMA* 285:2763–2773.

Dixon, C. W. (1962). *Smallpox*. London: J & A Churchill Ltd; p. 1460.

Emergent Biosolutions. (2002). Anthrax Vaccine-Adsorbed (BIOTHRAX™). Manufacturer's prescribing information. January 31. www.bioport.com/pdf/emergent_biothrax_us.pdf.

Evans, M. E., Gregory, D. W., Schaffner, W., & McGee, Z. A. (1985). Tularemia: a 30-year experience with 88 cases. *Medicine (Baltimore)* 64:251–269.

Fenner, F., Henderson, D. A., Arita, I., et al. (1988). *Smallpox and its eradication*. Geneva, Switzerland: World Health Organization; pp. 1–276. www.who.int/emc/diseases/smallpox/Smallpoxeradication.html.

Gill, D. M. (1982). Bacterial toxins: a table of lethal amounts. *Microbiol Rev* 46:86–94.

Henderson, D. A., Inglesby, T. V., Bartlett, J. G., et al. (1999). Smallpox as a biological weapon: medical and public health management. *JAMA* 281:1735–1745.

Huggins, J. W., Bray, M., Smee, D. F., et al. (2001). Potential antiviral therapeutics for smallpox, monkeypox, and other orthopox virus infections. Presented at The WHO Advisory Committee on Variola Virus Research, December 3–6, Geneva, Switzerland.

Inglesby, T., Dennis, D., Henderson, D., et al. (2000). Plague as a biological weapon: medical and public health management. *JAMA* 283:2281–2290.

Inglesby, T., O'Toole, T., Henderson, D., et al. (2002). Anthrax as a biological weapon, 2002: updated recommendations for management. *JAMA* 287:2236–2252.

Jernigan, J. A., Stephens, D. S., Ashford, D. A., et al. (2001). Bioterrorism-related inhalational anthrax: the first 10 cases reported in the United States. *Emerg Infect Dis* 7:933–944. www.cdc.gov/ncidod/EID/vol7no6/pdf/jernigan.pdf.

Johnson, E., Jaax, N., White, J., et al. (1995). Lethal experimental infections of rhesus monkeys by aerosolized Ebola virus. *Int J Exp Pathol* 76(4):227–236.

Lew, D. (1995). Bacillus anthracis. In: Mandell, G. L., Bennett, J. E., & Dolin, R., eds. *Principles and practice of infectious disease*. New York, NY: Churchill Livingstone Inc.; pp. 1885–1889.

Lichtblau, E., & Wade, N. (2008). F.B.I. details anthrax case, but doubts remain. *The New York Times*. August 18, 2008. www.nytimes.com/2008/08/19/us/19anthrax.html#.

Mina, B., Dym, J. P., Kuepper, F., et al. (2002). Fatal inhalational anthrax with unknown source of exposure in a 61-year-old woman in New York City. *JAMA* 287:858–862.

Morris, T. (2001). Tom Morris' last call. *TIME*. November 19, 2001. www.time.com/time/magazine/article/0,9171,1001251,00.html.

Muyembe-Tamfum, J. J., Kipasa, M., Kiyungu, C., & Colebunders, R. (1999). Ebola outbreak in Kikwit, Democratic Republic of the Congo: discovery and control measures. *J Infect Dis* 179 (suppl. 1):S259–S262.

Oyston, P. C. F., & Quarry, J. E. (2005). Tularemia vaccine: past, present and future. *Antonie van Leeuwenhoek* 87:277–281.

Pavlin, J. A. (1999). Epidemiology of bioterrorism. *Emerg Infec Dis* 5:528–530.

Pavlin, J. A., Kelley, P., Mostashari, F., et al. (2002). Innovative surveillance methods for monitoring dangerous pathogens. In: Knobler, S. L., Mahmoud, A. F., Pray, L. A., eds. *Biological threats and terrorism: assessing the science and response capabilities*. Washington, DC: National Academy Press; pp. 185–191.

Regaldo, A., Fields, G., & Schoofs, M. (2002). FBI makes military labs key focus on anthrax. *Wall Street Journal*. Retrieved Februrary 12, A4.

Rotz, L. D., Khan, A. S., & Lillibridge, S. R. (2002). Public health assessment of potential biological terrorism agents. *Emerg Infec Dis* 8:225–230.

Saslaw, S., Eigelsbach, H. T., Prior, J. A., et al. (1961). Tularemia vaccine study. I. Intracutaneous challenge. *Arch Intern Med* 107:121–133.

Shapiro, R. L., Hatheway, C., Swerdlow, D. L. (1998). Botulism in the United States: a clinical and epidemio-logic review. *Ann Intern Med* 129(3):221–228.

Shepard, C. W., Soriano-Gabarro, M., Zell, E. R., et al. (2002). Antimicrobial postexposure prophylaxis for anthrax: adverse events and adherence. *Emerg Infect Dis* 8:1124–1132. www.cdc.gov/ncidod/EID/vol8no10/02-0349.htm.

St. Louis, M. E., Peck, S. H., Bowering, D., et al. (1988). Botulism from chopped garlic: delayed recognition of a major outbreak. *Ann Intern Med* 108:363–368.

Stephenson, E. H., Larson, E. W., & Dominik, J. W. (1984). Effect of environmental factors on aerosol-induced Lassa virus infection. *J Med Virol* 14:295–303.

Tackett, C. O., Shandera, W. X., Mann, J. M., et al. (1984). Equine antitoxin use and other factors that predict outcome in type A foodborne botulism. *Am J Med* 76:794–798.

Thacker, S. B. (1994). Historical development. In: Teutsch, S. M., & Churchill, R. E., eds. *Principles and practice of public health surveillance*. New York, NY: Oxford University Press; pp. 8–9.

Torok, T. J., Tauxe, R. V., Wise, R. P., et al. (1997). A large community outbreak of salmonellosis caused by intentional contamination of restaurant salad bars. *JAMA* 278:389–395.

U.S. General Accounting Office. (2007). Actions needed to avoid repeating past problems with procuring new anthrax vaccine and managing the stock-pile of licensed vaccine. www.gao.gov/new.items/d0888.pdf.

U.S. General Accounting Office. (2008). Combating bioterrorism: actions needed to improve security at Plum Island-Animal Disease Center. Retreived July 29. www.gao.gov/atext/d03847.txt.,.

Wehrle, P. F., Posch, J., Richter, K. H., et al. (1970). An airborne outbreak of smallpox in a German hospital and its significance with respect to other recent outbreaks in Europe. *Bull World Health Organ* 43:669–679.

Weiss, R., & Snyder, D. (2002). 2nd leak of anthrax found at army lab. *Washington Post*. Retrieved April 24, B1.

Williamson, E. (2003). Ft. Detrick unearths hazardous surprises. *Washington Post*. Retrieved May 27, B1.

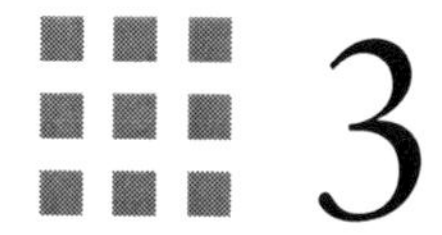

Bombings and Explosions

Objectives of This Chapter

- Describe recent trends in bombings around the world.
- List the various categories of blast injuries.
- Recognize the types of conditions that are considered primary blast injuries.
- Distinguish the differences between secondary and tertiary injuries.
- Provide examples of quaternary blast injuries.
- Describe several important healthcare facility lessons from tragedies like the 1995 Oklahoma City Bombing.
- Explain the difference between physical security and bomb incident plans.
- List three things that can reduce the morbidity and mortality impact of a car bomb.
- Describe how a threat may be identified and what an appropriate response of a targeted facility looks like.
- List several important factors and key differences of triaging victims of an explosion.

Introduction: 2005 London Terrorist Bombings

Anat was born and raised in Israel. After attending high school in Jerusalem and completing her service obligation with the Israeli army, she studied modern dance. This grew into a passion for the arts that led her to make a move to London where she immersed herself in the arts community and worked as an administrator with a children's charity. She was a petite 39-year-old. Her friends describe her as intelligent, warm, and witty. According to her boyfriend, she was hoping to return to Israel to visit her parents but was afraid to return because of the increasing number of suicide attacks. In a terrible ironic twist, her life was cut short by a suicide bomber on the streets of London near a popular statue of Ghandi at Tavistock Square. She was on the Number 30 bus talking with her boyfriend on the phone when a suicide bomber detonated a bomb in a backpack killing 13 (See Figure 3–1) (BBC News, 2005). This was just one of four suicide bomb attacks that day in London. Fifty-two people died in the four attacks and 770 were injured (Special reports, Depth: London attacks, 7 July bombings, *BBC News*, http://news.bbc.co.uk/2/shared/spl/hi/uk/05/london_blasts/what_happened/html/default.stm).

The use of bombs by terrorist organizations has increased dramatically in recent years. Though the number of bombing incidents only rose by about 4% between 2006 and 2007, the death toll increased by 30% and the number of suicide bombings increased by approximately 50% (National Counterterrorism Center, 2008). Although this dramatic rise in bombing morbidity and mortality is primarily attributed to the conflict in Iraq, there is a clear upward trend in bombings internationally, even when Iraqi attacks are excluded.

FIGURE 3–1 Ambulances at Russell Square, London after the July 7th bombings. Photo by FrancisTyers. http://en.wikinews.org/wiki/Image:Russell_square_ambulances.jpg.

There is also a growing desire to maximize the number of fatalities. This has not always been the aim of terrorist bombers. For many years, the Irish Republican Army detonated bombs throughout the United Kingdom to draw attention to their cause. Their bombs were crude and not very effective. In fact, the terrorists often accidentally detonated the bombs, killing themselves before they reached their objective (Rose, 1990). Civilian casualties from IRA bombings were usually low because their methods were low tech and their primary objective was media attention with the public and political pressure that it may generate. However, today's bombers often want to generate as many civilian casualties as possible and are acquiring the technologies and methods to reach that objective (See Table 3-2). As the materials and methods for bomb-making continue to proliferate, the likelihood of future incidents will continue to increase. This includes the possibility of suicide bombings in nations without a history of such events. Recent history supports the suggestion that these types of attacks are likely. There is a continual international growth in terrorism blast injuries. In fact, in the last decade, there has been a 10-fold increase in terrorist bombings making preparedness for this threat imperative (Fry and Berg, 2003).

Explosion Injury-Related Definitions

Barotraumas: Damage that results from a dramatic pressure change released from an explosion. These are injuries to the air-filled organs, such as the middle ear, sinuses, lungs, or digestive system.

Primary blast injury: Damage that occurs to the "hollow" organs of the body from the rapid overpressure generated by a blast wave. The most susceptible areas include the ears, lungs, and abdomen.

Quaternary blast injury: Any other damage that falls outside the first three categories including burns, subsequent respiratory problems from dust, or exposure to hazardous materials included in or released by the blast.

Secondary blast injury: Damage that results from the flying debris of an explosion including primary bomb fragments and secondarily propelled debris. Injuries include penetrating and blunt trauma.

Tertiary blast injury: Damage resulting from being thrown by the blast winds. These are primarily blunt trauma injuries but can also include penetrating injuries if a victim is propelled against something sharp.

Health Threat

Blast injuries are typically categorized into four categories. These include: primary, secondary, tertiary, and miscellaneous injuries (See Table 3-1). These broad categories have been established to aid in the understanding of the complex assortment of injuries associated with explosions and provide structure for the triage process.

Primary blast injuries result from the rapid overpressure or shock wave produced by an explosion (Benzinger, 1950). These injuries result from the dramatic changes in barometric pressure projected from the point of a detonation. Severe damage to the air-filled organs of the body result from this rapid pressure change and these injuries are referred to as barotraumas. Damage to air-filled organs includes the middle ear (otic barotraumas), the sinuses (sinus barotraumas), the lungs (pulmonary barotraumas), and the digestive system (gastrointestinal barotraumas). The most sensitive and most frequently injured hollow organ is the eardrum or tympanic membrane. A change in atmospheric pressure of no more than five pounds per square inch (psi) can rupture the eardrum while it usually requires nearly 10 times that amount (56–76 psi) of overpressure to damage other hollow organs in the body. However, there have been cases of bomb victims experiencing lung damage even when their eardrums are intact (Katz et al., 1989). Depending on the overpressure generated by a blast, the injury can extend into the middle and inner ear resulting in permanent hearing loss. The second most sensitive hollow organ is the lung, and blast damage to the lungs is the most common cause of life-threatening injury following an explosion (Leibovici et al., 1999). The change in pressure can result in lung contusions and rupture of alveolar capillaries. This can cause hemorrhage, pneumothorax (collapsed lung), hemothorax (internal bleeding in the chest), pneumomediastinum (air leaking from the lungs into the mediastinum), and subcutaneous emphysema (air leaking from the lungs under the skin of the chest) (Coppel, 1976; de Candole, 1967). There is little that can be done to protect against primary blast injuries. Even bomb technicians using the most advanced body armor cannot be protected from barotraumas during a blast. Although armor will offer some protection against projectile injuries, it cannot prevent primary trauma resulting from overpressure. (Mellor and Cooper, 1989). Other hollow organs affected by barotrauma include the large and small intestines and the eyes. They can be damaged or even ruptured from the pressure changes. In addition, the brain can experience a concussion or an acute air embolism (air bubbles in the blood) from sudden pressure changes. These injuries can lead to central nervous system damage and traumatic brain injury requiring years of rehabilitation.

Secondary blast injuries consist of the penetrating wounds caused by fragments flying from the blast. There are two categories of fragments. Primary fragments are those that are built into a weapon. For example, suicide bombers often pack nails or ball bearings around explosive charges to increase the lethality of the explosion. Secondary fragments are the other debris generated by an explosion, such as flying glass. In most mass casualty generating explosions, secondary blast injuries are responsible for the

FIGURE 3–2 The moment of a blast. A Japanese bomb hitting the flight deck of the U.S.S. Enterprise, August, 1942. This distinctive photo was taken by Robert Frederick Read, USN. Although the film in his camera was saved, he lost his life taking it.

Source: U.S. National Archives. http://www.archives.gov/exhibits/a_people_at_war/images/thumb_robert_fredric_read.jpg.

majority of morbidity and mortality. A conventional military explosive typically propels fragments up to 1500 minutes per second (over 3300 miles per hour) (See Figure 3–2) (Bellamy and Zajtchuk, 1991). These fragments reach much further and injure more people than the primary blast wave.

In attacks on large buildings with many windows, flying glass is a major cause of injury over a wide area. For example, the 1998 U.S. Embassy bombing in Nairobi sent flying glass up to 2 kilometer (over 1.2 miles) from the scene of the explosion and injured people even at that distance (Wightman and Gladish, 2001). These types of injuries can also be challenging for first responders to assess. A seemingly small wound may be an entry wound that is much more serious than it may appear (Nixon and Stewart, 2004).

The term most commonly used to describe primary fragments from a weapon is "shrapnel." This name comes from the inventor of the fragmentation bomb. Henry Shrapnel was a British Army Officer and inventor who created the fragmentation shell in the early 1800s.

Tertiary blast injuries are wounds sustained when an individual is thrown by the blast winds (Bellamy and Zajtchuk, 1991; Coppel, 1976; de Candole, 1967). A variety of injuries can result from rapid acceleration and deceleration including fractures and blunt traumatic injuries. Depending on the size of the explosion, the blast wave is often capable of rolling a car or collapsing a building.

Quaternary blast injuries are all the remaining injuries that may be observed among victims that are not due to the primary, secondary, or tertiary injuries. These injuries further complicate the triage process for healthcare providers with the introduction of patients with crush syndrome and compartment syndrome. Quaternary injuries include burns, crush injuries, breathing problems from dust, smoke, or chemical releases, and any other injury that falls outside the other categories.

Table 3–1 Types of Blast Injuries

Blast Injury Type	Mechanism of Injury	Health Impact
Primary	Rapid, crushing overpressure	Damage to "hollow" organs such as the ears, eyes, lungs, gastrointestinal tract
Secondary	Flying debris	Penetrating and blunt trauma injuries
Tertiary	Blast wind	Fracture and blunt trauma
Quaternary	Any complicating factor not in the other three categories	Burns, crush injuries, respiratory problems

FIGURE 3–3 Oklahoma City Bombing, Alfred P. Murrah Federal Building during recovery operations. Photo by SSGT Mark A. Moore, 4/21/1995.

Source: Department of Defense, DF-ST-96-00587.

An explosion is an extremely complex process with many variables that contribute to the impact on human morbidity and mortality. The type of delivery device used, amount and type of explosive material, distance from the explosion, barriers between the explosion and the victims, and whether it occurs indoors, outside, or in the water, are just a few of the major factors that make the outcome of an explosion difficult to predict. The remarkable energy released can rapidly dissipate in an open field while if it is released indoors or in an urban setting, it can accelerate down a hallway or alley. Although the blast injury presentation is consistent, the severity of injury is dramatically enhanced by indoor or enclosed detonation (Stein and Hirshberg, 1999).

1995 Oklahoma City Bombing

It started off as a typical day in Oklahoma City as people filed into the Alfred P. Murrah Federal Building. Some were going to work. Others were dropping off children at day care, stopping by the Federal Employees Credit Union, or taking care of personal business in one of the many federal offices. The faces included Dr. Charles Hurlburt and his wife Jean. She was a nurse and he was a retired professor from the University of Oklahoma. Working in his office, was Micky Maroney, a Secret Service Agent and former defensive end for the 1964 national champion University of Arkansas Razorbacks. From Tinker Air Force Base, Airman First Class Cartney McRaven had just dropped by the Social Security Office to report her upcoming name change. She was getting married in four days. Playing with over a dozen little friends at the America's Kids Child Development Center on the second floor was little Baylee Almon who just celebrated her first birthday the day before (Those who were killed, Oklahoma City National Memorial & Museum [Web page], www.oklahomacitynationalmemorial.org/secondary.php?section=1&catid=24). The faces at the federal building that day were a tapestry of our society. They were all different ages, races, and faiths. In an instant, those mentioned above were gone. A truck bomb ripped away the front of the building claiming the lives of 168 people and injuring over 800. Shortly after the dust began to settle, a Pulitzer Prize winning photo was captured by a loan specialist and amateur photographer, Charles Porter. It was Oklahoma City Fire Captain Chris Fields carrying the tiny broken body of a child from the scene. The child was Baylee Almon (Kurt, 2005). It was the first major terrorist attack on U.S. soil and was perpetrated by anti government, domestic terrorist, Timothy McVeigh who was put to death for the attack in the summer of 2001.

The Oklahoma City Bombing (Figure 3–3) was the impetus for many preparedness activities that are sustained today. The health and medical lessons learned form the basis for much of what is done in healthcare facilities across the nation to enhance preparedness for future events. Perhaps the most concise summary of the lessons learned are from the Oklahoma Hospital Association's Sheryl McLain (McLain, 1995):

- Lesson One: Never underestimate the importance of having disaster preparedness plans in place which are practiced regularly via mock disaster drills.
- Lesson Two: No amount of planning or disaster drills could have thoroughly prepared us for the thunderous blast which shattered us—physically and emotionally—on April 19, 1995.

- Lesson Three: Our medical community was prepared to handle mass casualties. Immediately following the blast, thousands of medical and nonmedical volunteers converged upon the downtown area. Within 3 minutes, seven staffed ambulances were in route to the scene; within 60 minutes, 66 ambulances were staffed and operational. A total of 47 ambulance services and 103 units were involved—many from outlying areas. More than 400 doctors and nurses rushed to the hospital located closest to the scene; therefore, there were two physicians caring for every one of the 202 patients taken there. Most of the medical response took place within the first 2 hours after the blast.
- Lesson Four: It is critical to involve phone companies in disaster-preparedness planning. Within the first 2 hours following the bombing, 12 million calls were attempted within the metropolitan area; the majority of the 1800 "911" (emergency) calls attempted during the first hour alone received busy signals. Our local phone company donated all resources to the rescue and recovery effort; 1500 phone lines were set up in and around the perimeter of the site. Long distance services were also donated so that rescue workers could call home. Wireless phone companies donated hundreds of phones and batteries. Three cells on wheels were brought in to add cellular capacity. One wireless phone company donated 1052 phones and contributed $4 million worth of equipment, time, and resources. When forming a community-wide disaster preparedness committee, do not overlook leaders from telephone or cellular companies.
- Lesson Five: It is imperative to frequently test communications systems, including backup systems. At approximately 10:30 a.m., all Oklahoma City hospital emergency rooms lost phone and radio contact with the local ambulance service. A backup radio system was in place in some, but not all, of the emergency rooms. Because of the bombing, each hospital's backup radio system has been evaluated and, if necessary, upgraded or replaced. Daily roll calls now occur between the ambulance service's central dispatch office and each Oklahoma City hospital.
- Lesson Six: Building relationships with other disaster-response agencies will lessen the confusion during a disaster. With phone lines overloaded, communication between agencies was a tremendous problem. Most of the pleas for medical personnel, supplies, and other needs, were disseminated by the news media to the masses, instead of agency to agency. Some of the needs were justified, but many were not. This resulted in a lot of confusion, poor distribution of medical personnel, and wasted supplies.
- Lesson Seven: It is important to have the ability to share basic patient information with and among hospitals, as soon as possible after a community-wide disaster. Because victims were treated in 18 hospitals, a centralized tracking system would have been helpful to hospitals, and to those traveling from hospital to hospital searching for their loved ones.
- Lesson Eight: Working with the international media for an extended length of time can be very challenging. Local media coverage at the bomb site and outside hospital emergency rooms began almost immediately. This live coverage continued nonstop—with no commercial interruptions—for

110 hours. However, the international media spotlight would shine on Oklahoma City for several weeks.

- Public relations staffing will be necessary 24 hours a day during the first several days following a disaster. Consider seeking help from outside volunteers, including experienced public relations professionals from other hospitals or your hospital's system.
- A public relations/media room should be set up to accommodate a large number of people, cameras, and other media equipment. This area should be easily accessible to the main entrance of the hospital, but away from the emergency room and patient care areas. Clearly marked directions to the media room are a must, as is an alert security staff. Parking lot signs directing the media and family members to designated parking areas are an additional help.
- To avoid duplicating tiresome patient interviews, consider pool interviews. A separate room for interviews with family members should also be set up, away from the patient's room.
- Do not forget the importance of your local news media when the national and international media come into your community. Make every effort to ensure local media representatives are treated as fairly and equitably as the national media. Decide, in advance, whether or not television tabloid or talk show hosts will be treated as equitably as "news" reporters.

- Lesson Nine: During a disaster, people need to do something to help. Because of the number of requests the Oklahoma Hospital Association received from healthcare workers across the United States, "Hospitals Helping the Heartland," disaster relief fund was established as a means

Table 3–2 Morbidity and Mortality of Selected Terrorist Bombings (1970–2000)

Date	Target/Location	Deaths	Injuries
1998	US Embassy, Nairobi, Kenya (U.S. Department of State, 1998)	213	>4000
1983	US Marine Barracks, Beirut, Lebanon (Frykberg et al., 1989)	241	105
1995	Murrah Federal Bldg, Oklahoma City, USA (Mallonee et al., 1996)	167	648
1969–72	Belfast, Northern Ireland (Hadden et al., 1978)	117	1532
1980	Bologna, Italy (Brismar and Bergenwald, 1982)	73	218
1975–79	Jerusalem, Israel (Adler et al., 1983)	26	340
1985–86	Paris, France (Rignault and Deligny, 1989)	20	248
1996	Khobar Towers, Dhahran, Saudi Arabia (Thach et al., 2000)	19	500
2000	Destroyer USS Cole, Aden Harbor, Yemen (Lambert et al., 2003)	17	39
1998	US Embassy, Tanzania (New York Times, 1998)	11	86
1988	Jerusalem, Israel (Katz et al.,1989)	6	52
1993	World Trade Center, New York, USA (Quenonmoen et al., 1996)	6	548
1972–80	Northern Ireland (Pyper and Graham, 1983)	>5	339
1996	Centennial Park, Atlanta, USA (Anderson and Feliciano, 1997)	2	111
1974	Birmingham, England (Waterworth and Carr, 1975)	2	80
1991	Victoria Station, London, England (Johnstone et al., 1993)	1	50
1996	Manchester, England (Carley and Mackway Jones, 1997)	0	208
1973	Old Bailey, London, England (Caro and Irving, 1973)	0	160
1974	Tower of London, England (Tucker and Lettin, 1975)	0	37

through which hospital workers *could* assist bombing victims in a meaningful way.

- Lesson Ten: It takes a lot of resources, and time, for a community to heal emotionally. Ours was one of 14 provider groups to endorse the publication and distribution of an informational booklet designed to assist individuals with emotional healing following the disaster. A centralized mental health entity—Project Heartland—was established shortly after the bombing by the Oklahoma State Department of Mental Health, with initial funding from the Federal Emergency Management Association, and later by the Department of Justice. Its goal was to coordinate the bombing victims' diverse mental healthcare needs.
- Of all the lessons we learned, the most important ones were those which reminded us of how fragile and precious life is, and the value of our loved ones; and those which showed the nation and the world that we are all much stronger when we are unified and work together toward a common goal.

Prevention

As with any health threat, prevention of explosion morbidity and mortality begins with planning. The planning for bomb and explosion incidents includes two distinct parts, including physical security planning and bomb incident planning. Physical security planning may be complete for high-risk buildings but often missing specific bomb threat considerations. Security planning needs to be regularly reviewed and updated with bombing scenarios in mind. The bomb incident plan provides a detailed description of the procedures that are followed when a threat is detected. This includes raising building occupant awareness of threat identification and orienting them on appropriate individual response. Both the physical security and bomb incident plans are distinct but interdependent. Physical security measures can be bypassed if threats are not quickly recognized by those at risk. Once a threat is recognized, an appropriate response is essential to limiting potential morbidity and mortality, particularly due to secondary devices that sometimes follow an initial explosion. The only way to ensure appropriate response is through coordinated planning, training, sustained awareness, and exercises (Figure 3–4). This holds true for critical facilities that must have procedures in place to take appropriate initial action and for first responders and those in healthcare that must manage the public health impact of an event. Adequate planning, training, and exercises will prevent avoidable injuries from a number of explosion scenarios.

Physical Security

It is impossible to fully protect any facility from a determined attacker, particularly a suicide bomber. However, the impact of an attack can be vastly reduced through enhanced physical security. Begin by simply seeking out and eliminating locations where improvised explosive devices (IEDs) could be hidden. One common location is a trash can near a building entrance. Trash cans near the road are also popular places for roadside bombs in the Iraq conflict. Other popular places for hiding bombs are vending machines, mailboxes, phone booths, and heavy underbrush or weeds. Any place a small package could be discreetly concealed should be considered a possible threat when developing a physical

FIGURE 3–4 Victims of a simulated fuel explosion walk around in a daze during a collaborative multi-agency medical exercise in Bethesda, MD, Dec. 7, 2006. (U.S. Air Force photo by Tech. Sgt. Cohen Young) (Released) 061201-F-3798Y-138.

security plan. It is also important to limit parking spaces in close proximity to the building and/or increase security for parking that is located within or adjacent to the building. Potential explosive device carriers such as cars and trucks should be kept at a safe distance. Establishing this distance between buildings and vehicle access moves the potential blast wave far enough away that the exponential overpressure decrease of a blast in open air is used as a buffer to protect the facility (Figure 3–5). If adequate distance cannot be established, barriers can be erected to deflect the blast wave away from a targeted facility. Solid walls of concrete, steel, and composite materials have been used. There are also lower tech solutions such as mounds of sand, dirt, or large containers of water. In addition, there are window laminates and treatments that can reduce the likelihood of secondary blast injuries due to flying glass. Many nations have invested tremendous resources in "hardening" facilities using these materials and work continues in developing the blast mitigation materials of the future that will be used as barriers or incorporated into the design of newer high-risk facilities (Figure 3–6). Additional security measures include a procedure for inspecting incoming visitors and mail; effective key control; keeping exits passable; and ensuring sufficient lighting inside, outside, and for emergencies. All these security measures should be tested periodically to identify and correct gaps. It is also helpful to invite local law enforcement in to review your plans and provide suggestions on ways to reduce vulnerability and increase the effectiveness of a response.

Threat Detection and Identification

There has been a proliferation of new explosives detection technology in recent years. The growing threat of terrorism has also raised the awareness of how to recognize a variety of related risks, including how to identify and report suspicious persons and packages. A threat can be detected in a variety of ways. In airports, X-ray machines are used to screen luggage and dogs are often used to identify explosive materials (Figure 3–7). Some

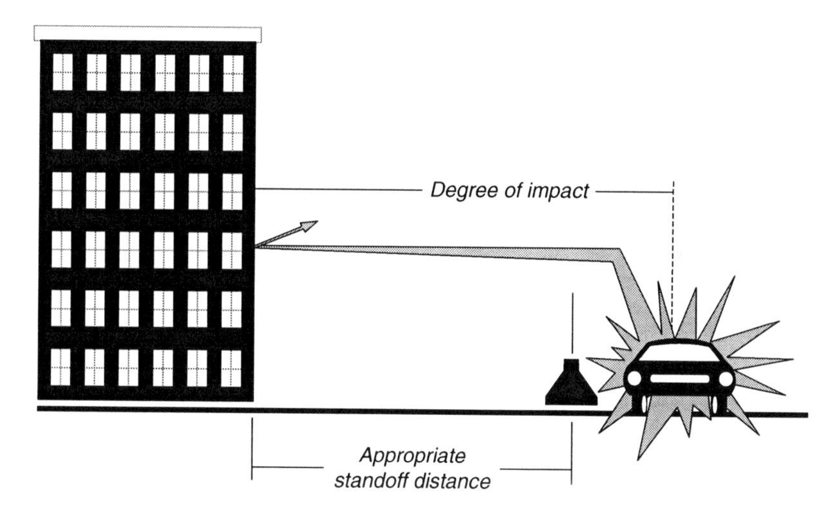

FIGURE 3–5 Stand-off distance.

Source: U.S. Department of Transportation. http://transit-safety.volpe.dot.gov/security/SecurityInitiatives/DesignConsiderations/CD/images/fig6_1.jpg.

FIGURE 3–6 A contractor installs a blast-resistant window frame in the outer facade of the Pentagon's E-ring.

Source: U.S. Department of Energy. http://www1.eere.energy.gov/femp/news/news_detail.html?news_id=7341.

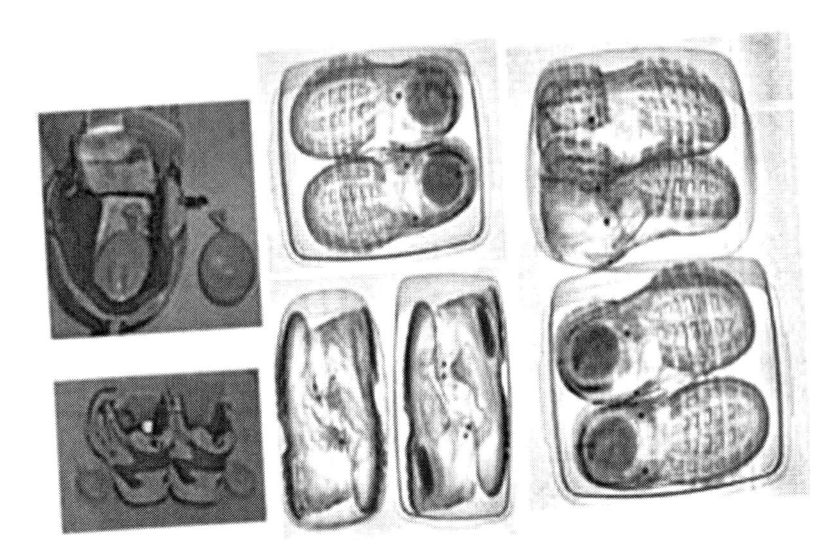

FIGURE 3–7 On left, images of shoes with hidden explosives. On right, the X-ray images of the shoe with explosives and the shoe without explosives. The dark spots on the X-rays indicate the explosives. http://www.tsa.gov/graphics/images/happenings_x-ray_shoes.jpg.

luggage is checked with trace explosives-detection devices (TEDDS) as well. As the threat has grown, this technology has accelerated to include electronic nose technology capable of literally sniffing out explosives using an assortment of sensors that respond to different odors. Each odor has a complex fingerprint or "smell print" that can be identified by this technology (Pearce et al., 2002). There are also new spray-on polymers that change color under ultraviolet light detecting specific types of explosive materials (Barras, 2008). These are among the many innovations emerging to identify threats.

Individual awareness is also essential but when training building occupants in recognizing threats, it is important to emphasize their limitations. They should be trained to recognize things that are out of place but should leave the final determination of the threat up to trained responders. A threat assessment should be accomplished with the help of local law enforcement to help make the decision on the appropriate course of action. Sometimes a threat is unannounced in writing or with a phone call. If a written threat is received, all materials, including the envelope should be secured and turned over to law enforcement. Depending on the overall objective of a bomber, the threat may be phoned in. This is sometimes done to reduce the number of fatalities. When a call is received, the individual taking the call should:

- Attempt to keep the caller on the line as long as possible.
- Get all the information possible.
- Ask for details on where the bomb is located and how it can be activated.
- Remind the caller that there are innocent people that could be hurt.
- Listen for background noises such as traffic or music that could indicate the location of the caller.
- Note the caller's voice (male/female, accents, and impediments) and tone (calm, angry, or excited).

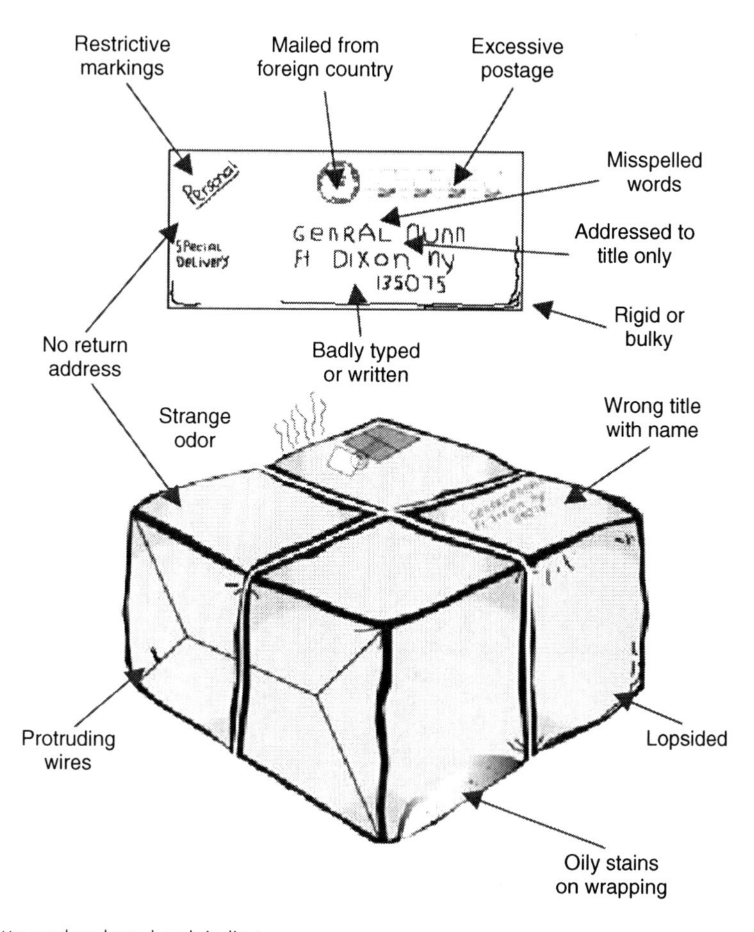

FIGURE 3–8 Letter and package bomb indicators.
Source: Los Alamos National Laboratory. http://www.lanl.gov/news/albums/security/Mail_Pkg_BombInd.jpg.

- Ask the caller to repeat as much as possible.
- Never hang up before the caller.
- Immediately contact law enforcement.

Recognizing Suspicious Packages

The use of a letter or a package to deliver an explosive device is a threat that should be well understood especially by those handling incoming mail and deliveries. There are many clues that should raise suspicion and demand action. If a letter or package is received that has one or more of the conditions displayed in Figure 3–8, procedures should be in place and quickly followed to isolate and report it. If things seem out of order with misspellings, a return address that does not match the postmark, restrictive markings such as "Confidential," "Personal," "Opened Only by Addressee," the package should not be handled and local authorities should be contacted immediately.

Table 3–3 Bomb Threat Evacuation Distances

Threat		Explosives Capacity[1] TNT Equivalent (in Pounds and Kilograms)	Building Evacuation Distance[2] (in Feet, Miles, Meters, and Kilometers)	Outdoor Evacuation Distance[3] (in Feet, Miles, Meters, and Kilometers)
	Pipe bomb	5 lbs, 2.3 kg	70 feet, 21 m	850 feet, 0.16 miles, 259 m, 0.26 km
	Briefcase, suitcase, or backpack bomb	50 lbs, 23 kg	150 feet, 46 m	1,850 feet, 0.35 miles, 564 m, 0.56 km
	Compact sedan	500 lbs, 227 kg	320 feet, 0.06 miles, 98 m	1,500 feet, 0.28 miles, 457 m, 0.46 km
	Sedan	1,000 lbs, 454 kg	400 feet, 0.08 miles, 122 m, 0.12 km	1,750 feet, 0.33 miles, 533 m, 0.53 km
	Passenger, cargo van	4,000 lbs, 1,814 kg	600 feet, 0.11 miles, 183 m, 0.18 km	2,750 feet, 0.52 miles, 838 m, 0.84 km
	Small moving van, delivery truck	10,000 lbs, 4,536 kg	860 feet, 0.16 miles, 262 m, 0.26 km	3,750 feet, 0.71 miles, 1,143 m, 1.14 km
	Moving van, water truck	30,000 lbs, 13,608 kg	1,240 feet, 0.24 miles, 378 m, 0.38 km	6,500 feet, 1.23 miles, 1,981 m, 1.98 km
	Semitrailer	60,000 lbs, 27,216 kg	1,500 feet, 0.28 miles, 457 m, 0.46 km	7,000 feet, 1.33 miles, 2,134 m, 2.1 km

1. These are estimates based on the estimated maximum quantity or weight of explosives (or the corresponding amount of TNT) that could realistically be placed in a device or vehicle.
2. The recommended safe distance from an outdoor explosive device for those in nearby buildings is difficult to accurately determine due to the number of variables, including building materials, type of structure, etc. These estimates are based on typical U.S. commercial construction materials. There are still risks for any individuals inside buildings that are within the Outdoor Evacuation Distance. Terrorist sometimes attempt to attract curious onlookers with a smaller explosion that draws them closer to the scene or to their windows and then follow it with a larger one to increase casualties.
3. The safe outdoor distance is estimated based upon how far fragments are likely to be thrown. Pipe, briefcase, and backpack bombs are often designed to throw fragments that may travel further, relative to the volume of explosive, than a vehicle bomb.

Source: U.S. National Counterterrorism Center, Bomb Threat Standoff Distances. http://www.nctc.gov/site/images/technical/bomb_threat.pdf.

Immediate Actions

If a threat necessitates an evacuation from a building, occupants should quickly collect personal items such as purses and briefcases so they are not among the possible threats needing to be assessed by bomb technicians. As with a building fire, elevators should not be used and there should be a rallying point away from the building for a head count. If a specific threat has been identified, it is imperative that everyone at risk be evacuated immediately to a safe distance. Refer to Table 3–3 for distances based upon the type of explosive threat. Everyone inside the "building evacuation distance" must evacuate immediately (both inside and outside buildings). Those who are outside the "building evacuation distance" but inside the "outdoor evacuation distance" should seek shelter immediately inside a building away from windows and exterior walls. There should not be anyone allowed outside unless they are beyond the "outdoor evacuation distance."

Health and Medical Response

It is important for all the first responders to be aware of risks when approaching a potential bomb scene. In addition to the threat posed by a suspect device, secondary devices are a common way to strike at responders. Even if an explosion has already occurred, it is important to consider the possibility of residual explosive material in the immediate area of the blast or an intentional device intended to be detonated when the response is underway (Suprun, 2004). A triage area is established upwind from the immediate blast area to minimize the possible exposure to inhalation hazards resulting from a blast. If it is a suspect terrorist explosion, these hazards include chemical, biological, or radiological substances introduced with the explosive device to complicate the response and inflict greater injury. There is also an assortment of hazardous building substances that

FIGURE 3–9 In South Baghdad, an explosion goes off from a second car bomb aimed at U.S. and Iraqi forces arriving to inspect the first car bomb detonated an hour earlier during Operation Iraqi Freedom. Photo by SPC Ronald Shaw Jr., USA, 4/14/2005, 050414-A-3240S-025.

can be released from an explosion. This includes particles such as asbestos, lead, silica, fiberglass, and a variety of chemical by-products from burning material.

According to the Centers for Disease Control and Prevention, there are several essential concepts that caregivers need to keep in mind concerning provision of care to those injured in explosions (Centers for Disease Control and Prevention, 2003):

- There will be unique injury patterns that are usually only seen in combat situations.
- The most common injuries will be penetrating and blunt trauma.
- The most common lethal injury among those that initially survive is "blast lung."
- Explosions in confined spaces of buildings or transportation systems have higher morbidity and mortality.
- Structural collapse following an explosion also results in higher morbidity and mortality.
- Half of all casualties will seek care in the first hour.
- The number of casualties seen in the first hour is a useful predictor of how many more to come.
- Expect an "upside-down" triage. The less injured will often arrive at the hospital before the more critical.

Local public health agencies also have an important role in long-term monitoring and follow-up of survivors. If any hazardous materials are released by an explosion, including chemical, biological, or radiological material, a public health agency may establish a registry to track the long-term mental and physical health effects associated with the event. An excellent example of a large health registry associated with a disaster is the World Trade Center Health Registry. It is monitoring the physical and emotional health impact of the World Trade Center collapse for the next 20 years by periodically collecting and analyzing information on over 71,000 enrollees (Brackbill et al., 2006). This provides a mechanism to communicate with the affected population and answer important questions about their health. It is also used to inform those enrolled about new treatments that may become available and facilitate research that provides healthcare providers with current information on how to identify and treat associated conditions.

Summary

Over the past several years, bombings have become more sophisticated and more lethal. If recent trends continue, it is likely that bombings will occur in nations that have rarely seen them. This creates tremendous challenges for the general public and response communities alike. High-risk buildings must consider the physical security measures in place and determine if they adequately address the threat of a bomb being mailed, carried, or driven into the building. Measures may be taken to enhance the ability to screen incoming persons and mail for suspicious materials and distance must be established between vehicle access and buildings. Everyone at risk needs to understand the proper approach to an identified threat and be able to carry out proper identification, reporting, and evacuation. Those working in the first responder and healthcare communities also must be aware of the differences between a bomb scene and other disasters. The risks are unusual, the patients are different, and the steps taken to properly manage the scene are distinctive. To reduce morbidity and mortality associated with these threats, it requires careful planning, training, and exercises.

Websites

California Emergency Survival Program, Bomb Threats: www.sdcounty.ca.gov/oes/docs/ESP_2008_Jul.pdf.

Centers for Disease Control and Prevention, Blast Injuries: Fact Sheets for Professionals: www.bt.cdc.gov/masscasualties/blastinjuryfacts.asp.

Centers for Disease Control and Prevention, Bombings: Injury Patterns and Care, Blast Injury Training: www.bt.cdc.gov/masscasualties/bombings_injurycare.asp.

Centers for Disease Control and Prevention, Explosions and Blast Injuries: A Primer for Clinicians: www.bt.cdc.gov/masscasualties/explosions.asp.

Centers for Disease Control and Prevention, In a Moment's Notice: Surge Capacity in Terrorist Bombings: www.bt.cdc.gov/masscasualties/pdf/surgecapacity.pdf.

Centers for Disease Control and Prevention, Predicting Casualty Severity and Hospital Capacity: www.bt.cdc.gov/masscasualties/capacity.asp.

Department of Education, Emergency Planning Resources: www.ed.gov/admins/lead/safety/-emergencyplan/index.html.

Department of Justice, Bureau of Alcohol, Tobacco, Firearms and Explosives (ATF), Arson and Explosives Publications: www.atf.treas.gov/pub/index.htm#arson.

Department of Justice, Federal Bureau of Investigation (FBI), Explosives Unit: www.fbi.gov/hq/lab/html/eu1.htm.

Department of Justice, Office of Justice Programs, A Guide for Explosion and Bombing Scene Investigation, June 2000: www.ncjrs.gov/pdffiles1/nij/181869.pdf.

Department of Transportation, Transit Security Design Considerations: http://transit-safety.volpe.dot.gov/security/SecurityInitiatives/DesignConsiderations/CD/ftasesc.pdf.

Department of Veterans Affairs Medical Center, Bomb Threat Standard Operating Procedure: www1.va.gov/emshg/apps/emp/7_2_sop_all/bomb_threat_sop.htm.

Federal Emergency Management Agency, Telephone Bomb Threats: www.fema.gov/hazard/terrorism/exp/exp_threat.shtm.

Kansas Emergency Management, Suspicious Package(s)/Envelope(s) and Airborne Hazards First Responders Guide, Version 2, December 10, 2001: www.kansas.gov/kdem/pdf/news/2001_suspicious_package_guideline.pdf.

National Counterterrorism Center, Bomb Threat Call Procedures: www.nctc.gov/site/images/technical/bomb_threat_call_procedures.pdf.

National Counterterrorism Center, Bomb Threat Standoff Distances: www.nctc.gov/site/images/technical/bomb_threat.pdf.

WebMD, e-medicine, Blast Injuries: www.emedicine.com/emerg/topic63.htm.

References

Adler, J., Golan, E., Golan, J., Yitzhaki, M., & Ben-Hur, N. (1983). Terrorist bombing experience during 1975–1979. Casualties admitted to the Shaare Zedek Medical Center. *Isr J Med Sci* 19:189–193.

Anderson, G. V., Jr., & Feliciano, D. V. (1997). The Centennial Olympic Park bombing: Grady's response. *J Med Assoc Ga* 86:42–46.

Barras, C. (2008). Glowing spray lets CSI operatives 'dust' for explosives. *NewScientist.com*. June 3, 2008. http://technology.newscientist.com/channel/tech/dn14048-glowing-spray-lets-csi-operatives-dust-for-explosives.html.

BBC News. (2005). Obituary: Anat Rosenberg. *BBC News*. August 3, 2005. http://news.bbc.co.uk/1/hi/england/london/4738127.stm.

Bellamy, R. F., & Zajtchuk, R., eds. (1991). *Conventional warfare: ballistic, blast, and burn injuries*. Washington, DC: Office of the Surgeon General of the U.S. Army; pp. 107–162.

Benzinger, T. (1950). Physiological effects of blast in air and water. In: *German aviation medicine World War II*. Vol 2. Washington, DC: U.S. Government Printing Office; pp. 1225–1259.

Brackbill, R. M., Thorpe, L. E., & DiGrande, L., et al. (2006). Surveillance for World Trade Center disaster health effects among survivors of collapsed and damaged buildings. *MMWR Surveill Summ* 55(2):1–18. www.cdc.gov/mmwr/preview/mmwrhtml/ss5502a1.htm.

Brismar, B., & Bergenwald, L. (1982). The terrorist bomb explosion in Bologna, Italy, 1980: an analysis of the effects and -injuries sustained. *J Trauma* 22:216–220.

Carley, S. D., & Mackway-Jones, K. (1997). The casualty profile from the Manchester bombing 1996: a proposal for the -construction and dissemination of casualty profiles from major incidents. *J Accid Emerg Med* 14:76–80.

Caro, D., & Irving, M. (1973). *The Old Bailey bomb explosion*. *Lancet* 1:1433–1435.

Centers for Disease Control and Prevention. (2003). Explosions and blast injuries: a primer for clinicians. May 9, 2003. www.emergency.cdc.gov/masscasualties/pdf/explosions-blast-injuries.pdf.

Coppel, D. L. (1976). Blast injuries of the lungs. *Br J Surg* 63:735–737.

de Candole, C. A. (1967). Blast injury. *Can Med Assoc J* 96:207–214.

Fry, K., & Berg, E. R. (2003). Medical management of disasters and mass casualties from terrorist bombings: how can we cope? *J Trauma* 53:201–212.

Frykberg, E. R., Tepas, J. J., III, & Alexander, R. H. (1989). The 1983 Beirut airport terrorist bombing: injury patterns and implications for disaster management. *Am Surg* 55:134–141.

Hadden, W. A., Rutherford, W. H., & Merrett, J. D. (1978). The injuries of terrorist bombing: a study of 1532 consecutive patients. *Br J Surg* 65:525–531.

Johnstone, D. J., Evans, S. C., Field, R. E., & Booth, S. J. (1993). The Victoria bomb: a report from the Westminster Hospital. *Injury* 24:5–9.

Katz, E., Ofek, B., Adler, J., Abramowitz, H. B., & Krausz, M. M. (1989). Primary blast injury after a bomb explosion in a civilian bus. *Ann Surg* 209:484–488.

Kurt, K. (2005). A lasting imprint: photo defined Oklahoma City tragedy for the world, but mother, photographer still search for meaning. *The Standard Times*. April 17, 2005. http://archive.southcoasttoday.com/daily/04-05/04-17-05/b01pe269.htm.

Lambert, E. W., Simpson, R. B., Marzouk, A., & Unger, D. V. (2003). Orthopaedic injuries among survivors of USS COLE attack. *J Orthop Trauma* 17(6):436–441. http://cat.inist.fr/?aModele=afficheN&cpsidt=14940796.

Leibovici, D., Gofrit, O. N., & Shapira, S. C. (1999). Eardrum perforation in explosion survivors: is it a marker of pulmonary blast injury? *Ann Emerg Med* 34:168–172.

Mallonee, S., Shariat, S., Stennies, G., Waxweiler, R., Hogan, & D., Jordan, F. (1996). Physical injuries and fatalities resulting from the Oklahoma City bombing. *JAMA* 276:382–387.

McLain, S. R. (1995) *The Oklahoma City bombing: lessons learned by hospitals*. Oklahoma: Oklahoma Hospital Association.

Mellor, S. G., & Cooper, G. J. (1989). Analysis of 828 servicemen killed or injured by explosion in Northern Ireland 1970–84: the hostile action casualty system. *Br J Surg* 76:1006–1010.

National Counterterrorism Center. (2008). 2007 Report on terrorism. April 30, p. 10. http://wits.nctc.gov/-reports/crot2007nctcannexfinal.pdf.

New York Times. (1998). New U.S. disclosures on Tanzania bombing. October 17, 1998. http://query.nytimes.com/gst/fullpage.html?res=9A03E7DC1E3AF934A25753C1A96E958260.

Nixon, R. G., & Stewart, C. (2004). When things go boom: blast injuries. *Fire Eng* 2. www.fireengineering.com/articles/article_display.html?id=204602.

Pearce, T. C., Schiffman, S. S., Nagle, H. T., & Gardner, J. W., eds. *Handbook of machine olfaction: electronic nose technology*. Weinheim: Wiley-VCH; 2002.

Pyper, P. C., & Graham, W. J. H. (1983). Analysis of terrorist injuries treated at Craigavon Area Hospital, Northern Ireland, 1972–1980. *Injury* 14:332–338.

Quenonmoen, L. E., Davis, Y. M., Malilay, J., Sinks, T., Noji, E. K., & Klitzman, S. (1996). The World Trade Center bombing: injury -prevention strategies for high-rise building fires. *Disasters* 20:125–132.

Rignault, D. P., & Deligny, M. C. (1989). The 1986 terrorist bombing experience in Paris. *Ann Surg* 209:368–373.

Rose, D. (1990). Devices reveal IRA know-how. *The Guardian (London)*. May 18, 1990.

Stein, M., & Hirshberg, A. (1999). Medical consequences of terrorism: the conventional weapons threat. *Surg Clin North Am* 79(6):1537–1552.

Suprun, S. C. (2004). Explosive events: EMS response to a bombing incident. *Emerg Med Serv* 33(4):61–65.

Thach, A. B., Ward, T. P., Hollifield, R. D., Cockerham, K., Birdsong, R., & Kramer, K. K. (2000). Eye injuries in a terrorist bombing: Dhahran, Saudi Arabia, June 25, 1996. *Ophthalmology* 107:844–847.

Tucker, K., & Lettin, A. (1975). The tower of London bomb explosion. *Br Med J* 3:287–290.

U.S. Department of State. (1998). Report of the accountability review boards: bombings of the US embassies in Nairobi, Kenya and Dar es Salaam, Tanzania, on August 7, 1998. n.d. www.state.gov/www/regions/africa/board_nairobi.html.

Waterworth, T. A., & Carr, M. J. T. (1975). Report on injuries sustained by patients treated at the Birmingham General Hospital following the recent bomb explosions. *Br Med J* 2:25–27.

Wightman, J. M., & Gladish, S. L. (2001). Explosions and blast injuries. *Ann Emerg Med* 37:664–678.

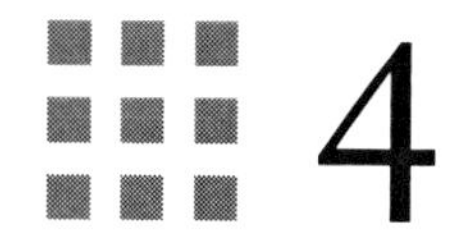

Chemical Disasters

Objectives of This Chapter

- Explain the dose–response relationship and how chemicals can enter the body.
- List the physical, physiological, and environmental factors that influence an exposure.
- Recognize the various systems used to classify and label chemical threats.
- Explain the difference in values of occupational and accidental chemical exposure guidelines.
- Describe the similarities and differences between chemical accidents and chemical terrorism.
- List the unique treatment requirements for chemical terrorism agents.
- Describe how a community can prepare for and reduce the impact of chemical emergencies.
- List the indicators that a chemical release may have occurred.
- Explain what an individual should do if they are notified of a chemical release in the community.
- Describe which populations are most vulnerable to chemical exposures and why.

Introduction (Case Study: Train Accident and Chlorine Gas Leak, Graniteville, South Carolina, 2005)

In the early morning hours of January 6, 2005, Graniteville, South Carolina experienced a catastrophic train derailment and chemical release. The night before, a local train engineer parked his locomotive and two train cars on a rail spur just off the main rail line that runs through town. Early the next morning a Norfolk Southern train with two locomotives and 42 rail cars departed Augusta, Georgia, bound for Columbia, South Carolina. Among the rail cars in tow were several tankers of pressurized cars filled with over 200 tons of liquid chlorine. As the train entered the small town of Graniteville, South Carolina, something went horribly wrong. A rail switch was locked in the wrong position directing oncoming trains off the main line and on to the spur where the stationary train sat from the night before. Realizing what was happening, the engineer applied the emergency brake but it was too late. The train cars began to derail, rupturing the tanks of chlorine and releasing a cloud of deadly green gas across the community (See Figure 4–1) (Hart and Wald, 2005; National Transportation and Safety Board [NTSB], Collision of Norfolk Southern freight train 192 with standing Norfolk Southern local train P22 with subsequent hazardous materials release at Graniteville, SC, Railroad accident report NTSB/RAR-05/04 (PB2005-916304)). The train engineer died at the scene and eight more people died from the initial release. Most of the fatalities were night shift workers at nearby Avondale Mill. Although some made it to their cars and sped away and others stayed inside the building, some tried to outrun the cloud on foot. A couple of them were found dead in a nearby wooded area. An additional 529 people sought care for possible exposures. Of those, 511 were seen in area emergency departments and 69 of them were hospitalized (Centers for Disease Control [CDC], 2005b). The tragic chlorine release in Graniteville accentuates the widespread vulnerability we all have to industrial chemical disasters. Though some in Graniteville died at or near the mill, the remaining fatalities included a truck driver sleeping in his cab, a man who drove through the cloud returning home from a convenience store, and even a nearby resident who never left his home. Thousands were evacuated and dozens were seen in local hospitals. Many continue to have chronic respiratory problems from this incident.

What Went Right with the Graniteville Accident Response and Recovery?

Officials in South Carolina quickly declared a state of emergency and evacuated several thousand residents. Local businesses and schools were closed and coordination with state and federal responders brought resources into the area to better manage the scene. A coordinated community assessment was carried out including an environmental assessment with the assistance of the Environmental Protection Agency and a rapid epidemiological assessment with the assistance of the Centers for Disease Control and Prevention. Good communication was established with the community and resources were developed ranging from fact sheets about reentering the home after the evacuation to tips on taking care of pets that were left behind (CDC, 2005b; U.S. Environmental Protection Agency, 2005).

Chlorine is a widely used industrial chemical. Applications include everything from water purification to making plastics. However, it has also been used as a chemical warfare agent. It was used on the battlefield in World War I and more recently has been

FIGURE 4–1 A collision of two freight trains in Graniteville, South Carolina on January 6, 2005, resulted in a release of chlorine gas to the atmosphere.

Source: U.S. Environmental Protection Agency.

used as a weapon by insurgents in Iraq who often placed cylinders of chlorine in car bombs (Cave and Fadam, 2007). In addressing preparedness and response to chemical disasters, it is important to note that preparing for toxic industrial chemical accidents is very similar to preparing for chemical terrorism. One release is accidental and the other is intentional but the resulting human suffering is the same.

There is no place where people live or work that is completely free from the risks of hazardous chemical exposures. Toxic industrial chemicals are stored and used in large quantities across highly populated areas. They are also transported in large quantities on trucks or trains through nearly every rural town on the map. Even many common household products can be dangerous if improperly stored or used. This chapter addresses chemical emergency preparedness on two levels. We will begin with chemical risks that originate outside the home and then focus on those that are in the home.

Chemical Disaster Definitions

Acute exposure guideline levels: Estimated concentrations of a chemical that a healthy person may be safely exposed to for 10 minutes, 30 minutes, 60 minutes, 4 hours, or 8 hours during an emergency. These values are used to make decisions about evacuation.

Acute poisoning: A one time, high-dose exposure during a short period of time resulting in a fast, severe response.

Biotoxin: A poisonous substance derived from a living organism.

Blister agent: Also called a vesicant; a substance that produces chemical burns and blisters when it comes in contact with the skin and mucous membranes.

Blood agents: Substances such as cyanide compounds that prevent normal oxygen transfer by the blood.

Carcinogen: A substance that initiates or promotes cancer.

Choking agents: Substances such as chlorine that cause severe damage to the respiratory tract and can result in suffocation.

Chronic poisoning: A prolonged exposure over a longer period of time resulting in illness that is often more difficult to identify than with an acute exposure.

Corrosive: A highly reactive substance, such as an acid or alkaline, that causes damage to living tissue.

Cumulative effect: A reaction that occurs over a longer period of time following multiple exposures.

Emergency response planning guidelines (ERPGs): Estimated concentrations of a chemical that a healthy person may be exposed to for up to 1 hour during an emergency.

Local effect: A reaction to an exposure observed at the specific site on the body exposed.

Nerve agents: Substances such as sarin that interfere with the proper function of the nervous system.

Oxidizer: A substance that reacts with other materials by providing an oxygen source. This can stimulate combustion in organic materials.

Synergistic effect: A reaction of two or more chemicals that is greater than the sum of the two individual exposures.

Systemic effect: A reaction in the body occurring away from the original site of exposure.

Basic Facts about Chemical Health Threats

Chemical toxicity is the ability of a chemical molecule or compound to damage susceptible sites or cells in the human body or in other living biological systems including plants, animals, or even ecosystems. The toxicity or harmful reaction from a chemical exposure is referred to as the "response." The amount of a chemical needed to result in that damage is called the "dose." Therefore, the amount of a substance that causes damage among biological systems is called the dose–response relationship. An illness-causing or lethal dose of a chemical can enter the body through inhalation, skin or mucous membrane exposure, ingestion, or injection.

Routes of Entry into the Body

1. Inhalation through the respiratory tract. This route is the most important in terms of potential severity and is the basis for exposure standards such as threshold limit values (TLVs) and permissible exposure limits (PELS).
2. Skin or mucous membrane absorption includes contact with the skin, eyes, nose, or mouth.
3. Ingestion through the digestive tract. This can occur through eating, smoking, or even applying lip balm or make-up with contaminated hands or in a contaminated work area.
4. Injection into the bloodstream by accidental needle stick or sharp object punctures of the skin.

Chemical Exposure Factors

The route and rate of an exposure determines the dose received. The results can include *acute poisoning* that is a rapid absorption resulting in an especially fast and severe response. Examples of typical acute exposures include substances such as cyanide and carbon monoxide. They are typically the outcome of a single oral or dermal exposure, multiple doses in a 24-hour period, or an inhalational exposure over several hours. There are also *chronic poisonings* that result from prolonged exposures over a longer period of time ranging from days to years. The symptoms of chronic poisonings are often far less apparent. This is especially true for chronic low level exposures to heavy metals such as lead and mercury that can accumulate in the body for years and cause slow, incremental damage.

Exposure Factors

1. Dose
2. Time of exposure
3. Route of exposure
4. Physical characteristics of chemical substance
5. Previous or simultaneous exposures to other chemical substances
6. Environmental conditions
7. Physiological characteristics of the exposed
   a. Genetics
   b. Age

c. Race
d. Health status
e. Lifestyle

There are a variety of factors that determine poisoning outcomes. Environmental conditions such as temperature and humidity influence exposure levels. The physical properties of a chemical are also important factors. The state of a substance (liquid, solid, or gas) influences the route of exposure. If it is a solid, the size determines where a substance will be deposited in the airway. Larger particles are trapped in the upper airways, and smaller particles are more likely to go deep into the lung. Previous chemical exposures can result in an increased sensitivity or even an additional tolerance. The existing state of health and current lifestyle has tremendous influence in the toxicity of some chemical exposures. Genetics, age, race, and gender, as well as smoking and the use of alcohol, have substantial influence in an individual's sensitivities to chemical exposures. Some exposures display a *local effect*. This is when a specific reaction is observed at the site of an exposure. This occurs in the respiratory tract with chlorine exposure or on the skin with corrosives. Some chemicals also have a *systemic effect*. This is when a chemical affects a system at a site other than the point of exposure, suggesting it was absorbed. For example, benzene can be absorbed through the skin or the lungs. Once it is in the bloodstream, it is converted to damaging metabolites in the liver and bone marrow. There is also a *cumulative effect* with many chemicals, particularly heavy metals, where the signs and symptoms are usually observed after numerous exposures over a long period of time. A *synergistic effect* occurs when two or more chemicals result in a response that is greater than the two of them individually. The classic example of this synergy is seen with drugs and alcohol where often 2 + 2 = 10 or more.

FIGURE 4–2 Environmental specialists carefully dispose of chemicals that Hurricane Rita spilled.
Source: Marvin Nauman/FEMA photo. http://www.photolibrary.fema.gov/photolibrary/photo_details.do?id=20446.

Categorization and Measurement of Chemical Health Threats

Historically, there have been a variety of ways various organizations across the United States and in other nations have categorized chemical threats to human health. The U.S. classification and communication of chemical hazards is a responsibility shared by multiple agencies. Although the physical characteristics of a chemical can be defined clearly and objectively with measurements such as vapor pressure or specific gravity, health hazards are far more difficult to categorize and define. It is difficult to develop concise measurements of things such as irritant properties or other highly subjective parameters. As a result, there is a myriad of far less precise titles and definitions for health hazards. Chemical threats are communicated through standardized labels, placards, and material safety data sheets (MSDSs). However, there have been different labels and warnings used by various nations. This sometimes causes confusion and increases the likelihood of accidental exposures or inappropriate responses to spills and accidents. As national economies continue toward globalization, these divergent labeling practices have introduced increasing safety challenges and difficulties in communicating chemical hazard information for imported and exported materials. After many years of work, the United Nations Economic and Social Council (ECOSOC) adopted the Globally Harmonized System of Classification and Labeling of Chemicals (GHS) in 2003 (Figure 4–3). This is an effort to get as many nations as possible to voluntarily standardize, or harmonize, the way in which the chemical hazards are classified and communicated (United Nations, 2007). Although this is part of the solution to the divergent descriptions and categories of chemical health hazards, there will continue to be a wide range of categories and descriptions embedded for years to come in key references of various U.S. agencies with chemical preparedness responsibilities, as well as those in other nations. Understanding these hazard communication approaches is a critical facet of protecting public health.

Globally Harmonized System of Human Health Classes

The globally harmonized system (GHS) includes 10 classes of chemical threats to human health. Each class has an established criteria with numeric cut-offs that place each substance in a category that conveys the level of hazard. For example, acute toxicity is based on how toxic a substance is when ingested, inhaled, or placed on the skin. An acute toxicity estimate (ATE) value is calculated based on the available toxicity data. The results are used to place a substance in a category ranging from 1 through 5. Category 1 agents are the most toxic and Category 5 agents are the least hazardous. A similar categorization is carried out for all 10 classes. However, not all classes have five categories. "Skin corrosion" includes three categories while "specific target organ systemic toxicity—repeated exposures" only has two categories. The latter includes Category 1 substances that are known to cause damage to specific organs and Category 2 that can possibly cause damage to specific organs (United Nations, 2007). This numeric hazard ranking is the opposite of the current U.S. labeling and placarding practice. The United States uses the National Fire Protection Association, 704M System. Those numbers increase as the hazard increases but conversely the GHS system decreases the numeric designations as the hazard increases. Although the implementation of this system is necessary given the globalization of industry, it will come with a substantial learning curve for nations that have firmly established practices.

Globally Harmonized System of Human Health Hazard Classes

1. Acute toxicity.
2. Skin corrosion/irritation.
3. Serious eye damage/irritation.
4. Respiratory or skin sensitization.
5. Germ cell mutagenicity.
6. Carcinogenicity.
7. Reproductive toxicity.
8. Specific target organ systemic toxicity—single exposure.
9. Specific target organ systemic toxicity—repeated exposure.
10. Aspiration hazard.

Source: Globally Harmonized System of Classification and Labeling of Chemicals (GHS), Program website: http://www.unece.org/trans/danger/publi/ghs/ghs_welcome_e.html.

The GHS is being implemented in the United States but requires a tremendous effort on the part of the Occupational Safety and Health Administration, Environmental Protection Agency, Consumer Product Safety Commission, Department of Transportation, and industries across the country. Although the initial goal set at the World Summit on Sustainable Development in 2002 was to have this system fully implemented by 2008, it will take several more years in the United States to make the necessary changes.

Improper HAZMAT Labeling—A Worst Case Scenario from Flight 592

Walter Hyatt was a songwriter and performed for many years with "Uncle Walt's Band" in Austin, Texas. Although they never experienced commercial success, their following grew in the 1970s and included people around the world and some noteworthy Americans such as Lyle Lovett, a young college student at the time, who would later experience huge commercial music success and become friends with the band members, including Walter. It was a clear day as Walter boarded a plane. He was on his way home after attending his daughter's wedding in Miami (Everitt, 2004). It was May 11, 1996, when he boarded ValuJet flight 592. It was a DC-9 passenger plane destined for Atlanta and had 105 passengers plus a crew of five. About 10 minutes after taking off from Miami, the pilot called in an emergency. There was smoke in the cabin and the cockpit. The flight was immediately cleared to return to Miami International Airport but never made it. They crashed in the everglades and there were no survivors. The loss of Walter and the other 109 people on that flight was attributed to poor labeling of a hazardous material. Poorly labeled, expired oxygen generators were being shipped in a cargo container on the flight with other materials. When the generators are in use, they are safely mounted above each passenger seat and are activated when the emergency masks drop from the ceiling. Although they produce heat and release oxygen, the generators activate safely when they are in place. If they are packaged in bulk for shipping, they present a very different hazard. This shipment of expired generators was identified as the cause of the tragic accident. According to the National Transportation Safety Board (NTSB) report, "...the Safety Board concludes that had a warning label or emblem clearly indicating the significant danger posed been affixed to each generator, personnel handling the generators, including the personnel in the shipping and stores who prepared them for shipment to Atlanta, might have been alerted to the need to determine how to safely handle and ship the generators. Had they done so, they might have learned of the need for (and acquired) safety caps and they might also have learned that the unexpended generators demand special packaging and identification requirements (and taken appropriate actions). Even if they did one of these actions, the accident would not likely have occurred" (National Transportation Safety Board, Aircraft accident report: in-flight fire and impact with terrain, Valujet Airlines Flight 592, DC-9-32, N904VJ, Everglades, near Miami, FL. NTSB/AAR-97/06, www.ntsb.gov/publictn/1997/aar9706.pdf).

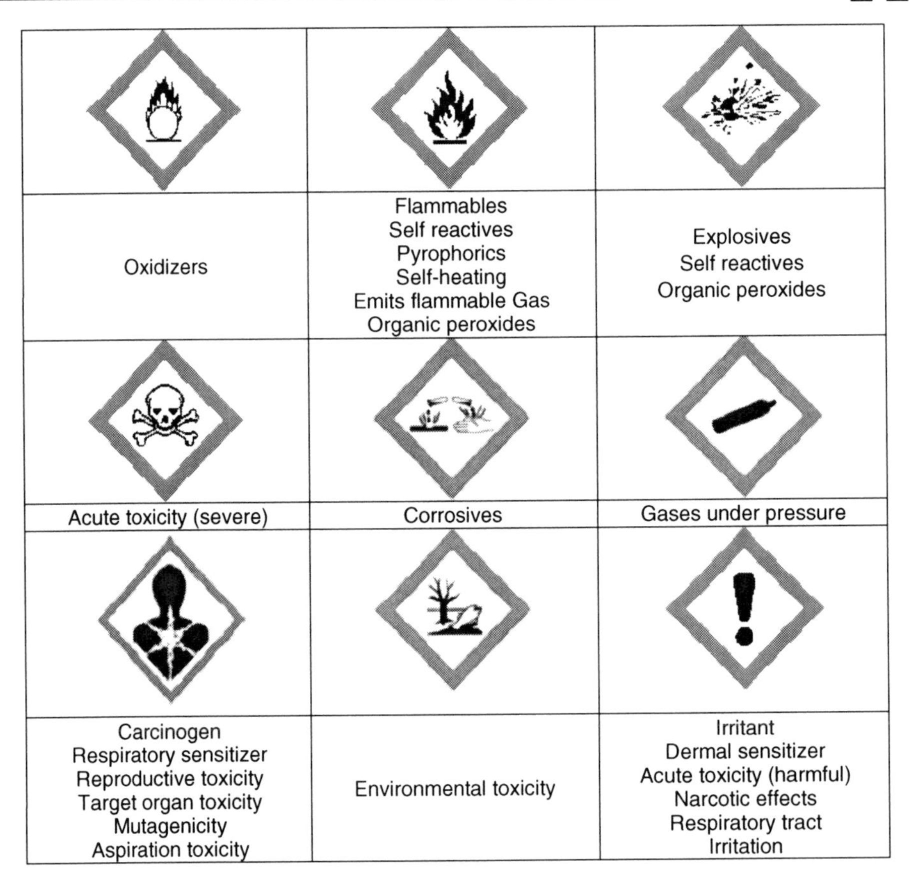

FIGURE 4–3 Globally harmonized system labels.

Source: Globally Harmonized System of Classification and Labelling of Chemicals (GHS), Annex 2, Classification and Labeling Summary Tables. http://www.unece.org/trans/danger/publi/ghs/ghs_rev01/English/06e_annex2.pdf.

National Fire Protection Association (NFPA) 704M System of Communicating Hazards

The individuals at greatest risk for hazardous materials exposure are first responders. Any time there is a transportation accident or damage at an industrial operation, it is essential that they have rapid, clear communication of the potential hazards that could be encountered to protect the responders as well as the affected community. One system currently used in the United States, NFPA 704M, was developed through the National Fire Protection Association (NFPA) (Figure 4–4). This is a simple placard placed on a container or facility to communicate the nature of the substances inside so first responders can take the proper precautions before handling the material or entering the facility during an emergency. The system uses a large diamond sign, called the "fire diamond," comprised of four smaller diamonds each having a different primary color. A large number is placed in one or more of the diamonds to indicate the nature and the level of risk. Red indicates flammable hazards, yellow is reactive, white is special hazards, and blue is health hazards. A number of 0 through 4 is placed in each diamond, except the white one that uses letters, to communicate the level of risk. For example, a blank blue diamond indicates that there are no substances that pose health hazards, and a four (4) indicates there are extremely toxic materials. A zero in the red diamond indicates that there is no flash point for the material inside. A one (1) in the red diamond indicates the presence of materials that have a flash point more than 200°F and a four (4) indicates a flash point less than 73°F. The yellow diamond communicates reactive materials and the degree to which they can explode when they are heated, disturbed by mechanical shock, or by coming in contact with water. Finally, if the white diamond at the bottom is blank, there are no specific or special health hazards present. If there are specific hazards, an abbreviation will be placed in the box such as CA for carcinogens, CO for corrosives, T for toxins, etc. (National Fire Protection Association, 2007).

Other U.S. Labeling Systems

The NFPA 704M System is only one of several systems that will eventually be incorporated into the Globally Harmonized System. The Hazardous Materials Identification System (HMIS) was developed by the National Paint & Coatings Association. It is similar to the 704M system but provides more detail. A problem with the NFPA 704M system

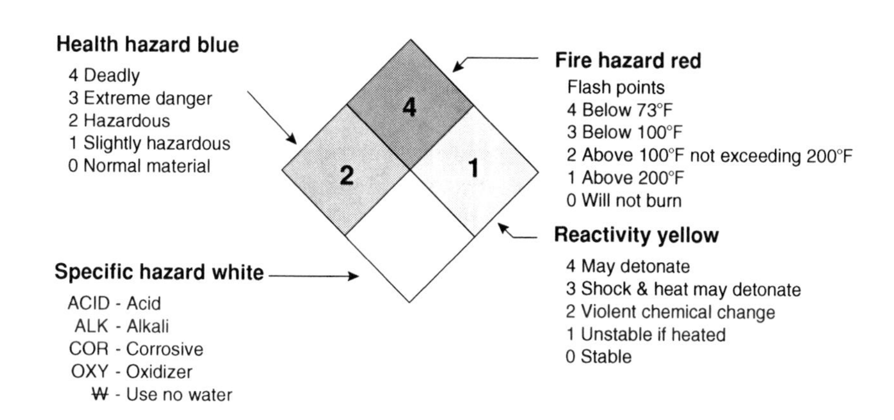

FIGURE 4–4 National Fire Protection Association, 704M System, "Fire Diamond."
Source: National Fire Protection Association.

is that it does not adequately denote occupational hazards or precautions. The HMIS system has similar numeric and color coded indicators but provides more detail, including a code recommending personal protective equipment. In addition, the American National Standards Institute (ANSI) adopted more detailed warning language developed by the Chemical Manufacturers Association (CMA) and the American Conference on Chemical Labeling but it has not been widely adopted. Finally, the Department of Transportation (DOT) hazard labeling system uses color coded diamonds to display the major hazard associated with material in transport (Figure 4–5). These placards use a combination of key words, color, and numeric classes to designate the nature of hazardous materials in transport. These divergent U.S. systems must all be incorporated directly or in tandem into the GHS implementation process.

Nomenclatures: Chemical Categorization of the Occupational Safety and Health Administration and Centers for Disease Control and Prevention Hazardous Chemicals

In addition to the differing placards and labels used in the United States and in other nations, there continue to be fundamental differences in the nomenclature of various hazards. Categories commonly used by American health and safety professionals often divide hazardous chemicals by their basic structure or composition, such as solvents or metals. There are also chemicals that are categorized based on their target organs or physiological damage. These include agents such as nerve, blister, and pulmonary agents. There is no definite pattern or trend that determines which term or label is most acceptable. The terms most commonly used today have emerged from various disciplines such as defense, emergency medicine, and the chemical industry over many years. No categorization list is exclusive. The Centers for Disease Control and Prevention use a combination of categorical titles that describe the health systems affected, the symptoms produced, and the chemical structure.

CDC Hazardous Chemical Categories

1. Biotoxins: Poisons derived from plants or animals.
2. Blister agents/vesicants: Chemicals that severely blister the eyes, respiratory tract, and skin.
3. Blood agents: Chemicals that affect the body by being absorbed into the blood.
4. Caustics (acids): Chemicals that burn or corrode the skin, eyes, and mucous membranes of the nose, mouth, throat, and lungs.
5. Choking/lung/pulmonary agents: Chemicals that cause severe irritation or swelling of the respiratory tract including the nose, throat, and lungs.
6. Incapacitating agents: Drugs that alter thinking and consciousness, sometimes leading to unconsciousness.
7. Long-acting anticoagulants: Poisons that prevent blood from clotting properly and lead to uncontrolled bleeding.
8. Metals: Poisons that consist of metallic agents such as lead, arsenic, and mercury.

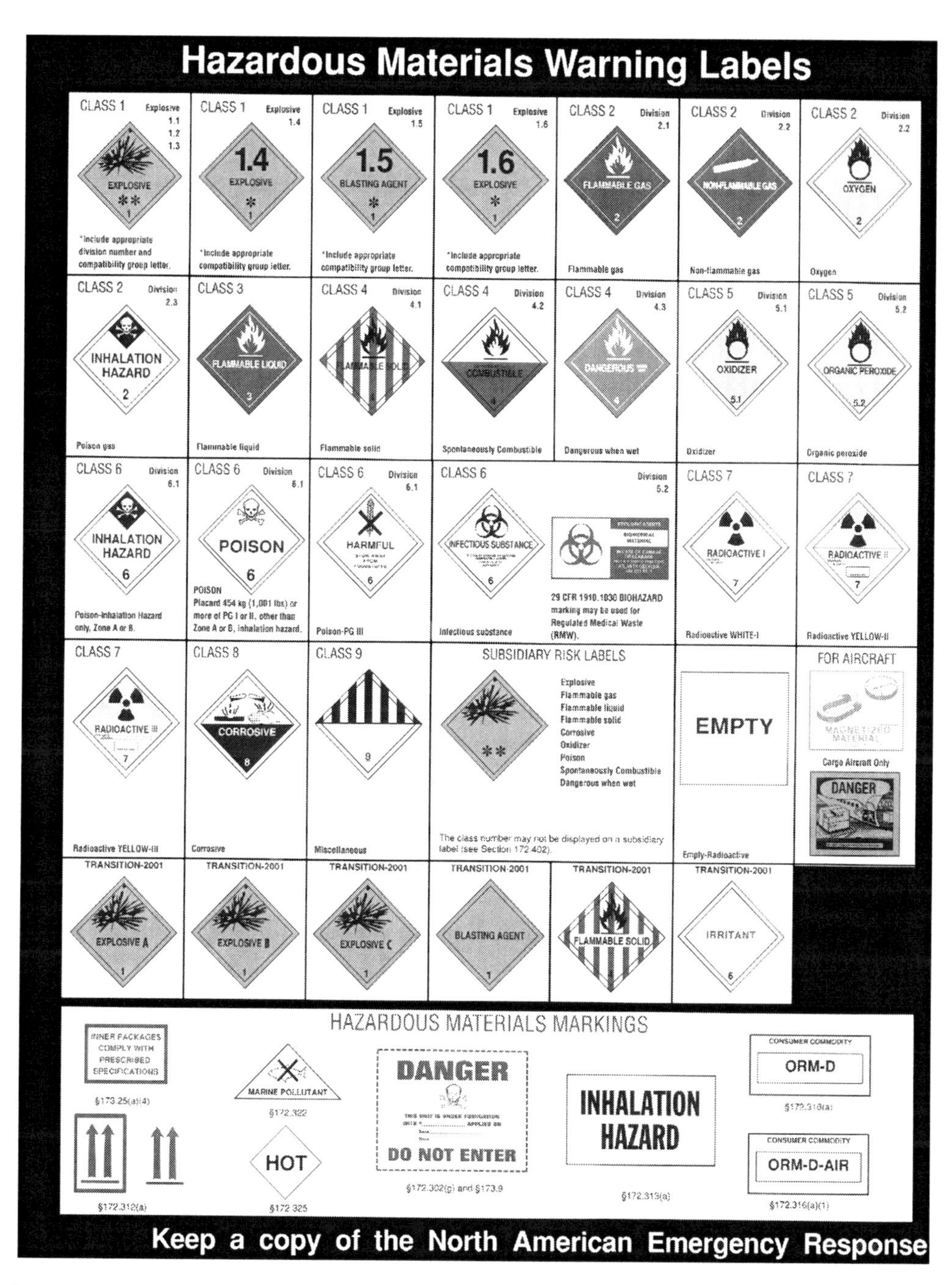

FIGURE 4–5 U.S. Department of Transportation, Hazardous Materials Warning Labels.
Source: U.S. Department of Transportation.

9. Nerve agents: Chemicals that prevent the nervous system from working properly.
10. Organic solvents: Chemicals that damage the tissues of living things by dissolving fats and oils.
11. Riot control agents/tear gas: Highly irritating, nonlethal chemicals used by law enforcement for crowd control, such as tear gas, or by individuals for protection, such as mace or pepper spray.

12. Toxic alcohols: Poisonous alcohols that can damage the heart, kidneys, and nervous system.
13. Vomiting agents: Chemicals that cause nausea and vomiting.

Source: Centers for Disease Control and Prevention; http://www.bt.cdc.gov/agent/agentlistchem-category.asp#vomiting

In the Code of Federal Regulations (29 CFR, 1910.1200, Appendix A), the Occupational Safety and Health Administration uses categorical descriptors related to workplace exposures such as carcinogens, irritants, and sensitizers. They also categorize chemical exposures according to specific organ systems, such as hepatotoxins that damage the liver or teratogens that produce birth defects. This is in contrast to the CDC nomenclature. The difference is a product of the particular focus of each agency. OSHA is focused on regulatory enforcement across a variety of industries, whereas CDC is sharing facts and reference materials for public health and healthcare organizations. When engaging in preparedness and response activities it is important to be mindful of the overall mission of each agency. Because the CDC has a public health and medical focus, they are more likely to have more current health information. OSHA is focused more on enforcement of standards to protect the workforce from long-standing occupational hazards and is less likely to have current health information for emerging public health issues.

Occupational Safety and Health Administration (OSHA), Health Hazard Definitions

1. Carcinogens: Chemicals suspected of or known to produce cancer. It includes:
   a. Acknowledged by the International Agency for Research on Cancer (IARC) to be a carcinogen or potential carcinogen.
   b. Listed as a carcinogen or potential carcinogen in the Annual Report on Carcinogens published by the National Toxicology Program (NTP) (latest edition).
   c. Regulated by the Occupational Safety and Health Administration (OSHA) as a carcinogen.
2. Corrosives: Chemicals that causes visible damage or destruction of living tissue by chemical action at contact site.
3. Highly toxic chemicals: A chemical that:
   a. Has a median lethal dose, LD_{50}, (killing 50% of test rats) of 50 milligrams or less per kilogram of body weight when administered orally.
   b. Has a median lethal dose, LD_{50}, of 200 milligrams or less per kilogram of body weight when administered by continuous contact for 24 hours or less.
   c. Has a median lethal concentration dose, LC_{50}, in the air of 200 parts per million by volume or less of gas or vapor, or 2 milligrams per liter or less of mist, fume, or dust, when administered by continuous inhalation for 1 hour.

4. Irritants: Chemicals that are not corrosive, but cause reversible inflammatory effects on living tissue by chemical action at the contact site.
5. Sensitizers: Chemicals that cause a substantial proportion of exposed people or animals to develop an allergic reaction in normal tissue after repeated exposure.
6. Toxic chemicals: A chemical that:
   a. Has a median lethal dose, LD_{50}, of more than 50 milligrams per kilogram but not more than 500 milligrams per kilogram of body weight when administered orally.
   b. Has a median lethal dose, LD_{50}, of more than 200 milligrams per kilogram but not more than 1000 milligrams per kilogram of body weight when administered by continuous contact for 24 hours or less.
   c. Has a median lethal concentration dose, LC_{50}, in air of more than 200 parts per million but not more than 2000 parts per million by volume of gas or vapor, or more than 2 milligrams per liter but not more than 20 milligrams per liter of mist, fume, or dust, when administered by continuous inhalation for 1 hour or less.
7. Chemicals with target organ effects: Chemicals that have an effect on specific organs, such as:
   a. Hepatotoxins produce liver damage.
   b. Nephrotoxins produce kidney damage.
   c. Neurotoxins produce nervous system damage or effects.
   d. Agents that act on the blood or hematopoietic system and decrease hemoglobin function or deprive the body tissues of oxygen.
   e. Agents that irritate or damage pulmonary tissue.
   f. Reproductive toxins that affect the reproductive capabilities including chromosomal damage (mutations) and effect fetuses (teratogenesis).
   g. Cutaneous hazards are chemicals that affect the dermal layer of the body.
   h. Eye hazards are chemicals that affect the eye or visual capacity.

Source: Occupational Safety and Health Administration, 29 CFR, 1910.1200, Appendix A http://www.osha.gov/pls/oshaweb/owadisp.show_document?p_table=STANDARDS&p_id=10100.

Chemical Exposure Measurements: Occupational versus Emergency Values

The potential exposure an individual has to a substance can be measured in several ways. Air sampling may be conducted, particularly in occupational settings. The results are compared to TLVs established by the American Conference of Governmental Industrial Hygienists (ACGIH). These are guidelines that help to make decisions on safe levels of chemical exposures in the workplace. They are based on inhalation of a volume of chemical in the air expressed as parts per million (ppm) or milligrams of a chemical per cubic meter of air (mg/m^3). TLVs are not legal standards but reflect a level of exposure a typical healthy worker can experience during 8-hour workdays, for 40-hour work weeks, over a life time without increased risk of disease or injury.

The components of TLVs include time-weighted average (TWA) concentrations that are a measurement of chemical exposures in a worker's breathing zone for an entire work day. A TWA established by ACGIH is simply the average amount of exposure that should not be exceeded in an 8-hour work day. Ceiling values are the levels that should never be exceeded by anyone at any time during the work day. Short-term exposure limit (STEL) values are the levels of exposure that should never be exceeded in a 15-minute period, even if the average TWA is below the recommended level. The STEL values are for contaminants that have known short-term hazards. None of these measures are quantitative risk estimates and they cannot be used as acceptable levels for public exposure during an emergency. With the growing numbers of vulnerable individuals in our communities, response activities must be based on the assumption that susceptible individuals could be injured by exposures that are below the established TLVs. The Occupational Safety and Health Administration (OSHA) also has PELs. Although many of these limits are exactly the same as TLVs, the OSHA PELs are regulatory limits and not just recommendations. There are over 500 PELs established in 29 CFR, 1910.1000 Table Z-1, Limits for Air Contaminants and Table Z-2, Toxic and Hazardous Substances. In reality, the most prudent approach is to review the TLV and PEL and the use of whichever standard is most protective. Although these standards do not apply to allowable public exposures, they become important during the recovery process after a disaster. The importance of enforcing these recommendations and standards during recovery is highlighted by the continuing health problems being experienced by those who worked at Ground Zero in 2001 and 2002 following the collapse of the World Trade Center towers in New York City on 9/11. Had the TLV recommendations and PEL standards been enforced during the recovery operations at Ground Zero, many of the chronic respiratory problems experienced by those workers would have been avoided.

Acute Exposure Guideline Levels (AEGLs) are exposure descriptions that can assist in decision-making when a toxic chemical exposure potential exists for the general public (See Figure 4–6). These recommendations are developed by the U.S. Environmental Protection Agency (EPA) and the National Research Council's Committee on Toxicology (NRC COT) through an open review and consensus process using a National Advisory Committee for AEGLs and stakeholders. AEGLs have been developed for hundreds of chemical agents and others continue to be added. They include recommended exposure limits at five exposure time intervals including 10 minutes, 30 minutes, 1 hour, 4 hours, and 8 hours. When these guideline levels are used with air modeling predictions, they can be very helpful in planning the best course of action to protect the public following an accidental industrial chemical release or a chemical terrorism attack (U.S. Army Center for Health Promotion and Preventive Medicine [CHPPM], 2003).

Using a chlorine release, such as the one in Graniteville, South Carolina, below is a breakdown of the AEGLs. According to the data, most members of the population, including the majority of susceptible individuals will be allright, aside from some irritant effects, if the airborne concentrations are kept below 0.50 ppm. Given a scenario of a nursing home downwind from a chlorine release, if they are sheltering in place and the chlorine levels are predicted to be below 0.50 ppm, it may be best to plan on keeping them sheltered in place because the levels outside are higher than inside and the levels inside are tolerable. However, if an evacuation plan is being considered, an informed decision can be made with estimated outcomes using the AEGLs.

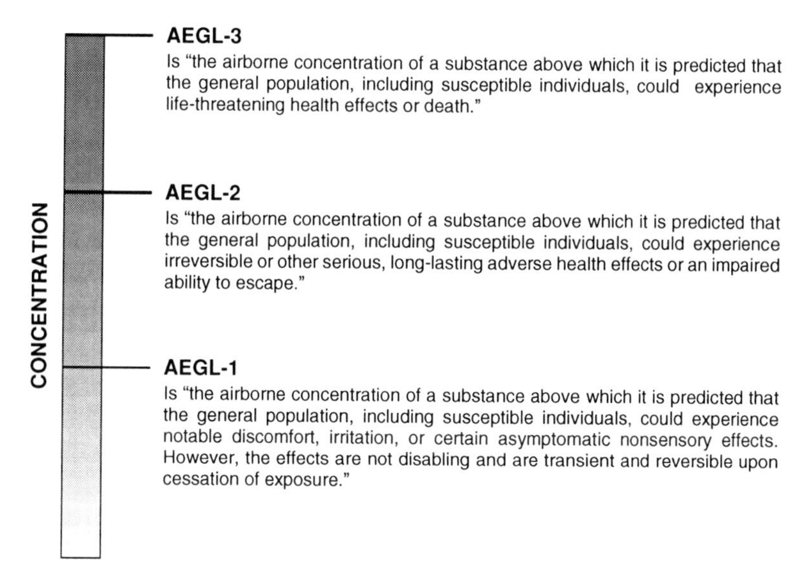

FIGURE 4–6 Acute Exposure Guideline Levels (AEGLs).

Source: National Oceanic and Atmospheric Administration, Office of Response and Restoration http://response.restoration.noaa.gov/topic_subtopic_entry.php?RECORD_KEY%28entry_subtopic_topic%29=entry_id,subtopic_id,topic_id&entry_id(entry_subtopic_topic)=662&subtopic_id(entry_subtopic_topic)=24&topic_id(entry_subtopic_topic)=1.

TIME	AEGL-3 (ppm)	AEGL-2 (ppm)	AEGL-1 (ppm)
10 minutes	50	2.8	0.50
30 minutes	28	2.8	0.50
1 hour	20	2.0	0.50
4 hours	10	1.0	0.50
8 hours	7.1	0.71	0.50

ERPGs are estimated concentrations that are also developed from published experimental data and apply to most individuals but do not address vulnerable populations (See Figure 4–7). There are just over 100 chemicals with established ERPGs. These are planning values developed for a 1 hour maximum exposure scenario. They can be used as a health-based guideline for a single exposure to a chemical for up to 1 hour to develop response plans and mitigation strategies (American Industrial Hygiene Association, 2006). The development of ERPGs is carried out by the American Industrial Hygiene Association, ERP Committee. These recommendations should never be used for occupational exposure estimates or environmental exposures lasting more than 1 hour. They can be used for chemicals that do not have an established AEGL and when there is a chemical release with a short duration.

ERPGs have a close correlation to AEGLs. However, AEGLs have lower values than ERPGs as they consider vulnerable populations. For example, the 1-hour AEGLs for chlorine compared to ERPGs are as follows:

Comparison of Chlorine AEGLs and ERPGs	
AEGL-3 = 20 ppm	ERPG-3 = 20 ppm
AEGL-2 = 2.0 ppm	ERPG-2 = 3.0 ppm
AEGL-1 = 0.50 ppm	ERPG-1 = 1.0 ppm

With the endless lists of chemical substances used in industry, many do not have an established AEGL or ERPG. Work is continuing toward expanding the lists of available values. For many chemicals that do not have these values established, a Temporary Emergency Exposure Limit (TEEL) may be used. As the name suggests, these are temporary guidelines that may be used to estimate the potential impact on the health of the general public to different chemical concentrations. These values are established by the U.S. Department of Energy, Subcommittee on Consequence Assessment and Protective Actions (SCAPA). The way the TEELs are defined is the same as the ERPGs. In other words, a TEEL-1 is defined exactly the same way as an ERPG-1. The differences are that this is a temporary recommendation calculated as the peak 15-minute Time Weighted Average (TWA) concentration.

AEGLs, ERPGs, and TEELs are all values or measurements of chemical substance in the air. The values represent the expected physiological response of the majority of a population at a given exposure level. Although these values are helpful for chemical emergency response planning, they should never be used as a tool for workplace exposure assessments. For chemical preparedness purposes, an AEGL is the first choice. If there is not an AEGL available, an ERPG should be used. If there is no AEGL or ERPG value available, a TEEL may be used. Keep in mind that the ERPGs and TEELs do not factor in vulnerable populations such as elderly, very young, etc. When carrying out recovery operations, occupational values should be used for those working in the area of potential exposure. It is recommended that you review the ACGIH TLVs and OSHA PELs and use whichever is most protective. There is one final value that

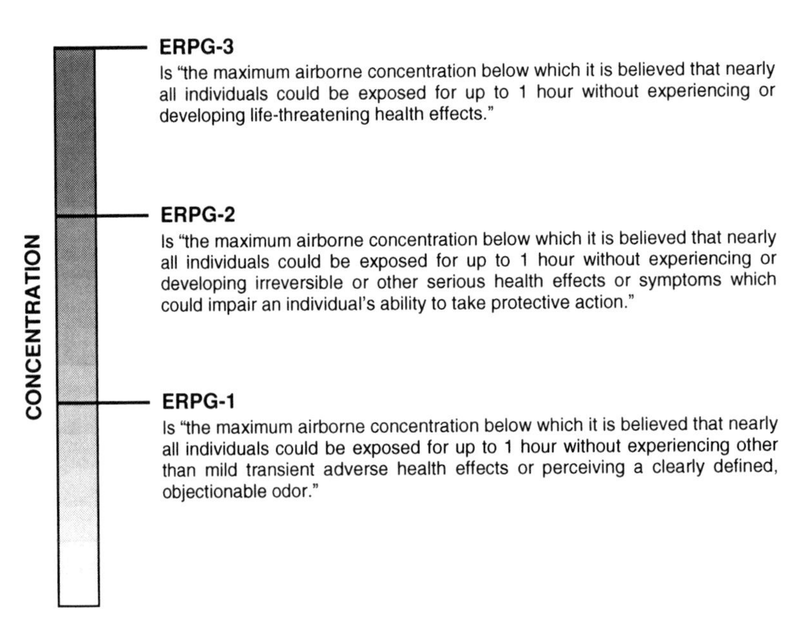

FIGURE 4–7 Emergency Response Planning Guidelines (ERPGs). Source: National Oceanic and Atmospheric Administration, Office of Response and Restoration. http://response.restoration.noaa.gov/topic_subtopic_entry.php?RECORD_KEY%28entry_subtopic_topic%29=entry_id,subtopic_id,topic_id&entry_id(entry_subtopic_topic)=663&subtopic_id(entry_subtopic_topic)=24&topic_id(entry_subtopic_topic)=1.

is a last resort in preparedness planning if there are no available AEGL, ERPG, or TEEL values. If there are no public exposure guidelines available for a chemical substance, the final option is using 1/10TH of the Immediately Dangerous to Life and Health (IDLH) value. Although the IDLH value is a workplace exposure limit, it is established by the National Institute for Occupational Safety and Health (NIOSH) and has been used in the past by many U.S. agencies for hazard planning. Using the chlorine example once again, if the IDLH is 10 ppm, one-tenth of that is 1 ppm. That is the same value as the ERPG-1 level. Always keep in mind that this is not applicable to vulnerable populations.

Chemical Accidents versus Terrorism

In recent years, much attention has been focused on the increasing prospect of chemical terrorism. The U.S. Department of Homeland Security (DHS), 15 National Planning Scenarios include four chemical incidents. Two of these are industrial or transportation disaster threats including a chlorine tank explosion and a toxic industrial chemical release. Although these scenarios are more likely to occur as accidents, they may also be instigated by terrorists. In fact, a chemical attack with a chlorine tank or other toxic industrial chemical is far more likely than the other two DHS chemical scenarios. Those are attacks with chemical warfare agents such as blister and nerve agents. In terms of specific federal, state, and local actions, what is the difference between preparing for an accidental chemical release compared to an intentional chemical attack? As far as the healthcare and public health challenges, the differences are few. Although law enforcement challenges increase exponentially with an intentional incident, the health and safety

Table 4–1 Summary of Occupational and Emergency Exposure Guidelines

Developers/Guideline	Use	Exposure Duration
ACGIH TLV and OSHA PEL	Occupational	Occupational exposure for an 8-hour day, 40 hours per week.
ACGIH STEL	Occupational	Occupational exposure for a 15-minute maximum time limit.
NIOSH IDLH	Occupational	Highest concentration from which escape is possible with no adverse health effects within 30 minutes.
EPA and NRC-COT AEGL	Public emergency	Three-tier emergency response recommendations for 10 minutes, 30 minutes, 1 hour, 4 hours, and 8 hours.
AIHA ERPG	Public emergency	Three-tier emergency response recommendations for 1 hour.
FEMA, EPA, and DOT 1/10 IDLH	Public emergency	Highest concentration considered safe for most of the public for 30 minutes.

issues are nearly identical. Any of these scenarios may require mass decontamination, mass care, evacuation, or sheltering in place. Therefore, with a few exceptions, the preparedness activities for accidental and intentional chemical incidents are the same.

Tokyo Subway Nerve Agent Attack by the Aum Shinrikyo Cult

The Aum Shinrikyo cult was founded by Shoko Ashara who claimed to be "enlightened" after spending time meditating in the Himalayan Mountains. The cult he established in Japan espoused the belief that Ashara was destined to be a great ruler but in order to usher in his reign, an apocalyptic social collapse would be required. To that end, they pursued chemical and biological weapons (Kaplan and Marshall, 1996). After several unsuccessful attempts at using biological weapons and one somewhat effective chemical attack in Matsumoto City with seven casualties, they devised a plan to release sarin nerve agent in the Tokyo Subway System. On March 20, 1995, cult members boarded five trains converging at a busy central terminal. The sarin nerve agent was place in plastic bags, wrapped in newspapers, and laid on the floor of each train. When it arrived at the station, the cult members each ruptured the bags with the sharpened tip of an umbrella and exited the trains. The number of people affected was approximately 3800 with about 1000 requiring hospitalization and 12 fatalities (Olson, 1999).

There are many lessons to be learned from this attack. It sets a precedent and displays a willingness among terrorists to inflict mass casualties using chemical weapons. This also highlights the difficulties in protecting the public from a diverse number of attack scenarios. The need for rapid diagnosis and treatment is crucial to limiting the impact of an attack on public health. This requires first responders and healthcare providers to understand how to recognize and react quickly to a variety of threats. It also requires a medical surge capacity that is exercised often enough to sustain effectiveness. In the case of sarin and several other chemical agents, there are unique antidotes that need to be maintained. Plans must be established and exercised to quickly deploy these resources during a crisis.

Chemical Terrorism

One difference between chemical terrorism and accidents is that some agents have no use in industry. Should an incident occur, it is reasonable to assume that it is an intentional release. The most important examples of these are classic chemical warfare nerve and blister agents (See Table 4–2). Nerve agents include two primary categories, V-Series and G series. The most popular of the V-series agents is VX. The G agents include tabun (GA), sarin (GB), and soman (GD). The most widely used has been sarin. The "G" designation stands for German. Each of the agents in this series were developed in the 1930s and 40s by Germany. Tabun was discovered first and called German agent A or "GA," sarin (GB) was second, and soman (GD) was third. The GC designation was skipped because it is already a widely used abbreviation for gonorrhea. A lesser known GF agent was also discovered

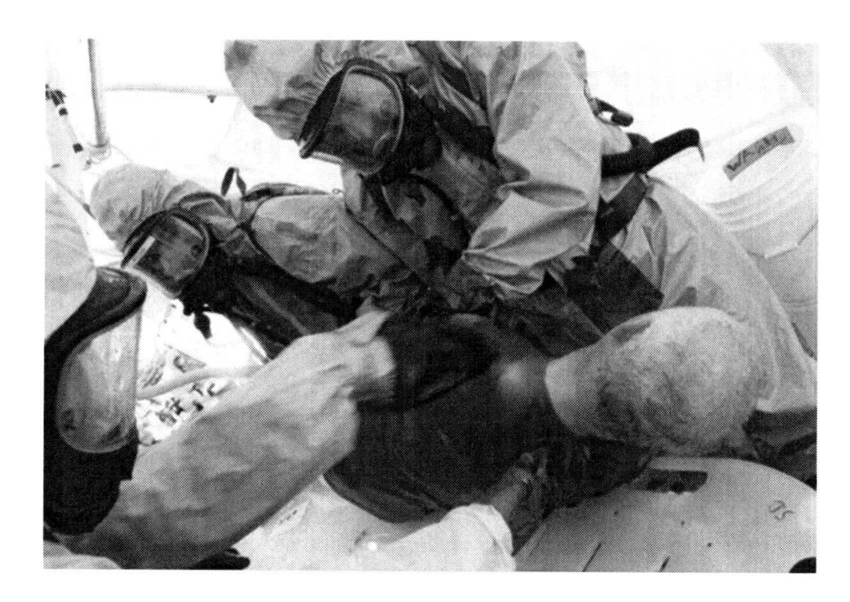

FIGURE 4–8 U.S. Soldiers from the Virginia Army National Guard decontaminate a role-player during a joint chemical, biological, radiological, nuclear and high-yield explosive enhanced response force exercise at Fort Pickett, Va., June 26, 2008. DoD photo by Sgt. Teddy Wade, U.S. Army.

(cyclohexyl methylphosphonofluoridate) (Organisation for the Prohibition of Chemical Weapons, Nerve agents: lethal organo-phosphorous compounds inhibiting cholinesterase (a FAO briefing book on chemical weapons), www.opcw.org/index.html). Again, an abbreviation was skipped due to other uses. GE is a common abbreviation for gastroesophageal. Blister agents or vesicants fall into one of four categories including, lewisite (L), nitrogen mustard (HN), phosgene oxime (CX), and sulfur mustard (H). These nerve and blister agents were designed for the battlefield and serve no industrial purpose. However, there are still demilitarization activities under way around the world by nations who signed the Chemical Weapons Convention in 1997. This includes the deactivation, emptying, cleaning, and destruction of stockpiled chemical weapons including bombs, missiles, and land mines. As long as these activities are incomplete, there will be a slight risk for an industrial accident that could release a nerve or blister agent. Outside that remote scenario, any discovery of a nerve or blister agent release is highly likely to be an act of terrorism.

The most likely chemical terrorism scenarios are also the most likely industrial accident scenarios. Ubiquitous toxic industrial chemicals such as chlorine, ammonia, and cyanide compounds are constantly being moved in large quantities across our highways and rail lines. It is far more likely that a terrorist will attack an existing target of opportunity rather than establishing the necessary infrastructure to develop chemical weapons.

Chemical Accidents

There are two types of chemical incidents that can have a widespread impact on public health in nearly every community, fixed facility, and transportation accident. Much can and has been done in the United States to prepare for each type of accident. There are mandates requiring fixed facilities to notify local and state authorities with details of hazardous materials they produce, store, or use. This allows for detailed planning in advance on how to best manage a release. Even though there are also laws regulating the

Table 4–2 Likely Chemical Terrorism Agent Symptoms and Treatments

Category	Agents	Symptoms	Treatment
Blister agents	Nitrogen mustards HN-1, HN-2, HN-3 Sulfur mustards H, HD, HT	Skin burns, burning and swelling of the eyes, blindness, respiratory damage.	Decontaminate and offer supportive care. No antidotes exist for most blister agents (Agency for Toxic Substances and Disease Registry [ATSDR], 2002b).
	Lewisites L, L-1, L-2, L-3 Mustard lewisite (HL) Phosgene oxime (CX)		British anti-lewisite is effective against lewisite if given quickly (ATSDR, 2002d).
Blood agents	Arsine (SA) cyanide Cyanogen chloride (CK)	General weakness and confusion, headache, dizziness, shortness of breath.	Decontaminate and offer supportive care.
	Hydrogen cyanide (AC) Potassium cyanide (KCN), sodium cyanide (NaCN)	Exposed individuals with light skin may appear pink or red. This is due to the buildup of oxygen in the blood.	Arsine: no specific antidote (ATSDR, 2002a). Cyanides: cyanide antidote kit contains amyl nitrite perles, sodium nitrite and sodium thiosulfate (ATSDR, 2002c). Newer treatments for cyanides are emerging, such as hydroxycobalmin (Cyanokit®).
	Sodium monofluoroacetate (Compound 1080)	Advanced symptoms include convulsions, low blood pressure, and respiratory failure.	Sodium monofluoroacetate: No antidote.
Choking agents	Ammonia Bromine (CA) Chlorine (CL) Hydrogen chlorine Methyl bromide Methyl isocyanate	Eye irritation, airway irritation, stridor (high pitched airway sounds), chest tightness, sore throat, pulmonary edema.	Decontaminate and offer supportive care.
	Osmium tetroxide Phosgene Diphosgene (DP) Phosgene (CG) Phosphine Phosphorous Sulfuryl fluoride		Priority is given to airway management and treatment of eye injuries.

(Continued)

Table 4–2 (*Continued*)

Category	Agents	Symptoms	Treatment
Nerve agents	G agents	Miosis (small pupils) is a hallmark of exposure.	Decontaminate and offer supportive care.
	Tabun (GA)	An easy way to remember common symptoms is the acronym "SLUDGE": salivation, lacrimation, urination, diarrhea, gastroenteritis, emesis.	Antidotes include atropine and pralidoxime chloride (2-PAM CL) (ATSDR, 2002e).
	Sarin (GB) Soman (GD) V agents VX	Advanced symptoms of higher exposures include seizures, muscle fasciculations, flaccid paralysis, apnea, and death.	

safe transportation of chemicals, transportation accidents pose unique challenges as there is such a wide array of chemicals being transported and so many response variables of how, when, and where an accident may occur.

Most Common Hazardous Materials Releases (Binder, 1989)

1. Natural gas
2. Chlorine
3. Gasoline
4. Sulfuric acid
5. Hydrochloric acid
6. Ammonia
7. Sodium hydroxide
8. Diesel oil
9. Corrosive liquids
10. Nitric acid/phosphoric acid

During a 10-year period, 1998–2007, there were over 169,000 transportation accidents in the United States involving hazardous materials (hazmat). Over 86% of these accidents occurred on highways compared with only 8% by air, 5% by rail, and less than 1% on the water (Figure 4–9). It is interesting to note that among the over 2800 resulting injuries and illnesses, 55% were from highway accidents and 39% from rail accidents. Rail accidents occur far less frequently but when they do occur, they have a much larger footprint and a greater impact on public health per incident. However, the vast majority of the 141 hazmat-related fatalities during this period are attributed to highway incidents. During this timeframe, highway accidents account for 88% of hazmat fatalities with 12% due to rail accidents (Pipeline and Hazardous Materials Safety Administration, 2008). Given the massive quantities of hazardous industrial chemicals transported and used each day, the total number of individuals who ever become injured, ill, or suffer a lethal exposure from a chemical accident is relatively low. However, there is very little tolerance on behalf of the general public for accidents involving hazardous materials. Chemical accidents are preventable and with the knowledge, tools, and regulations that exist in the United States today, these incidents should be exceptionally rare.

Morbidity and Mortality of U.S. Hazardous Materials Accidents: A 10-Year Summary (1999–2007)

- Total Hazmat accidents = 169,076
  - Highway = 146,584 (86.7%)
  - Air = 13,563 (5.1%)
  - Rail = 8652 (2.1%)
  - Water = 277 (<1%)
- Total Hazmat injuries = 2824
  - Highway = 1546 (54.7%)
  - Rail = 1104 (39.1%)
  - Air = 154 (5.5%)
  - Water = 20 (<1%)
- Total fatalities = 141
  - Highway = 124 (88%)
  - Rail = 17 (12%)

Prevention

Preparing for the full range of possible chemical emergencies requires action at every level of government and in every home. Federal, state, and local governments all play a role in developing and implementing measures that reduce the likelihood of chemical accidents. The activities of emergency managers, planners, and others are important but cannot reach into homes across the community where many more chemical hazards exist. It is not enough to simply follow labeling instructions for use. The way household chemicals are stored and where they go when disasters such as earthquakes or hurricanes disturb them, can also pose significant chemical hazards.

Community Prevention Activities

To reduce the inherent public health risks of having chemicals throughout industry and crossing the network of highways, railways, and waterways, a variety of local and state agency activities are sustained. These public agency activities begin with regulations such as the Emergency Planning and Community Right-to-Know Act (EPCRA), Superfund Amendments and Reauthorization Act of 1986 (SARA), and chemicals listed under

FIGURE 4–9 Overturned tanker truck.
Source: U.S. Department of Transportation. http://www.itsdocs.fhwa.dot.gov/JPODOCS/REPTS_TE/13884/flipedgas.jpg.

section 112(r) of the Clean Air Act (CAA). The Environmental Protection Agency has compiled the "Consolidated List of Chemicals Subject to the Emergency Planning and Community Right-To-Know Act (EPCRA) and Section 112(r) of the Clean Air Act," also referred to as the "List of Lists." This document lists the chemicals required to be reported by industry to local and state officials (United States Environmental Protection Agency, 2001). Industry identifies the reportable quantities of chemicals on the list and reports them to state and local planners. The planners use this industry reported information to determine worst case chemical release scenarios, develop their plans, and carry out training and exercises to better prepare for the most likely scenarios in their region.

The Local Emergency Planning Committees (LEPCs) were established by the Emergency Planning and Community Right-to-Know Act to carry out preparedness activities at the community level. The LEPC is required to include local first responders, emergency management, public health, environmental, transportation, industry officials, and the local media. When a local industrial operation reports the presence of a substantial quantity of potentially hazardous chemicals to the LEPC, the members develop an emergency response plan. This plan is reviewed and updated as needed, at least annually. The information and plans they compile are open to the community and are shared with the state. Each state maintains a State Emergency Response Commission (SERC) that appoints members and establishes jurisdictional areas for each LEPC. Should there be an accidental release of a hazardous chemical, the industry responsible for it is required to report the details, including estimates on what was released, what the potential public health impact may be, and what community precautions should be instituted. This process was instituted in the United States shortly after the 1984 disaster in Bhopal, India where a Union Carbide facility experienced an accidental release of methyl isocyanate resulting in thousands of deaths and tens of thousands of long-term injuries.

The Worst Industrial Accident in History: Bhopal, India

In the early morning hours of December 3, 1984, an accidental release occurred at a Union Carbide facility in Bhopal, India. Tons of methyl isocyanate leaked throughout the morning resulting in over 3800 deaths and thousands more chronic illnesses that continue decades later to contribute to premature death among those exposed. The facility was manufacturing pesticides and struggling to stay in business. Local officials knew there were safety issues but did not want to do anything that may increase the financial burdens of the facility and risk losing a large employer in their community (Shrivastava, 1987). As a massive malfunction of their facility began releasing methyl isocyanate, thousands of mostly impoverished local residents began collapsing in the streets. There was a lack of community awareness of what the facility produced and a lack of community preparedness, including in local healthcare facilities. Local hospitals did not know what chemical or chemicals had been released and had no idea how to treat the exposed who managed to make it to the hospital (Fortun, 2001). Debates raged for years concerning how many people were actually exposed and how many died. The true numbers may never be fully known. A legal settlement was reached several years after the accident and payments were made to survivors and their families. Compensation was awarded to over 500,000 people who claimed they were injured and to over 15,000 surviving family members of those who perished. The total settlement was $470 million dollars paid by Union Carbide to the Government of India who in turn paid out an average of $2,200 to each survivor. Thousands of Bhopal residents require daily medical care more than two decades after the release (Kumar, 2004).

Many difficult and important lessons were learned from the Bhopal tragedy. There has never been an event in modern history that had a greater impact on industrial safety. The lessons include the need for communities to know in advance what potential hazards are introduced through their local industry. In the United States, the Emergency Planning and Community Right-to-Know Act (EPCRA) of 1986 was the regulatory response. The challenge is finding the balance between informing the public and providing too much information to those who may use it to formulate a crude chemical attack by targeting the facilities with toxic materials. Although the EPCRA legislation led to the establishment of State and Local Emergency Planning Committees in the United States, similar requirements are needed internationally to prevent companies from building dangerous operations in highly populated areas of nations with lax regulation. Local governments must exercise caution in where they allow facilities with large quantities of dangerous materials to be sited. Prevention activities in every nation need to begin with safety legislation and facility design. In addition, risk reduction activities need to be required in all industrial facilities with large quantities of hazardous chemicals (Bertazzi, 1999).

Reducing Building Vulnerability

High-risk facilities like hospitals, government buildings, and public venues need to complete a comprehensive vulnerability assessment of their air handling unit (AHU) and ensure a detailed risk reduction plan is developed and implemented. For other buildings, there are several simple steps that can be taken to reduce risks associated with chemicals accidentally or intentionally being introduced into the AHU. One of the most important vulnerabilities is the positioning of the outdoor air intakes. Many are located on ground level in an easily accessible area. This poses two problems. If it is easy to access, the building is more susceptible to an attack. A person with a simple hand sprayer can easily deliver material into the air handling system that can affect occupants. Many toxic chemicals are heavier than air and stay close to the ground. Air intakes positioned near the ground are also more likely to draw in chemicals from an industrial release or transportation accident. To reduce these threats, several steps may be taken. Physical security of the outdoor air intake, as well as limiting access to the building maintenance areas, is the first step. If mechanically and financially feasible, it is better to relocate low lying outdoor air intake vents to less accessible, more elevated areas. There are also ways to build intake extensions that draw the air in at a higher level without disturbing the function of the building's AHU. An intake that is elevated above 12 feet (3.7 meters) with a 45° sloped top, will keep it out of reach for those who may want to tamper with it and will also limit the risk of drawing in toxic chemicals that are heavier than air (See Figure 4–10) (U.S. Department of Health and Human Services, National Institute for Occupational Safety and Health, 2002). It is also important for building managers to develop procedures for quickly shutting down the AHU when there is an emergency and have a plan on managing building occupants safely if the need arises to shelter in place.

Home Prevention Activities

In the home, many household products are accumulated that individually or combined can pose a risk to residents. These risks can result from improper usage or from the inadvertent mixing of chemicals resulting from other disasters. Following any major disruption of a home, an assessment must be made of the chemical substances stored inside. Harmful concentrations of the by-products of mixed chemicals can pose a risk to

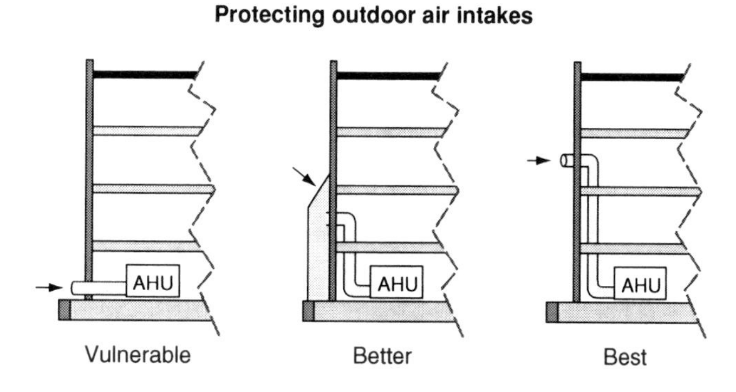

FIGURE 4–10 Reducing the vulnerability of Air Handling Units (AHUs) by repositioning the outdoor air intake.

those entering buildings after a disaster. Harmful chemicals can be found in household products such as:

- Paints and solvents.
- Indoor and outdoor pesticides.
- Automotive fluids.
- Kitchen and bathroom cleaners.
- Health and beauty products.
- Lawn and garden products.
- Swimming pool supplies.
- Barbeque products.
- Mercury thermometers.

It is far more likely that you will experience an emergency due to chemicals in the home than from industrial or transportation accidents. Before a disaster strikes, make sure that unnecessary chemical materials are properly disposed off and no needless chemical materials are stored in the home. Buy only what you need and never store hazardous chemicals in unlabeled containers. Shelves where hazardous chemicals are stored should be mounted securely and labeled. Incompatible chemical products should be identified and segregated. A metal storage cabinet is deal for storing hazardous chemicals.

If a home is near a major highway or railroad, or if there are industrial operations in the community that store and use hazardous chemicals, several items may be added to the household disaster kit described in Chapter 1. In 2003, the U.S. Secretary of Homeland Security, Tom Ridge, was met with cynicism when he suggested that Americans should purchase items such as plastic sheeting and duct tape to protect themselves from chemical terrorism (Lichtblau and Drew, 2004). Practically speaking, if these items are added to a home disaster kit and used correctly during an industrial disaster such as the Union Carbide catastrophe in Bhopal or a chemical terrorism incident, it could save lives. The part of the message that failed to reach much of the public in 2003 was how and when these materials are used.

Immediate Actions

There are a variety of ways that a chemical release may be detected. The sensory clues depend upon the chemical. A strong odor may be recognized in the environment or from exposed individuals. Common symptoms may be shared among the exposed. Chemical exposure symptoms include watery eyes or nose, burning of the eyes or skin, difficulty breathing or tightness in the chest, nausea, vomiting, headache, or a number of other

Indicators of a Chemical Release:

1. Syndromic
   a. Dermatological
   b. Respiratory
   c. Neurological
   d. Gastrointestinal

2. Epidemiological
   a. An abnormal number of patients with similar symptoms
   b. A cluster of illnesses among patients with something in common, such as a shared food or water source
   c. A rapid onset of symptoms among patients with a common exposure
   d. A specific pattern of symptoms or a syndrome suggesting a chemical exposure
3. Environmental
   a. Odors
   b. Unusual plume, cloud, vapor, or droplets
   c. Rapid onset of illness or death among wild or domestic animals
   d. Changes in the appearance or color of plants.

symptoms that comprise various chemical exposure syndromes (CDC, 2003). With some chemical exposures, these clues may be easy to miss in individual patient diagnosis. The defining clues may be in a cluster of patients with similar symptoms. There may also be clues in the environment such as dead insects, birds, or other animals. Plants may die or experience changes in appearance or color. Although a chemical release can have large scale, dramatic effects that are impossible to miss, some chemical releases can have a much more subtle progression (See Figure 4–11).

The clues that a release is occurring depend upon the mechanism of release. There may be obvious signs from an industrial facility or transportation accident including fire, explosion, or a plume release. If a spill has occurred, there may be little evidence of

FIGURE 4–11 Ninth Ward, New Orleans, LA, 9-16-05. This neighborhood remains flooded 2 weeks after the storm came through. The foul-smelling flood water is contaminated with petrol chemicals, household chemicals, and biological hazards. Marvin Nauman/FEMA photo. http://www.photolibrary.fema.gov/photolibrary/photo_details.do?id=15826.

a release for those who do not observe it. Vapors emitted from a spill may be invisible, have no odor, and travel a substantial distance making it impossible to recognize for those downwind. If a home or other structure is structurally damaged by disasters such as floods, earthquakes, or hurricanes, there may also be spilling and mixing of household chemicals posing a risk to those entering. Chemical contamination may also be intentionally introduced into food, water, or over-the-counter pharmaceuticals.

Individual Response

If a major chemical release occurs, you may be notified directly by first responders. They will sometimes go door-to-door, drive through the affected community using loudspeakers, contact residents by phone, use television and radio warnings, or use a combination of any or all these approaches. Depending on the type and quantity of chemical released, it may be more hazardous to evacuate than to simply stay inside until it passes. Response officials assess the circumstances and make the decision on what is best for the surrounding community. They will then provide information to the public on who, what, where, when, why, and how.

- Who is at risk of exposure.
- What the hazard includes.
- Where they should or should not go for information, healthcare, or evacuation.
- When they should take action.
- Why these recommendations are important.
- How to carry out suggestions that have more detail, such as sheltering in place.

Residents who are sheltered in place should immediately get all family members and pets inside. Shut down heating, ventilation, and air conditioning systems. Close all openings including windows, doors, and fireplace dampers. Here is where the plastic and duct tape become important. No building is air tight. To minimize what can slip through the cracks and expose you in the home, plastic and tape may be used to seal around windows, doors, vents, and outlets. Because many hazardous chemicals are heavier than air, everyone should take shelter on the highest floor of the building in a room with the fewest windows. Damp towels may be placed under the door of the room as well. With the home disaster kit described in Chapter 1, the family is prepared to shelter in place until an "all clear" message is announced. This is typically less than a couple hours. Otherwise, evacuation would be recommended (Federal Emergency Management Agency, 2006).

Residents asked to evacuate during a chemical emergency should pick up their disaster kit and follow the instructions immediately. Advice on which routes to follow or avoid should be included in the evacuation order. If there is time, all windows, doors, and vents in the home should be shut down. Also, if there are vulnerable neighbors nearby, offer as much assistance as you can. Those who have infants or who are disabled or elderly may not be able to evacuate without support. As you drive in areas near a chemical release, keep vehicle windows closed and keep the vents, air conditioning, and heating systems off until you are clear of the hazards. Everyone should take essential items with them such as medications, personal care items, and a change of clothes. If you evacuate to a shelter, keep in mind that most shelters will only provide minimal support including a cot, blanket, and meal (American Red Cross, 1994). Any other personal necessities should be brought.

Healthcare Response

The medical response for many chemical scenarios is similar. The primary challenges are diagnosis and decontamination of the exposed. This was an enormous challenge in Tokyo following the 1995 sarin release in the subway system. For every person seeking care, there were about five more who sought care even though they did not need it. The "worried well" can complicate the health and medical response to a chemical release. Unlike traumatic injuries where it is easy to conduct triage based on the initial presentation of a patient, chemical exposure is much more difficult to triage. Although some chemical exposures will have acute effects that are obvious, many exposures have delayed or chronic effects that are not easy to evaluate during the triage process. The perception that some victims may have that they were exposed makes it challenging to distinguish the worried well from those with delayed effects. As a result, it is necessary to establish a case definition and use case classifications such as "suspected," "probable," and "confirmed." Establishing the case definition can be complicated by the exposure factors previously mentioned in this chapter and further convoluted by exposure to more than one chemical substance (CDC, 2005a). Healthcare providers must work closely with poison control and local public health agencies in establishing the case definition and carrying out proper requests for resources and case reporting.

When patients from a suspected chemical incident begin to arrive, they should be held outside until the emergency department is prepared (Centers for Disease Control [CDC], Emergency room procedures in chemical hazard emergencies: a job aid, www.cdc .gov/nceh/demil/articles/initialtreat.htm). Healthcare staff should don personal protective equipment and ensure that screening for contamination is conducted and appropriate patient decontamination accomplished before the entry is allowed (See Figure 4–12). Again, this was a challenge after the sarin release in Tokyo. Many of the exposed entered the emergency departments and spread contamination in the facility, exposing healthcare providers as well. If patients are ambulatory, they should remove their own clothing and do as much self decontamination as possible. If they are not able to do so, hospital staff may be required to cut them out of their clothing and decontaminate them. There are several decontaminants options available. There are two primary methods to decontaminate a chemically exposed patient including physical removal and chemical deactivation. Profuse amounts of water or dry absorbent materials may be used for removal. Applying flour, followed by wiping it with a wet tissue paper has been described as being effective against nerve and blister agents (United States Army Medical Research Institute of Chemical Defense [USAMRICD], 2000). Chemical deactivation includes hydrolysis and oxidation. Hydrolysis options include mildly acidic solutions, alkaline solutions, or soap and water. Oxidizing options include chlorine- or peroxide-based solutions. Patient decontamination needs to be trained and exercised frequently enough to ensure all staff on all shifts can carry it out rapidly and effectively.

For most chemical disasters, the primary medical goals are patient evacuation, decontamination, and supportive care. However, unique antidotes are required for nerve and cyanide agents. Nerve agent antidotes include atropine and 2-PAM chloride. The biggest challenge healthcare providers may encounter is acquiring enough antidotes for a mass casualty event. The Centers for Disease Control and Prevention have been working with state and local public health and healthcare organizations to field "Chempacks." This program places dozens of containers storing nerve agent antidote at secure locations in each state they are available at the quickly during a crisis at the community level.

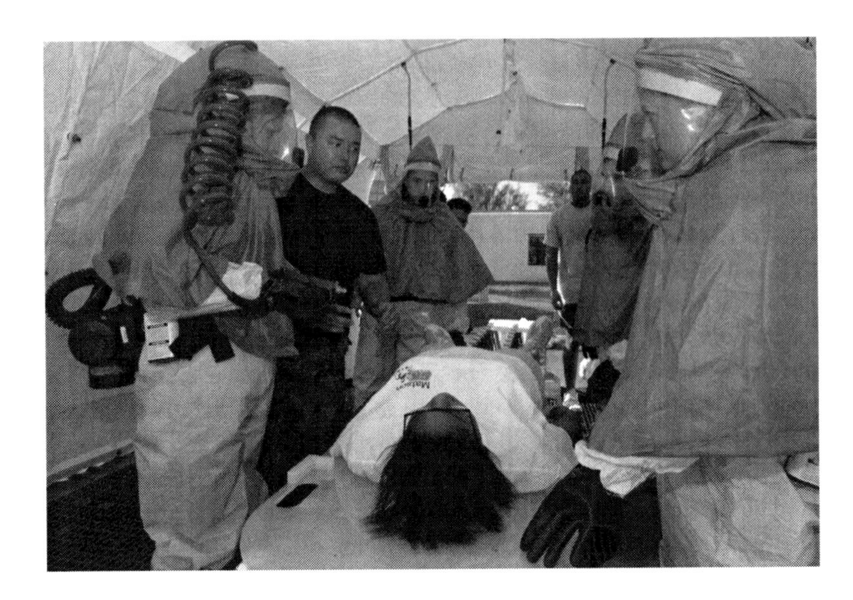

FIGURE 4–12 U.S. Air Force healthcare providers at Andersen Air Force Base receiving training on a new in-place patient decontamination system. Photo by Staff Sgt. Vanessa Valentine. http://www.andersen.af.mil/shared/media/photodb/photos/080416-F-8078V-129.JPG.

When a community is threatened by a nerve agent release and lives are at risk, hospitals may utilize the resources from Chempacks they host and request more resources from other hosting organizations if they exceed their capacity. However, the Chempacks only provide antidote for nerve agents. The other agents that require unique drug therapy are cyanide compounds. This is also a more likely scenario than nerve agents due to the large amounts of cyanide used in industry and the possibility of being exposed to cyanide by-products during structural fires. There are currently two methods available to treat cyanide poisoning. The method that has been around for several decades is a multistep process using amyl nitrite perles, sodium nitrite IV, and sodium thiosulfate IV. Another treatment option approved for use in the United States by the Food and Drug Administration in 2006 is hydroxocobalamin. This is a precursor of vitamin B12 that has been used with great success for smoke inhalation patients in Europe (Davis, 2006). The problem with available cyanide treatments is a lack of antidote surge capacity similar to what now exists for nerve agents treatment through the Chempack Program.

Public Health Response

Much of the medical surge capacity that may be needed during a chemical disaster is facilitated through public health agencies. Accessing the stockpiles of antidotes is coordinated through state and local public health agencies (LPHAs) and if more medical staffing is needed, additional healthcare professionals may be requested through LPHAs as well. LPHAs can assist in coordinating local healthcare or shelter volunteers by sending formal requests to the state for additional state or federal resources. Yet their role in chemical disaster response goes far beyond resource acquisition and coordination. Depending on the nature of a chemical release, there may not be a clear emergency scene or enough patient surge to indicate anything unusual has occurred. It is often through public health surveillance or epidemiological investigation that exposures are detected

FIGURE 4–13 A team member of the Miami-Dade County Hazardous Materials Crime Unit gathering evidence on suspicious drums. http://www.miamidade.gov/mdpd/BureausDivisions/IGB/cis.asp

and characterized. These activities are also the key to long-term follow-up for residual effects. Public health also develops health-risk communication messages for the public in times of crisis and shares state-of-the-art diagnosis and treatment information with healthcare organizations and providers. Although environmental chemical samples are not typically taken or handled by public health professionals, clinical specimens sometimes are. The Centers for Disease Control and Prevention and many State Public Health Labs have established the capacity to analyze blood, serum, and urine specimens to detect and measure exposures to nerve agents, blister agents, ricin, cyanide, and others. If the chemical release under investigation appears to be intentional, public health officials will work closely with local, state, and federal law enforcement officials. All clinical and epidemiological data may be used as evidence in a future criminal case.

There are unique chemical risks associated with most nonchemical disasters as well. When a structure is damaged by a flood, hurricane, tornado, or earthquake, chemical substances are often spilled and mixed with other chemicals. This can pose substantial risks to first responders conducting search and rescue or damage assessment operations and to individuals returning to their property. Environmental health professionals play a key role in assessing the range of chemical health risks produced by disasters and providing recommendations on chemical risk reduction.

Recovery Actions

Once a chemical emergency is over, residents of the affected area should not reenter the area until local authorities announce that it is safe to return. When building owners and residents return to the area, they should thoroughly ventilate the structure by opening all windows and turning the air handling systems on. If at any point they begin to detect the presence or feel the effects of a chemical exposure, they should immediately leave the facility and contact authorities. Although structures offer some protection to chemicals outside and can be used to shelter in place, once the hazard clears away in the outdoor environment there is sometimes a residual, low-level build-up of chemical vapors inside. If you come into contact with any suspicious liquid or solid material in or near the structure, quick action should be taken to decontaminate anyone exposed. Place contaminated clothes in a sealed bag or container, decontaminate by taking a shower, seek medical attention immediately for any symptoms, notify authorities of the material, and advise others in the surrounding area. When cleaning up a spill use the personal protective equipment recommended by the product labeling and place the rags, paper towels, or other clean-up debris in a plastic bag for disposal.

A variety of other safety issues must be considered. Avoid any potentially contaminated food or water and watch for household hazardous materials. If the disaster is not a chemical incident but one that causes structural damage, it is possible that hazardous chemicals have been spilled and possibly mixed with others to create a potentially toxic exposure. If there are widespread issues such as this, public agencies may establish drop- off locations to bring hazardous materials in for proper disposal (Figure 4–14). If there is structural damage, make sure that utilities have been turned off and use a battery powered flashlight when entering the building. Never smoke or initiate any open flames until an evaluation of household chemicals has been completed. Even though there may be no apparent odors or obvious evidence of a flammable risk, a flame can result in a fire or explosion. Turn on the flashlight before entering the building. If there is natural gas present, even the small spark generated by a flashlight could cause an explosion. If gas is smelled, immediately exit the building and contact authorities. Do not make the call inside

FIGURE 4–14 The Environmental Protection Agency (EPA) has set up a hazardous waste material collection site for the disposal of toxic and otherwise hazardous materials. The site is located in the center of Cameron, LA which was severely affected by Hurricane Rita. Robert Kaufmann/FEMA. http://www.photolibrary.fema.gov/photolibrary/photo_details.do?id=21306.

the house and do not spend any time gathering items. Get out of the structure and move upwind of the residence to make the emergency call on a cell phone or neighbor's phone.

If the building has been flooded and water remains inside or if there has been a structural fire, do not enter it until a qualified inspector or engineer declares it safe to enter. Make sure that all important local emergency numbers are readily available, including the Poison Control number. The National Poison Control Number is 800-222-1222 (Centers for Disease Control [CDC], Tips to prevent poisoning: safety tips for you, your family, and friends, www.cdc.gov/ncipc/factsheets/poisonprevention.htm.).

If a power outage is being experienced when residents return to their homes, some will use generator power. This introduces the risk of carbon monoxide (CO) poisoning. Running a generator or vehicle in a garage or other unventilated area can quickly produce enough carbon monoxide to result in illness or death. CO is colorless and odorless making it very difficult to detect without a CO detector. Burning in a fireplace with a blocked flue, or using charcoal grills or other fuel burning appliances inside can also generate CO (Underwriters Laboratories, 2008).

Vulnerable Populations

Chemical exposures have a dramatically different impact on vulnerable populations throughout an impacted community. Children have more respirations per minute than adults and have breathing zones closer to the ground. As a result, the release of sarin, chlorine, or other toxic chemicals with greater vapor density will be more concentrated in a lower breathing zone. Children will take more respirations in a more highly concentrated area of the release (Bearer, 1995). Infants also have more permeable skin and a greater surface-to-mass ratio making them more vulnerable to dermal exposures (Slater, 1997). This can also result in rapid heat loss and place infants at greater risk for hypothermia during decontamination operations. The other decontamination challenge is the handling of infants by workers in protective clothing. The protective gear limits vision, mobility, and dexterity making it difficult to effectively handle small patients (Tucker, 1997). Children are not "small adults." They respond differently to chemical exposures (Landrigan et al., 1998). For example, a newborn child is up to 164 times more susceptible to organophosphates and 65 times more sensitive to diazinon (Furlong et al., 2006). Between the ages of 1 and 6 years old, their rapid development can be interrupted by chemical exposures and impose challenges to proper cell development, keeping vital organs from developing properly.

According to the 2007 American Community Survey of the Census Bureau, nearly 15% of Americans over 5-year-old also have at least one disability. That is 41 million people who have some sensory, physical, mental, self-care, or employment disability (U.S. Census Bureau, 2007). These groups each pose unique challenges to chemical preparedness and response. Some disabled individuals will not understand instructions during a crisis and others may not be capable of following them. Local emergency planners need to work closely with disability advocacy groups to better characterize the vulnerable populations in their community. These individuals need to be identified in advance and their needs addressed by local planners. This includes consideration of personal care equipment, mobility aids, adaptive feeding devices, service animals, specially equipped vehicles, and medical equipment requiring power. During a chemical emergency, evacuation of these populations will be challenging and decontamination of the disabled and their equipment must be addressed prior to an emergency.

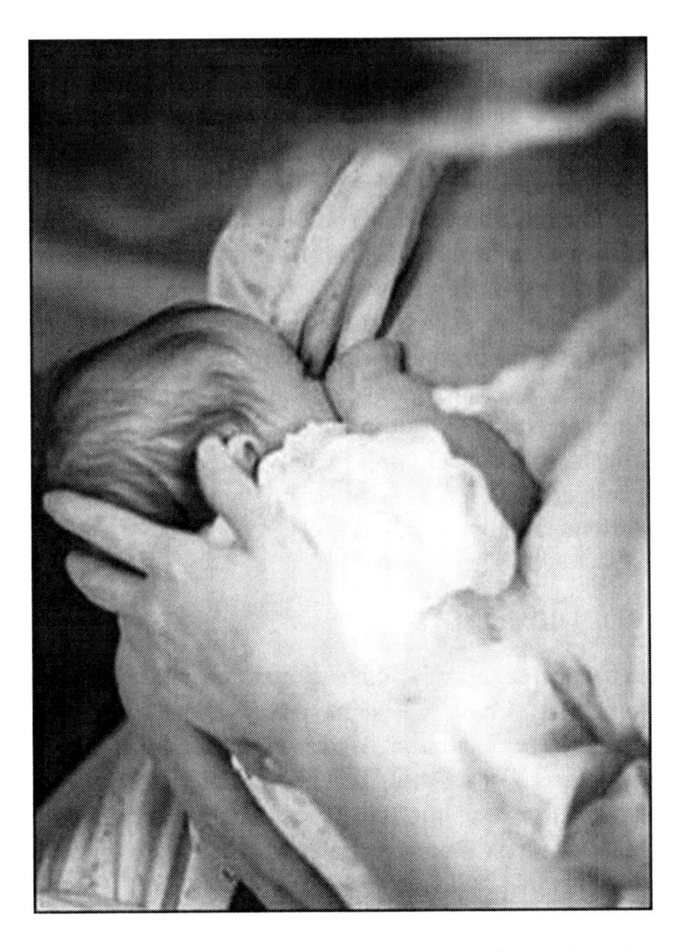

FIGURE 4–15 Infants are particularly susceptible to chemical exposures. http://www.cdc.gov/Features/MedicationUse/MedicationUse_250px.jpg.

There are special public health challenges with pregnant and nursing mothers during a chemical release. Chemical exposures can permanently damage the developing fetal organs. Priority evacuation, treatment, and decontamination should be offered to pregnant victims whenever possible. Following a chemical release, nursing mothers with newborns can unknowingly increase the potential exposure of their babies to certain chemicals as well. A variety of environmental chemicals have been found in human breast milk including mercury and methyl mercury, lead, cadmium, polychlorinated biphenyl (PCBs), chlorinated dibenzo-*p*-dioxin (CDDs), and many others (U.S. Department of Health and Human Services, Agency for Toxic Substances and Disease Registry, Interaction profiles for persistent chemicals found in breast milk (chlorinated dibenzo-*p*-dioxins, hexachlorobenzene, *p,p´*dde, methylmercury, and polychlorinated biphenyls), www.atsdr.cdc.gov/-interactionprofiles/IP-breastmilk/ip03.pdf). Following any chemical incident, recommendations should include a description of any exposure risks associated with exposed women nursing their babies.

Summary

As the threats of chemical terrorism persists and as hazardous industrial chemicals continue to be produced, stored, and transported across our highways, railways, and waterways, the need for enhancing our chemical emergency preparedness remains crucial. The differences between preparing for intentional and accidental chemical releases are few for those with public health and healthcare responsibilities. Unique public health challenges that result from chemical incidents span across every facet of the emergency management process. Exposures are mitigated by continual industrial process safety and enhanced security. There are also important mitigation steps that must begin in each home through reducing the hazardous chemicals present in the home, organizing a home preparedness kit, making a plan, and following the instructions of officials. Consistent, uniform labeling and nomenclature continues to develop and emergency exposure standards continue to be refined through lessons learned exercises and responses. The focus of these activities must cover the full lifecycle of chemical products, from industry to homes. The response to a major chemical release may include rapid diagnosis and treatment, mass care, mass fatality management, and long-term plans to monitor exposed survivors. Much progress has been made in these areas in recent years and much more is needed. Sustained improvements require a public–private partnership. There is little that public agencies can do to enhance their response and recovery without the cooperation of industry and citizens before, during, and after a disaster.

Websites

Agency for Toxic Substance and Disease Registry: www.atsdr.cdc.gov/.

Agency for Toxic Substance and Disease Registry, Medical Management Guidelines (MMGs) for Acute Chemical Exposures: www.atsdr.cdc.gov/MHMI/mmg.html.

American Industrial Hygiene Association, Emergency Response Planning Committee: www.aiha.org/-Content/InsideAIHA/Volunteer%2bGroups/ERPcomm.htm.

American Red Cross Guide to Home Chemical Safety and Emergency Procedures: www.redcross.org/static/file_cont157_lang0_67.pdf.

Centers for Disease Control and Prevention, Case Definitions for Chemical Poisoning: www.cdc.gov/mmwr/preview/mmwrhtml/rr5401a1.htm.

Centers for Disease Control and Prevention, Emergency Preparedness & Response Chemical Emergencies: www bt.cdc.gov/chemical/.

Centers for Disease Control and Prevention, Public Health Emergency Response Guide for State, -Local, and Tribal Public Health Directors: www.bt.cdc.gov/planning/pdf/cdcresponseguide.pdf.

Department of Energy, Subcommittee on Consequence Assessment and Protective Actions (SCAPA): www.orise .orau.gov/emi/scapa/default.htm.

Department of Transportation, Emergency Response Guidebook (ERG2008): http://hazmat.dot.gov/pubs/erg/gydebook.htm.

Department of Transportation, Pipeline and Hazardous Materials Safety Administration, Hazmat Safety Community: www.phmsa.dot.gov/hazmat.

Environmental Protection Agency, Acute Exposure Guideline Levels: www.epa.gov/oppt/aegl/.

Environmental Protection Agency, Computer-Aided Management of Emergency Operations (CAMEO) -Program: www.epa.gov/emergencies/content/cameo/index.htm.

Environmental Protection Agency, Pesticide A-Z Index: www.epa.gov/pesticides/a-z/index.htm.

Food and Drug Administration, Food Defense and Terrorism: www.cfsan.fda.gov/~dms/defterr.html.

National Institute for Occupational Safety and Health (NIOSH) Guidance for Protecting Building Environments from Airborne Chemical, Biological, or Radiological Attacks: www.cdc.gov/niosh/docs/2002-139/-default .html.

National Institute for Occupational Safety and Health (NIOSH) Emergency Preparedness for Business: www.cdc .gov/niosh/topics/prepared/prepared_facility.html.

National Institute for Occupational Safety and Health (NIOSH) Pocket Guide to Chemical Hazards: www.cdc.gov/niosh/npg/default.html.

National Library of Medicine, Toxicological Data Network—TOXNET: toxnet.nlm.nih.gov/.

National Pesticide Information Center: npic.orst.edu/emerg.htm.

Protective Action Criteria (PAC) with AEGLs, ERPGs, and TEELs: www.orise.orau.gov/emi/scapa/teels.htm.

United Nations, Globally Harmonized System of Classification and Labeling of Chemicals (GHS) Program: www.unece.org/trans/danger/publi/ghs/ghs_welcome_e.html.

References

Agency for Toxic Substances and Disease Registry (ATSDR) (2002a). *Managing hazardous materials incidents, vol 3: medical management guidelines for arsine*. Atlanta, GA: U.S. Department of Health and Human Services. www.atsdr.cdc.gov/MHMI/mmg169.pdf.

Agency for Toxic Substances and Disease Registry (ATSDR) (2002b). *Managing hazardous materials incidents, vol. 3: medical management guidelines for blister agents: nitrogen mustard (HN-1, HN-2, and HN-3)*. Atlanta, GA: U.S. Department of Health and Human Services. www.atsdr.cdc.gov/MHMI/mmg164.pdf.

Agency for Toxic Substances and Disease Registry (ATSDR) (2002c). *Managing hazardous materials incidents, vol 3: medical management guidelines for hydrogen cyanide*. Atlanta, GA: U.S. Department of Health and -Human Services. www.atsdr.cdc.gov/MHMI/mmg8.pdf.

Agency for Toxic Substances and Disease Registry (ATSDR) (2002d). *Managing hazardous materials incidents, vol 3: medical management guidelines for lewisite and mustard lewisite mixture*. Atlanta, GA: U.S. Department of Health and Human Services. www.atsdr.cdc.gov/mhmi/mmg163.pdf.

Agency for Toxic Substances and Disease Registry (ATSDR) (2002e). *Managing hazardous materials incidents, vol 3: medical management guidelines for nerve agents: tabun (GA), sarin (GB), soman (GD), and VX*. Atlanta, GA: U.S. Department of Health and Human Services. www.atsdr.cdc.gov/MHMI/mmg166.pdf.

American Industrial Hygiene Association (2006). *ERP committee procedures and responsibilities*. Fairfax, VA: AIHA.

American Red Cross (1994). Your guide to home chemical safety and emergency procedures. www.redcross.org/static/file_cont157_lang0_67.pdf.

Bearer, C. F., (1995). How are children different from adults? *Environ Health Perspect* 103:7–12.

Bertazzi, P. A. (1999). Future prevention and handling of environmental accidents. *Scand J Work Environ Health* 25:580–588.

Binder, S. (1989). Death, injuries, and evacuations from acute hazardous materials releases. *Am J Public Health* 79(8):1042–1044. www.ajph.org/cgi/reprint/79/8/1042.pdf.

Cave, D., & Fadam, A. (February 21, 2007). Iraqi militants use chlorine in 3 bombings. *The New York Times*. www.nytimes.com/2007/02/21/world/middleeast/21cnd-baghdad.html.

Centers for Disease Control (CDC) (2003). Recognition of illness associated with exposure to chemical agents—United States, 2003. *MMWR* 52(39):938–940. www.cdc.gov/mmwr/preview/mmwrhtml/mm5239a3.htm#tab.

Centers for Disease Control (CDC) (2005a). Case definitions for chemical poisoning. *MMWR* 54(RR01):1–24. www.cdc.gov/mmwr/-preview/mmwrhtml/rr5401a1.htm.

Centers for Disease Control (CDC) (2005b). Public health consequences from hazardous substances acutely released during rail transit—South Carolina, 2005; selected states, 1999–2004. *MMWR* 54(3):64–67. www.cdc.gov/mmwr/preview/mmwrhtml/mm5403a2.htm.

Davis, R. (2006). FDA gives green light to cyanide treatment. *USA Today*. December 28, 2006. www.usatoday.com/news/health/2006-12-27-cyanide_x.htm.

Everitt, R. (2004). *Falling stars: air crashes that filled rock and roll heaven*. Augusta, GA: Harbor House.

Federal Emergency Management Agency (2006). What to do during a hazardous materials incident. www.fema.gov/hazard/hazmat/hz_during.shtm.

Fortun, K. (2001). *Advocacy after Bhopal*. Chicago, IL: University of Chicago Press.

Furlong, C. E., Holland, N., Richter, R.J., et al. (2006). PON1 status of farmworker mothers and children as a predictor of organophosphate sensitivity. *Pharmacogenet Genomics* 16:183–190.

Hart, A., & Wald, M. L. (January 8, 2005). Cloud rising from train wreck, then death and a ghost town. *The New York Times*. www.nytimes.com/2005/01/08/national/08train.html.

Kaplan, D. E., & Marshall, A. (1996). *The cult at the end of the world*. New York: Crown Publishers, Inc.:85.

Kumar, S. (2004). Victims of gas leak in Bhopal seek redress on compensation. *BMJ* 329:366. www.bmj.com/cgi/reprint/329/7462/366.pdf.

Landrigan, P. J., Carlson, J. E., Bearer, C.F., et al. (1998). Children's health and the environment: a new agenda for prevention research. *Environ Health Perspect* 106(Suppl 3):787–794.

Lichtblau, E., Drew, C. (December 4, 2004). Ridge's record: color alerts and mixed security reviews. *New York Times*. www.nytimes.com/2004/12/01/politics/01home.html.

National Fire Protection Association (2007). *NFPA 704: standard system for the identification of the hazards of materials for emergency response*. www.nfpa.org/aboutthecodes/aboutthecodes.asp?docnum=704&cookie%5Ftest=1.

Olson, K. B. (1999). Aum Shinrikyo: once and future threat? *Emerg Infect Dis* 5(4):513–516.

Pipeline and Hazardous Materials Safety Administration (2008). *Ten year hazardous materials incident data*. U.S. Department of Transportation. http://hazmat.dot.gov/pubs/inc/data/tenyr.pdf.

Shrivastava, P. (1987). *Bhopal: Anatomy of a Crisis*. Cambridge, MA: Ballinger Publishing.

Slater, M. S., & Trunkey, D. D. (1997). Terrorism in America: an evolving threat. *Arch Surg* 132:1059–1066.

Tucker, J. B. (1997). National Health and Medical Services response to incidents of chemical and biological terrorism. *JAMA* 278:362–368.

Underwriters Laboratories (2008). Product safety tips: CO alarms. www.ul.com/consumers/co.html.

United Nations (2007). *Globally harmonized system of classification and labeling of chemicals (GHS)*. 2nd revised ed. United Nations: UNECE. www.unece.org/trans/danger/publi/ghs/ghs_rev02/02files_e.html.

United States Army Medical Research Institute of Chemical Defense (USAMRICD) (2000). *Medical management of chemical casualties handbook*. 3rd ed. Aberdeen Proving Grounds, MD: United States Army Medical Research Institute of Chemical Defense (USAMRICD), Chemical Casualty Care Division. www.gmha.org/bioterrorism/usamricd/Yellow_Book_2000.pdf.

United States Environmental Protection Agency (2001). *Consolidated list of chemicals subject to the Emergency Planning and Community Right-To-Know Act (EPCRA) and Section 112(r) of the Clean Air Act*. EPA. www.epa.gov/earth1r6/6sf/pdffiles/title_iii_list_of_lists.pdf.

U.S. Army Center for Health Promotion and Preventive Medicine (CHPPM) (2003). *Basic Questions Regarding Acute Exposure Guideline Levels (AEGLs) in Emergency Planning and Response*. www.osha.gov/SLTC/emergencypreparedness/chemical/pdf/tier_2-aegls_-basic_usachppm1_03.pdf.

U.S. Census Bureau (2007). S1801 disability characteristics: 2007 American community survey 1-year estimates. www.factfinder.census.gov/servlet/STTable?_bm=y&-geo_id=01000US&-qr_name=ACS_2007_1YR_G00_S1801&-ds_name=ACS_2007_1YR_G00_&-_lang=en&-redoLog=false&-format=&-CONTEXT=st.

U.S. Department of Health and Human Services, National Institute for Occupational Safety and Health (2002). *Guidance for Protecting Building Environments from airborne Chemical, Biological, or Radiological Attacks*. CDC. www.cdc.gov/niosh/docs/2002-139/default.html.

U.S. Environmental Protection Agency (2005). Norfolk southern Graniteville derailment. www.epa.gov/-region4/graniteville/index.htm.

5 Earthquakes

Objectives of This Chapter

- List the factors that contribute to earthquake morbidity and mortality.
- Recognize the scales used to measure the intensity and magnitude of earthquakes.
- Describe the unique challenges associated with recognizing crush injuries.
- Explain the difference between crush syndrome and compartment syndrome.
- Provide examples of hazards to look for when assessing a home for earthquake risks.
- List the steps individuals may take to maximize their safety during an earthquake.
- Explain what an earthquake feels like and what initial actions persons should take when they experience it.
- Describe where the greatest opportunities are to reduce earthquake related morbidity and mortality.
- Recognize the populations at greatest risk during a seismic event.
- List the key partners necessary to carry out a public awareness campaign for earthquake preparedness.

"We learn geology the morning after the earthquake."

Ralph Waldo Emerson

Introduction—1906 San Francisco Earthquake

Lloyd was a young boy living in San Francisco with his parents and baby sister. It was just after 5 a.m. and he was sleeping soundly on that early Wednesday morning when the shaking woke him. So frightened he could not scream, he buried his face under his pillow and held tightly to keep from being shaken out of bed. As he peeked out across his room from under his pillow, the knick-knacks on his bureau danced vigorously and he could hear dishes and other loose items in the house rattling loudly. As soon as the shaking slowed, he bolted to his parents' room. His baby sister sat up quietly in her crib as his parents sat in stunned silence. As soon as the shaking subsided, the family moved quickly to the backyard. When an extended moment of calm arrived, they ventured back into the house to see the damage. As they cleaned up the broken dishes, vases, and other breakables, they looked out the front window to see their shocked neighbors lining the sidewalks in bewilderment.

As the day progressed, the family ran back and forth, leaving the house every time the shaking resumed until they finally decided to build a makeshift camp site in the backyard. As they began cooking over an open fire, the ash from the fires raging in San Francisco began to fall on them (Figure 5–1). They stayed in their improvised tents for several days as those who were fleeing from the raging fires evacuated past their home to camps with clever names like Camp Grateful and Camp Thankful (Head, 1906).

This family is among the lucky ones. Hundreds died from collapsing structures that day across San Francisco and tens of thousands became homeless. The final casualty numbers became a source of controversy. Although an initial Army report estimated about 500 deaths in San Francisco, just over 100 deaths in San Jose, and 64 deaths in Santa Rosa, more recent estimates place the actual number of fatalities beyond 3000 (Greely, 1906; Hansen and Condon, 1989). The recovery efforts that followed the earthquake were impressive. Military units began immediate response with no formal activation or call up. In spite of the massive damage, civil order was quickly restored. A quarter of a million homeless residents were provided shelter through enormous tent cities and small modular housing, the forerunner of today's FEMA trailers. A safe food and water supply was quickly established, and toilet facilities and other key sanitation necessities were set up. Aid poured in from around the nation and rebuilding efforts began almost immediately.

The public health impact of this disaster was felt for over a year. In addition to the injuries and fatalities resulting from collapsed structures and fires, an associated outbreak ignited nearly a year later. Although local public health authorities worked with public works, military, and volunteers to reinstate healthy living conditions, there was one threat that was not fully addressed. At the beginning of the twentieth century, bubonic plague had been brought into San Francisco by ships arriving in port carrying infected rats. Local officials initially ignored the 1900 plague outbreak among poor citizens living in Chinatown. In spite of a racially motivated, feeble public health response, the outbreak

FIGURE 5–1 San Francisco, California, Earthquake, April 18, 1906. Downtown San Francisco showing residents watching fire after the 1906 earthquake. Photo by Ralph O. Hotz. http://libraryphoto.cr.usgs.gov/htmllib/btch476/btch476j/btch476z/btch476/hpe00104.jpg.

FIGURE 5–2 San Francisco, California, Earthquake, April 18, 1906. Ruined house on Shotwell Street. Photo courtesy of U.S. Geological Survey. http://libraryphoto.cr.usgs.gov/htmllib/batch01/batch01j/batch01z/batch01/ggk02894.jpg.

was brought under control. After the massive 1906 earthquake, the conditions were ripe for the reemergence of plague. The organism that causes the disease, *Yersinia pestis*, is carried by rats and transmitted to humans through the bite of fleas that have fed on infected rats. One year after the earthquake, a large plague outbreak spread across San Francisco. Between May and September of 1907, there were 160 cases with 77 fatalities (Todd, 1909). These cases occurred among predominantly white affluent citizens.

There were several important differences in the public health response to the 1907 San Francisco, post-earthquake plague outbreak compared to the Chinatown outbreak of 1900. More was understood 7 years later about the causes of plague and effective control measures. In addition, the city had just been through such a calamity that those rebuilding worked together with a greater sense of purpose than ever before. A massive effort was undertaken to clean up the city to exterminate rats and reduce their available food sources (See Figure 5–4). Businesses, schools, and households were provided with detailed instructions on simple measures to reduce the rat problems across the city (See Figure 5–3). This effort is considered to be one of the most effective public health responses to an outbreak in American history (Risse, 1992).

Earthquake-Related Definitions

Aftershock: An earthquake of similar or lesser intensity that follows the main earthquake. They are usually smaller than the initial shock and may occur days, weeks, months, or even years later. Larger earthquakes are usually followed by larger aftershocks over a longer period of time.

Crush injury: An injury that occurs when a body part experiences a high degree of pressure, usually under the weight of a heavy object. The resulting injuries include fractures, bruising, bleeding, lacerations, and compartment and crush syndromes.

GENERAL COMMITTEE

Homer S. King, Chairman
L. M. King, Secretary

E. H. Rixford	R. H. Swayne
Gustave Brenner	Capt. H. W. Goodall
T. C. Friedlander	James McNab
H. H. Sherwood	A. W. Scott, Jr.
Frank J. Symmes	L. M. King
Chas. C. Moore	Walter Macarthur
Henry M. Sherman, M. D.	John Gallwey, M. D.
Martin Regensburger, M. D.	Langley Porter, M. D.
John M. Williamson, M. D.	H. C. Moffitt, M. D.
George H. Evans, M. D.	P. M. Jones, M. D.
Charles G. Levison, M. D.	E. N. Ewer, M. D.
James H. Parkinson, M. D.	N. K. Foster, M. D.

CITIZENS HEALTH COMMITTEE

OF SAN FRANCISCO

HEADQUARTERS · 1233 MERCHANTS EXCHANGE BUILDING

TELEPHONE KEARNY 2183

EXECUTIVE COMMITTEE

Chas. C. Moore, Chairman
Gustave Brenner
Walter Macarthur
Geo. H. Evans, M. D.
Frank J. Symmes

SAN FRANCISCO, CAL.,

Important To Our Customers

Preservation of your business and the health of every person in San Francisco demand your instant and continued assistance in *exterminating RATS* and cleaning up the entire city.

The undersigned Associations are fully advised that a *quarantine* will be placed on the city *unless sanitary conditions* are made satisfactory to the federal government. Hence you are urged to give your personal attention to this matter.

Your business future depends upon the work you do to help the doctors.

San Francisco must be rid of rats within sixty days.

RATS SPREAD BUBONIC PLAGUE.

Fleas leave sick or dead rats and carry the plague to human beings.

Rats in San Francisco are infected with bubonic plague. There is no question about it, and every man, woman and child in the city must help the federal and city officials clean up.

Every line of trade must assist.

The Citizens Health Committee, advised by Dr. Rupert Blue of the United States Public Health and Marine Hospital Service, is directing the work.

GET BUSY and see that no open garbage cans are in your own or your neighbor's home or place of business. Don't leave food where rats can get it.

If quarantine is ordered you go out of business.

Enclosed are instructions prepared by the Health Committee and government inspectors.

Work, not complaints, is demanded. Notify Dr. Blue, 401 Fillmore Street; L. M. King, Secretary Citizens Health Committee, 1233 Merchants Exchange, or L. R. Levy, Secretary of the Beer Bottlers' Association, 524 Gough Street, of any insanitary condition in your neighborhood.

GET BUSY with rat traps and poison.

Make your scavenger do his work cleanly and often.

(Signed) Brewers' Protective Association.
S. F. Beer Bottlers' Board of Trade.
S. F. Soda Water Manufacturers' Association.
Bottlers' Protective Association.

TEN THOUSAND OF THESE CIRCULARS WERE DISTRIBUTED TO THE SALOONS AND OTHERS OF THEIR TRADE BY THE ORGANIZATIONS SIGNING.

FIGURE 5–3 A 1907 flyer distributed to saloons and other alcohol serving establishments in the San Francisco area asking for assistance in bringing the post-earthquake, 1907 plague outbreak under control. Image Courtesy of San Francisco Department of Public Health.

FIGURE 5–4 Eight public health service workers holding buckets filled with rat bait working to bring the 1907 San Francisco, CA, plague outbreak under control. Photo courtesy of the National Library of Medicine.

Earthquake: Ground shaking caused by an abrupt slipping or movement of a portion of the earth's crust, accompanied and followed by a series of vibrations.

Epicenter: The place on the earth's surface directly above the point on the fault where the earthquake begins. Once fault slippage begins, it expands along the fault during the earthquake and can extend hundreds of miles before stopping.

Fault: The fracture across which displacement has occurred during an earthquake. It may also push up, down, or sideways. The slippage may range from less than an inch to more than 10 yards in a severe earthquake.

Intensity: The measured effect of an earthquake on people, buildings, or land. The two scales used to measure intensity are the Richter Scale and Modified Mercalli Scale.

Liquefaction: The behavior of loose soil when it becomes saturated. Soil deposits near rivers and beaches are particularly susceptible. As the resistance of the soil is compromised by water and vibration, structures are more likely to shift or collapse as the ground loses strength and acts more like a liquid.

Magnitude: The amount of energy released during an earthquake, which is computed from the amplitude of the seismic waves. A magnitude of 7.0 on the Richter Scale indicates an extremely strong earthquake. Each whole number on the scale represents an increase of about 30 times more energy released than the previous whole number represents. For example, an earthquake measuring 6.0 is about 30 times more powerful than one measuring 5.0. With a logarithmic scale, that means a 7.0 is about 900 times more powerful than a 5.0 (30×30).

Modified Mercalli scale: This is a scale that has been modified for North American conditions that uses Roman numerals I through XII to describe the effects observed by an earthquake. The lower numbers describe how an event is felt by people in the area and the larger numbers describe the expected damage. This is an arbitrary qualitative ranking based on observations rather than a mathematically derived scale.

Richter scale: A logarithmic measurement of earthquake intensity. Each whole number increase on this scale represents a 10-fold increase in force. In other words, a 6.0 earthquake has 10 times more wave amplitude than a 5.0 as it is detected on a seismogram and it releases 30 times more energy.

Seismic hazard: The damage potential from seismic events. It depends on a variety of geological factors such as magnitude of an earthquake, distance from the epicenter, and type of soil and ground materials in the affected area.

Seismic waves: Vibrations that travel outward from the earthquake fault at speeds of several miles per second. Although fault slippage directly under a structure can cause considerable damage, the vibrations of seismic waves cause most of the destruction during earthquakes.

Earthquakes

One of the most powerful and potentially devastating forces of nature is the energy contained in the Earth's surface. Enormous tectonic plates comprising the Earth's crust constantly slide together. There are points where stress accumulates. The violent shaking of an earthquake results from the sudden release of this energy and can collapse buildings, fracture roadways, and create tsunamis. Fatalities can easily reach into the thousands from a single event and hundreds of thousands of lives are impacted. The intensity of these events is measured in two ways. The Richter scale is a logarithmic measurement of earthquake intensity (Table 5–1). Each whole number increase on this scale represents a 10-fold increase in force. In other words, a 6.0 earthquake is actually 10 times greater in magnitude than a 5.0. This scale was not designed to estimate consequential damage from a seismic event. The Modified Mercalli scale uses Roman numerals I through XII to describe the effects observed by an earthquake. The lower numbers describe how an event is felt by people in the area and the larger numbers describe the expected damage. Both scales are rough estimates because there has not been consistent methods used to gather the data on which they are based. Although there is not a concise correlation between the two measures, both the U.S. Geological Survey and Federal Emergency Management Agency have approximated the relationship of the two scales.

There are an array of variables that influence the morbidity and mortality resulting from a seismic event. On Good Friday, March 27, 1964, the second most powerful earthquake ever recorded in history hit the Prince William Sound area in Alaska (Figure 5–5). The earthquake had a moment magnitude of 9.2 on the Richter scale and lasted for nearly 4 minutes. It was followed by nine aftershocks greater than 6.0 on the same day and many more in the subsequent weeks and months. It caused landslides, a tsunami, and direct damage to buildings in many major Alaskan cities but the death toll was only 131, including 115 in Alaska and 16 along the coastlines of California and Oregon (Christensen, D., The great Alaska earthquake of 1964, Alaska Earthquake information center, www.aeic.alaska.edu/quakes/Alaska_1964_earthquake.html). This is in stark contrast to the 1976 Tangshan, China earthquake which registered just over 7.5 and killed about a quarter million people (Peek-Asa, 2003). The differences between the two disasters underscores the risk factors for increased morbidity and mortality. How can an earthquake that is nearly 20 times greater in magnitude and hundreds of times more powerful, generate less than 1% of the casualties of another one? It is a product of variables including individual

Table 5–1 Comparison of Scales Used to Measure the Earthquake Intensity with Damage Levels and Frequency

Modified Mercalli Scale		Damage Levels	Richter Scale	Annual Estimated Frequency
I	Slightly Detectable	Seldom felt or noticed; instrumentation detects occurrences	1.0–3.0	Thousands per day
II	Sensed by some	Felt by some people, especially on upper floors of structures. Negligible damage with small, unstable objects displaced; vibration similar to a large passing truck	3.0–3.9	50,000
IV–V	Felt by everyone	Slight damage with windows, dishes, glassware broken; furniture moved or overturned; weak plaster and masonry cracked	4.0–4.9	6,000
VI–VII	Minor damage	Slight to moderate damage in well-built structures; considerable in poorly built structures; furniture and weak chimneys broken; masonry damaged; loose bricks, tiles, plaster, and stones will fall	5.0–5.9	800
VIII	Destructive	Considerable structural damage, particularly to poorly built structures; chimneys, monuments, towers, elevated tanks may fall; frame houses moved; trees damaged; cracks in wet ground and steep slopes	6.0–6.4	120
IX	Ruinous	Severe structural damage; some will collapse; general damage to foundations; serious damage to reservoirs; underground pipes broken; conspicuous cracks in ground; liquefaction	6.5–6.9	
X	Disastrous	Most masonry and frame structures/ foundations destroyed; some well-built wooden structures and bridges destroyed; serious damage to dams, dikes, embankments; sand and mud shifting on beaches and flat land	7.0–7.4	20
XI	Very disastrous	Few or no masonry structures remain standing; bridges destroyed; broad fissures in ground; underground pipelines completely out of service; rails bent; widespread landslides	7.5–8.0	
XII	Catastrophic	Damage nearly total; lines of sight and level distorted; objects thrown into the air; damage for hundreds of miles	>8.0	1

Source: Adapted from U.S. Geological Survey and Federal Emergency Management Agency sources.

human characteristics and behaviors, population density, time of day, weather, structural standards, distance from the epicenter, and geology. These and other factors all work together in a complex, dynamic process that determines the resulting morbidity and mortality. For example, a sturdy structure may be sitting in an area prone to liquefaction and collapse or a vulnerable structure may be located on solid ground and remain standing. In both cases, the occupants may react in such a way that they increase their likelihood of injury. If an earthquake occurs at night, there are likely to be a smaller number of injuries in suburban and rural areas since fewer people will be in multistory and other vulnerable structures. This could be reversed in older urban areas where a large number of unstable multistoried residences could prove to be more dangerous locations than some of the newer office buildings.

Earthquake Morbidity and Mortality Variables

1. Geological factors: Earthquake magnitude, local geological composition, and aftershocks.
2. Geographical factors: Distance from epicenter, secondary emergencies generated in the immediate area such as fires and hazardous materials releases.
3. Timing: The time of day, day of the week, time of year.
4. Weather: Extreme cold or heat, excessive rain, and wind.
5. Individual characteristics: Demographics such as age, sex, physical and mental limitations, and previous earthquake experiences and preparedness awareness and training.
6. Individual behavior: The actions taken by those at risk prior to, during, and immediately after an earthquake.
7. Built environment: Type of building materials, number of stories, age and condition, mitigation history for building contents as well as structural retrofitting.
8. Community variables: Previous awareness and preparedness activities, social networks, medical and first responder surge capacity.

The dynamic nature of these variables makes outcome predictions for various sized events very difficult. Perhaps, the best effort to date in constructing a useful modeling tool is the Multi-Hazard Loss Estimation Methodology Earthquake Model used in the HAZUS-MH earthquake Model. This is a valuable planning, training, and exercise tool developed by the Federal Emergency Management Agency for federal, state, and local governments. Although it has many limitations, it can provide rough estimates of morbidity and mortality that are extremely valuable in preparedness training and exercises.

For preparedness initiatives, the two most influential determinants of morbidity and mortality are the built environment and human behavioral response to seismic events. Although the geological factors, timing, weather, and other factors cannot be changed, the built environment and human behavioral responses can be strengthened through a

FIGURE 5–5 Alaska Earthquake, March 27, 1964. This highway embankment fissured and spread as a result of liquefaction. The road was built on thick deposits of alluvium and tidal estuary mud. Image courtesy of U.S. Geological Survey. /htmllib/batch07/batch07j/batch07z/batch07/aeq00065.jpg.

variety of awareness programs and related efforts. Much can be done to mitigate the impact of an earthquake in the built environment before, during, and after a seismic event. If an individual is in a relatively stable building but makes poor decisions regarding immediate actions, the survivability of the structure may not be enough to prevent serious human injury. For example, if individuals do not take cover but choose to run from the building while the seismic vibrations continue, they are at greater risk of being struck by falling debris from the façade of the building than to be injured from a building collapse. If the building is not stable, individuals who take appropriate cover along an inside wall or under a sturdy piece of furniture are more likely to remain in a survivable void space and be rescued later from the collapsed structure.

It is not enough to put building codes into place across areas at increased risk for earthquake activity. For example, a 1999 earthquake centered near Izmit caused the collapse of buildings across seven Turkish provinces in an area larger than 150 miles (>250 km). The earthquake registered 7.4 on the Richter scale and left about a half million people homeless, tens of thousands injured, and over 15,000 dead (Mid-America Earthquake Center Research Brief, Earthquake in Turkey, http://mae.ce.uiuc.edu/research/research_briefs/Turkey_earth.html). Turkey has always been a region prone to destructive seismic activity. Its history includes a 1939 earthquake that killed over 30,000 and at least 18 other twentieth-century seismic events that measured 7.0 or greater (National Geospacial Data Center, Significant earthquake database, www.ngdc.noaa.gov/nndc/struts/form?t=101650&s=1&d=1). As a result, building codes were developed and instituted in Turkey over the course of several decades to reduce the vulnerability of new structures and make older structures less susceptible to collapse during an earthquake. Unfortunately, it was observed by engineers responding from around the world to assist in 1999 earthquake recovery efforts that building codes, though in place, were seldom followed. As a result, many died who otherwise could have survived. Building codes and other public safety requirements are useless without enforcement.

Top 10 Most Lethal Earthquakes:

Rank	Year	Location	Fatalities	Magnitude
1	1556	China, Shensi	830,000	~8
2	1976	China, Tangshan	255,000	7.5
3	1138	Syria, Aleppo	230,000	9.1
4	2004	Sumatra	227,898	9.1
5	856	Iran, Damghan	200,000	Unknown
6	1920	China, Haivuan, Ningxia	200,000	7.8
7	893	Iran, Ardabil	150,000	Unknown
8	1923	Japan, Kanto	142,800	7.9
9	1948	Turkmenistan, Ashgabat	110,000	7.3
10	1290	China, Chihli	100,000	Unknown

Source: U.S. Geological Survey (http://earthquake.usgs.gov/regional/world/most_destructive.php).

One uncontrollable but extremely influential variable is the weather. Although weather has no influence on what goes on deep below the Earth's surface where earthquakes originate, it plays a major role in the outcomes of those who survive the initial destruction and the success of earthquake response and recovery efforts. This is a variable that is often overlooked during the planning and exercise process. Most full-scale preparedness exercises are conducted in the spring and summer months. This may result in an underestimation of the complications that may be faced in the aftermath of an earthquake. The distribution of fault lines does not favor areas with mild weather and there is an equal chance of an earthquake in desert, arctic, or tropical conditions. In earthquake response efforts, weather seriously influences outcomes by inflicting additional environmental burdens on survivors and delaying the arrival of rescue teams and relief supplies as it did in the aftermath of several earthquakes in Pakistan.

A 6.4 magnitude earthquake in southwestern Pakistan occurred on October 29, 2008. It killed several hundred and left tens of thousands homeless. Many of the homes across the region were constructed of mud brick which quickly crumbled during the shaking. In the devastated hill town of Ziarat, the district's health officer, Ayub Kakar, described the terrible living conditions of the survivors to the international media. The most immediate need was for shelter. Displaced persons including vulnerable small children and elderly were sleeping on the streets in extremely cold temperatures. The shelter problems were compounded by the lack of safe water and food across the region. According to Kakar, "Due to the cold, hundreds of children are being treated for pneumonia, abdominal diseases, diarrhea, and chest problems ..." (Mansoor, 2008). If the disaster had occurred several weeks later, in the middle of winter, it could have been far worse. These post-earthquake shelter problems in Pakistan follow just 3 years after the

FIGURE 5–6 A family in Kashmir standing in front of an earthquake damaged home. Photo courtesy of USAID. http://www.usaid.gov/locations/asia_near_east/images/south_asia_quake/sae-in7-usaid-hi.jpg.

massive Kashmir earthquake that registered at a 7.6 magnitude and killed over 70,000, injured tens of thousands, and destroyed the homes of millions. In the aftermath of that recent disaster, the Relief Commissioner, C.B. Vvas, told the media that the nation's response efforts had failed (USA Today, 2005a). He said that they had an immediate need for 23,000 tents but could only come up with about 4,000. Several days later, Prime Minister Shaukat Aziz told reporters that, "We need tents, tents, tents, and prefab housing ..." (USA Today, 2005b). It is understandable that the massive 2005 earthquake response would fall short on needed resources. It was the worst earthquake in the nation's history. However, it is difficult to understand how 3 years later, when similar issues arose with a much smaller earthquake, it took at least 4 days to get shelters in place. Lessons learned from each catastrophe must be incorporated into future response operations.

Effects on the Human Body

Earthquakes seldom cause direct injuries. The injuries are caused by the ensuing falling debris, collapsing structures, fires, and other hazards set in motion by seismic activity. A variety of traumatic injuries may result from an earthquake, ranging from minor cuts and bruises to serious burns, fractures, and crush injuries. There are also pulmonary challenges including the possibility of airway obstruction or asphyxiation from the large quantities of dust and debris generated by collapsing structures. Though traumatic injuries and airway management are well understood and readily managed by trained emergency medicine professionals, the earthquake-related injuries that pose the most unique challenges to healthcare providers and emergency medical personnel are crush injuries. Crush injuries were first recognized in early twentieth century earthquakes but were not clearly described and patient management guidelines were not established until World War II (Better, 1997). These injuries are most commonly observed in earthquake responses and rarely observed any other time. The fact that these injury types are so infrequent poses a preparedness challenge for clinical providers and rescue workers. To reduce crush injury morbidity and mortality, extensive care may be needed at the rescue site. To effectively prevent renal and cardiac failure, treatment must be started before and continued during extrication of a patient from a collapsed structure (Ron et al., 1984). Crush injuries include two distinct syndromes, crush syndrome and compartment syndrome.

Crush syndrome is the systemic manifestations of crushed muscle tissue and associated cell death. When a person with a crush injury is quickly extracted without any supportive care prior to and during the extraction, rapid death can result. Releasing the pressure on a crushed victim causes reperfusion of the blood carrying a large "dose" of damaged cells and toxic by-products to vital organs. It is sometimes referred to as the "Grateful Dead syndrome." This is due to the often quick and unexpected death of victims who moments earlier were expressing gratefulness toward rescuers for finally being extracted from entrapment. The sustained crushing force provides some protection for the victim by keeping the toxic substances produced by the injury from being circulated throughout the body. As soon as the crush pressure is released, a rush of toxic substances enters the bloodstream.

Compartment syndrome results from pressure in a confined space of the body. The fascia is a tough tissue covering muscle groups throughout the body. It does not easily expand. As a result, when a fascia encased area is affected by trauma such as burns, contusions, fractures, and other injuries that cause swelling, it is tightly contained in the

---

The Greta Hanshin-Awaji earthquake hit Kobe, Japan, in 1995 killing thousands. A review of 372 patients admitted to local hospitals with crush injuries showed that 74% of the crush injuries involved the lower extremities, 10% in the upper extremities, and 9% involved the trunk. Fifty of those crush injury patients (13%) died within several days of the earthquake. The cause of death was hypovolemia (decreased blood volume) and hyperkalemia (elevated blood potassium) (Oda et al., 1997).

---

compartment by the fascia. As internal pressure builds from the swelling, muscle tissue, nerves, and blood vessels are compressed. This reduces oxygen flow to the tissue and the muscle and nerve cells begin to die. The most influential factor in patients with bad outcomes from a compartment syndrome injury is a delayed diagnosis (Matsen and Clawson, 1975; McQuillan and Nolan, 1968; Rorabeck, 1984). One easy way to remember the keys to proper diagnosis is to use the "five Ps" (Olson and Glasgow, 2005). These include:

- Pain: Most common sign.
- Paresthesia: Loss of feeling below the injury.

- Passive stretch: Severe pain when the affected muscle is stretched.
- Pressure: Obvious tenseness in the affected area.
- Pulselessness: Less reliable sign.

When extricating a trapped patient, it is often extremely difficult to ascertain the level of crush injury risk. The patient may show no signs at all until after being extracted from entrapment (Michaelson, 1992). Managing a patient trapped in rubble experiencing a crush injury runs contrary to the usual "scoop and run" approach that is so much a part of emergency medicine. That is why it is critical that emergency medical professionals working in areas known to have seismic risks should receive periodic training in the proper management of crush injuries. If their response is approached with the same mentality and procedures, it will increase morbidity and mortality.

Demographic information on those injured and killed by earthquakes indicates that there are fundamental population characteristics that influence survival outcomes. The risk of earthquake death increases with age and with the presence of disabilities for both

---

On December 16, 1811, about 2 o'clock a.m., we were visited by a violent shock of an earthquake, accompanied by a very awful noise resembling loud but distant thunder, but more hoarse and vibrating, which was followed in a few minutes by the complete saturation of the atmosphere, with sulfurious vapor, causing total darkness. The screams of the affrighted inhabitants running to and fro, not knowing where to go, or what to do—the cries of the fowls and beasts of every species—the cracking of trees falling, and the roaring of the Mississippi—the current of which was retrogade for a few minutes, owing as is supposed, to an irruption in its bed—formed a scene truly horrible.

From that time until about sunrise, a number of lighter shocks occurred; at which time one still more violent than the first took place, with the same accompaniments as the first, and the terror which had been excited in everyone, and indeed in all animal nature, was now, if possible doubled. The inhabitants fled in every direction to the country, supposing (if it can be admitted that their minds can be exercised at all) that there was less danger at a distance from, than near to the river. In one person, a female, the alarm was so great that she fainted, and could not be recovered. Eyewitness account of the 1811 New Madrid, Missouri, earthquake by Eliza Bryan (1849).

Although there are no details available on the woman who "fainted, and could not be recovered," there is some interesting information that could be related and it came from a study of LA County Coroner records from the 1994 Northridge earthquake. There were 24 sudden cardiac deaths the day of the earthquake. They normally only have four or five each day. That means that an earthquake can literally scare some people to death! (Leor et al., 1996)

---

adult women and men (Tanida, 1996). The risk also increases as age decreases below age 16, especially in less developed nations. Those with a lower socioeconomic status are at increased risk of death and the data suggests that if their dwelling is collapsed, those with a lower economic status are less likely to survive it. Those with mental or physical disabilities, as well as those with nonmental illnesses are also less likely to survive a serious earthquake (Boyce, 2000; Morrow, 1999; Osaki and Minowa, 2001).

On January 17, 1995, a 6.9 magnitude earthquake struck Kobe, Japan. It is also known as the Great Hanshin Earthquake. The demographic trends that emerged from that disaster were shocking. Over 6000 people lost their lives. More than 50% of these fatalities were older than age 60. Among that cohort of elderly fatalities, there were twice as many female than male casualties. The death rate of those beyond age 80 was more than six times higher than people under age 50 (Tanida, 1996). There is typically a priority given in response to elderly, children, and other vulnerable populations in Japan. The problem was that the disaster was so large that those who normally look out for those groups were busy taking care of their own families and could not attend to the needs of vulnerable populations.

As with many disasters that impact essential infrastructure, earthquakes can directly and indirectly pose countless threats to human health. For those relying upon medical devices and procedures, the loss of power can literally cut off life lines. Those in need of dialysis and other life sustaining procedures are at risk. Many other vulnerable populations are at increased danger simply through the loss of basic environmental controls like heating in cold climates and air conditioning in hot climates. The loss of power may also compromise perishable food supplies and increase the likelihood of foodborne disease outbreaks. Damage to water and sewer lines can contribute to increased health risks from poor sanitation. History has clearly shown that many opportunistic infections and vectors can pose a growing threat to public health in the weeks and months following a major earthquake.

In April 2008, a 5.2 magnitude earthquake hit West Salem, Illinois. Immediately following that seismic event, the supervisor of the University of Iowa Hygienic Lab sent a message to state public health authorities warning that water from private wells may have a color change to black or yellow for up to a week following this earthquake in their neighboring state. This warning was based on the 2002 observation of private wells in Northeastern Iowa experiencing "black water events" following a November 2002 earthquake in Alaska. The same thing happened in Iowa after the 1964 Good Friday earthquake in Alaska (University of Iowa, 2008). There is clearly a risk of private well contamination following seismic events and following any moderate seismic event, neighbors in the surrounding area of several hundred miles need to be reminded to watch for problems with their private water supply. As the magnitude of the seismic event increases, the area included in such a warning should be expanded as well.

Although outbreaks of infectious disease do not normally occur after earthquakes, it is a possibility that cannot be dismissed. A plague outbreak followed the 1906 San Francisco earthquake. More recently, a rotavirus gastroenteritis outbreak occurred in Pakistan following the 2005 Kashmir earthquake (Karmakar et al; 2008). It is foolish to dismiss the possibility of an infectious disease outbreak following any major disaster that impacts essential infrastructure. Federal, state, and local public health officials should diligently watch for occurrences of opportunistic infectious disease outbreaks following an earthquake or any other disaster.

Prevention

The most critical actions an individual can take to reduce the risk of injury during a seismic event are carried out prior to the event. Although the tendency of many is to believe that preparedness for such a catastrophic event is not possible, earthquake preparedness is not only possible but essential. It takes a multiagency, multidisciplinary approach. Prevention of earthquake associated morbidity and mortality starts with minimizing construction projects in areas known to have high risk of earthquake damage. These areas have been well delineated and characterized. We should not be building on known fault lines. Detailed information on these zones is available from local planning and emergency management officials. For necessary construction projects near fault lines, building codes and engineering solutions may be incorporated that reduce the risk of collapsing structures in areas known to have seismic activity. When structures collapse, there are void spaces that occur where trapped individuals may survive for hours or days before rescue. These void spaces can be maximized during the design process. In addition, the risk of suffocation from dust inhalation can be minimized by using materials that produce less dust during a collapse. If a structure cannot be hardened to prevent collapse during a seismic event, measures should be instituted that make collapsing structures as survivable as possible for those inside.

Enhancing individual and home preparedness requires an effective public education initiative. Much of this activity can be facilitated through working closely with the local media. Both print and broadcast media can be engaged to raise awareness across the community concerning earthquake risks and mitigation measures. Engaging the media can be extremely beneficial by promoting stories on locating hazards and safe spots in structures, shutting off utilities, assisting vulnerable neighbors and family members, and where to find more detailed preparedness information. This includes general preparedness material available through the Red Cross and local, state, and federal emergency management organizations. Details on shutting off utilities may be obtained and promoted through utilities companies. A variety of public and private organizations, including businesses, public and private schools, faith-based organizations, advocacy organizations for vulnerable populations, and other local organizations should be recruited to support the development and sharing of essential earthquake preparedness information in regions known to have seismic activity potential. The challenge is that there is no single set of rules for the most appropriate actions to take during an earthquake. There are stories of individuals running out of a building during an earthquake and being killed by falling debris from the façade of the structure. There are also stories of those in similar situations who choose to stay inside and perish in a building collapse. Rather than training people to do something that is black and white or specific, we need to train people to stay calm and think during an earthquake. If those at risk are trained to look for certain types of hazards before and during a seismic event, the risk of injury or death can be greatly reduced.

Hazard Hunts

Minimizing earthquake injury risks in a structure begins with a search for any items that can fall or break and cause injuries. This includes heavy items that can fall and injure an occupant, hazardous materials that can be released from spilled containers or ruptured pipes, electrical hazards from broken or fallen wires, and fire risks from gas lines and appliances. When assessing a structure for potential hazards, go room by room. Start with the ceiling and consider suspended and even recessed light fixtures. If they are not securely anchored, they could pose a hazard. Look for heavy items hanging on the walls that could fall. Pay particular attention to items hanging near beds, couches, chairs, and anywhere else where people may lie or sit. Heavy items hanging over these areas must be moved or braced so they will not fall and strike someone during an earthquake. Heavy items on shelves should be placed on lower shelves and tall furniture such as shelving units should be secured to wall studs. Breakable items should be placed in cabinets with strong latches. Any hazardous materials should be segregated by chemical compatibility and stowed inside latched cabinets. Water heaters should be strapped to wall studs. A water heater poses particular risks during a disaster. If it is knocked over during an earthquake, it cannot only cause serious water damage, it can also cause a gas leak leading to an explosion and fire. The water heater should be located between 1 and 12 inches from a wall. Using flexible steel straps (plumber's tape), the water heater should be secured to wall studs on both sides of the water heater. At least two straps should be

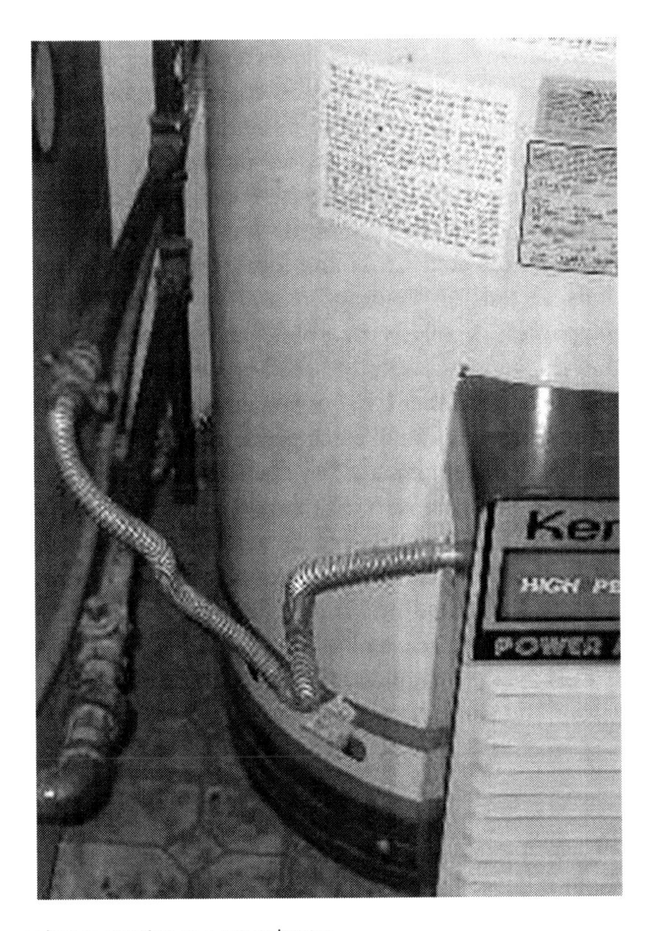

FIGURE 5–7 Flexible gas line connection on a water heater.

secured to the water heater. The top strap should be about 6 inches from the top of the water heater and the bottom strap should be about 6 inches from the bottom. In high-risk areas, this preventive measure may be doubled at the top and bottom with one strand of plumber's tape wrapped clockwise and the other counter clockwise at the top and the bottom sections of the water heater. The straps should be securely mounted to wall studs. The rigid gas connections on water heaters should be replaced with flexible connectors (see Figure 5–7) to minimize the risk of a broken gas line.

Top Ten Earthquake Hazard Mitigation Actions

1. Anchor overhead light fixtures.
2. Brace shelves securely to walls.
3. Avoid hanging heavy items such as pictures and mirrors over beds, couches, or anywhere where people sit.
4. Anchor heavy furniture and appliances to reduce movement.
5. Put heavier objects on lower shelves.
6. Secure water heaters by strapping them to wall studs.
7. Install flexible gas lines to water heaters and appliances.
8. Repair defective electrical wiring and gas connections to reduce fire risks.
9. Store breakable items in closed cabinets with strong latches.
10. Store household hazardous materials, such as flammables, pesticides, and cleaning products securely in closed cabinets with latches and on bottom shelves to avoid breaking, spilling, and mixing.

Individual Preparedness

Once a structure has been made as safe as it can be, the individual actions to be taken during an earthquake must be determined and practiced in advance. A safe spot should be selected in each room of a home, workplace, or anywhere that you spend a considerable amount of time. It may be under a large piece of furniture or simply along an inside wall that has no heavy objects hanging on it. The key to finding the safest place in the room is to look above each potential spot and see what can break and fall. Anything heavy or containing glass should be avoided. If there is a sturdy piece of furniture that can be used as a shield, it should be used for cover. For example, the fallen light fixtures in Figure 5–8 could cause serious injury to children in the classroom if the children have not been taught in advance to take cover under their desks during an earthquake. A plan for immediate action should include practicing the "drop, cover, and hold on" procedure. A safe spot should also be identified outside away from buildings, power lines, and other hazards. Individuals should be trained on basic first aid, the use of fire extinguishers, procedures for shutting off utilities, and communication planning using out of town contacts.

FIGURE 5–8 A classroom without braced light fixtures damaged by the 2001 Nisqually earthquake in the Puget Sound region of Washington. Photo courtesy of U.S. Geological Survey.

Top Ten Individual Preparedness Actions for Earthquakes

1. Identify the types of earthquake risks you face where you live, work, go to school, or travel.
2. Take action to mitigate all identified risks in your home and any place else you spend time.
3. Identify safe places in each room of your home and a safe location outside your home.
4. Establish a communication plan with your family.
5. Identify vulnerable friends and neighbors that may need assistance.
6. Store essential supplies in a home disaster kit.
7. Keep a smaller preparedness kit in your vehicle.
8. Plan transportation options if crossing bridges or using mass transit in daily commutes.
9. Learn how to use a fire extinguisher and maintain at least one in your home.
10. Get trained in first aid and be prepared to help others during a crisis.

Immediate Actions

When an earthquake begins, it will have a variety of initial presentations for different people depending on the magnitude of the seismic activity, how far you are from the epicenter, and where you are (e.g., inside, outside, in a car, etc.). If you are near the epicenter of a large earthquake it will feel like an abrupt jolt. It may knock you down if you are standing and even make it difficult to stay on your bed if it hits at night. It may be followed by violent shaking that could last up to several minutes. If it is a small earthquake nearby, it may feel like a smaller jolt followed by several stronger shakes that dissipate quickly (U.S. Geological Survey, FAQ—earthquake effects & experiences, www .earthquake.usgs.gov/learning/faq.php?categoryID=8&faqID=121). The most important immediate action is to stay calm.

A familiar phrase memorized by children on actions to take if their clothing catches fire is, "STOP, DROP, AND ROLL." A similar phrase promoted by the Federal Emergency Management Agency should be as familiar for those living in earthquake prone regions and that is "DROP, COVER, and HOLD." As soon as the shaking begins you should immediately drop to the floor, take cover under a sturdy table or other piece of furniture, and hold on. If there is not a sturdy piece of furniture to use for cover then move to an inside wall. Historically, many partially collapsed structures lose external walls while the internal sections remain intact. Stay away from tall, heavy furniture that could fall and from heavy items hanging on walls that could fall and cause injuries. Avoid

FIGURE 5–9 Loma Prieta, California, Earthquake, October 17, 1989. San Francisco. Cars crushed by a collapsed brick facade near 5th and Townsend. Five people were killed at this locality while leaving work.
Source: U.S. Geological Survey Open-File Report 90-547. http://libraryphoto.cr.usgs.gov/htmllib/btch136/btch136j/btch136z/btch136/mce00001.jpg.

glass from windows, pictures, and mirrors. Stay inside until the shaking stops. Many injuries occur when those inside a building attempt to move around inside or attempt to exit. When exiting the building, avoid using elevators and be aware that the electricity may go out or the fire alarms or sprinkler systems may activate.

On Wednesday, April 18, at 5:14 a.m., I was wakened by the crash of falling furniture, and a rocking, heaving house. Jim was sleeping in the next room, but I am used to slight earthquake shocks, so I lay still. The meteorological record has announced later that there were seven shocks in 10 minutes, 17 in the whole day, and numerous slight shocks every day since. I felt very calm, paralyzed perhaps, but I thought, "This is the worst thing I ever knew, and we may be going to be killed, and I want to die together." It was as much as I could do to walk across the floor, because it heaved so, and it made me very sea-sick. Jim always wakes slowly and dazed, and when I opened the door he thought the walls were falling in. I asked if we ought to get out of the house, but he said that we were in the safest place, under a cross beam in a frame house, which was true, as the streets were full of bricks, and our house was less damaged than most. So I got into bed with Jim, and he held me in his arms till the severe shocks were over. Excerpt from a personal letter written by Eleanor Watkins, wife of a San Francisco physician describing the 1906 San Francisco Earthquake (Watkins, 1981).

If the shaking begins when you are outside, make sure you stay outside but quickly move away from overhead electrical lines and tall trees or buildings. Once the shaking stops, stay away from buildings. Most injuries during earthquakes occur when people are running into, out of, or who are simply near buildings that collapse or drop debris during or immediately after an earthquake. If driving a car, you should stop on the side of the road as soon as it is safe to do so and stay in your vehicle until the shaking stops. Do not stop on or under bridges. Also avoid stopping near power lines, multistory buildings or large road signs. Once the shaking stops you should proceed slowly and avoid bridges or ramps that may have been damaged. This advice may have prevented some of the 33 fatalities that occurred during the 1994 Northridge earthquake in Los Angeles County. Of the 33 fatalities reported, 26 were injured indoors and 6 outdoors. Five of the six who died outdoors perished in car accidents and the other was electrocuted (Peek-Asa, 1998).

Response and Recovery Challenges

Once a seismic event is over, the immediate actions taken will be dictated by your surroundings. The first thing you should do is ensure that you are not injured and then take immediate action to keep it that way. That means shutting down utilities to avoid fire and electrocution risks. As soon as the area is made as safe as possible, attention should

FIGURE 5–10 Mexico City Earthquake, September 19, 1985. The top floors of this eight-story building collapsed because of pounding against surrounding buildings. Available at: http://libraryphoto.cr.usgs.gov/htmlorg/lpb030/land/cel00012.jpg.

turn to vulnerable friends and neighbors. The vast majority of those rescued from collapsed structures are not rescued by professional first responders and urban search and rescue team members but rather by untrained individuals nearby who are uninjured and make the decision to assist others. Those engaged in an initial rescue effort should be aware of potential aftershocks as they turn off utilities to damaged structures and seek out injured individuals. The use of anything that generates a spark or flame should be avoided and sturdy shoes should be worn by anyone participating in initial rescue and recovery activities. Particular attention should be paid to electrical hazards and hazardous materials that could have been spilled or released during the earthquake. Those who live near the coast line should also be aware of tsunami risks and avoid being near the beaches. This is not a statement condoning "freelance" rescue workers but rather a statement of fact that able bodied survivors are often willing to intervene on behalf of those around them that may be trapped and are the source of the majority of rescues. Waiting for the professionals to arrive will cost valuable time and cost lives that could otherwise be saved. This underscores the need for better community preparedness training for those spontaneous volunteers in high-risk areas. Nearby survivors are typically the ones who rescue those trapped in collapsed structures and technical rescues by professional teams follow a day or two later.

Those responsible for provision of medical care to the acutely injured should plan on 3–5 days of periodic patient surge. Those caring for the chronically ill should ensure they have taken measures to maintain continuity of care. This includes prescriptions, dialysis, home care, and other medical necessities. Public health authorities should step up surveillance activities. This includes documenting acute injuries, including when, where, and how injuries occur, as well as communicable diseases that may emerge due to compromised infrastructure.

Media advisories are an integral part of the healthcare and public health response. Appropriate warnings for physical and environmental hazards should be shared as well as advice for injury prevention. These are not messages that should be developed exclusively during the postdisaster phase. Others have been through these circumstances before and you should not reinvent the wheel. Prior to any major seismic event, public health and medical professionals should already have templates established to share essential post-earthquake information. Although there are regional risk communication issues that require content changes, the majority of the core information will remain the same regardless of the earthquake location.

The transition from rescue to recovery is one of the most difficult and controversial issues facing decision-makers when dealing with collapsed structures. An urban search and rescue operation is a very dangerous and costly process. As long as there is a reasonable hope that there still may be survivors, it is a moral necessity to do all you can to carry out extractions. The key question is how long it is reasonable to assume the possibility of survivors when you are placing rescue workers in dangerous circumstances and focusing resources away from those who survived. There is a point at which a declaration must be made that there are no survivors. In the aftermath of the September 11, 2001, terrorist attacks and collapse of the World Trade Center in New York City, Mayor Rudy Guiliani had to make the decision on when to issue death certificates for those not found and transition from Ground Zero rescue operations to recovery operations. Although the collapse was not initiated by an earthquake, the challenge is the same. The decision in that case was 14 days after the collapse (Cooper, 2001).

Although it is rare that survivors are found 2 weeks from the time of a disaster, it has been seen. All morbidity and mortality variables need to be considered when making this decision. Depending upon the conditions, the declaration of a transition from a response to a recovery operation may be sooner than 14 days but will rarely exceed that time.

> An interesting phenomenon occurs in many regions impacted by a major disaster like an earthquake. There is often a unique post disaster camaraderie among response organizations and citizens alike leading to enhanced teamwork for future preparedness endeavors. Perhaps, it is the shared disaster stress experience or realization of the common threats and benefits of teamwork. Regardless of the exact stimulus, a major disaster often has a unifying effect on responders and following a disaster, turf issues and other barriers to preparedness frequently give way to better collaboration that pays dividends for future preparedness efforts.

Summary

Earthquakes are one of the most destructive and unpredictable forces of nature. Fortunately, there are a variety of opportunities to reduce associated morbidity and mortality. Although the day and time of earthquakes cannot be predicted, the majority of locations with the greatest potential for devastating seismic activity are well defined. Like many other catastrophic disasters, minimizing earthquake associated injuries and death requires a multidisciplinary and multiagency approach. It includes adequate and enforced building codes. Healthcare preparedness initiatives must include training on the management of crush injuries and the coordination of mass care. Exercises should incorporate as many challenging variables as possible, such as poor weather and damaged transportation routes.

In all preparedness initiatives, a special focus should be placed on vulnerable populations. Family members and home healthcare providers should be engaged to offer tips to vulnerable disabled or ill individuals in private homes under their care. Initial actions and evacuation procedures should be discussed and practiced. Beds should be relocated to the safest area of the home away from windows and outer walls and without heavy objects or unanchored light fixtures over them. Large facilities such as hospitals and nursing homes should have comprehensive emergency plans that go beyond fire evacuation or immediate sheltering for a tornado. Evacuation plans should be developed and coordinated with emergency management professionals in the community. Unfortunately, there are plans that appear to be good on paper but lack proper coordination. If they have not been coordinated with others in the community, they will often not be successful during an emergency. For example, if a long-term care facility plans on using a specific bus service for resident evacuation, they must ensure that the bus service has an agreement in place with them, that there is not a competing agreement with other facilities that could delay or prevent evacuation of the facility due to other commitments, and that the bus company has a plan in place to ensure they can provide the service during a crisis. An ideal emergency plan may be in place for a facility but come unraveled due to an unanticipated external weakness.

The greatest public health challenges following earthquakes include provision of support for the initial patient surge, identification and assistance of vulnerable populations, epidemiological monitoring of injuries and illnesses resulting from the disaster, and sustaining safe environmental health conditions, including safe food and water, through the recovery of critical infrastructure.

Community awareness and education initiatives offer the greatest opportunities to reduce earthquake associated morbidity and mortality. The simple and inexpensive measures to reduce the risks of falling objects, broken glass, fire, and electrocution as well as individual self protection measures during and immediately following an earthquake are all topics that should be regularly addressed with populations at risk. This can be done through working closely with schools, community and advocacy organizations, and the local media.

Websites

American Red Cross, Earthquake Preparedness, (In English and Spanish): www.redcross.org/services/-disaster/0,1082,0_583_,00.html.

Centers for Disease Control and Prevention, Earthquake Preparedness: www.bt.cdc.gov/disasters/-earthquakes/prepared.asp.

City of Berkeley, CA, Earthquake Hazard Hunt for Renters and Property Owners: www.ci.berkeley.ca.us/uploadedFiles/Fire/Hazard%20HuntRenters-7.pdf.

City of Los Angeles, CA, Department of Building Safety, How you can strengthen your home for the next big earthquake in the Los Angeles area: www.cert-la.com/BAS-How-You-Can-Strengthen-Your-Home.pdf.

City of Los Angeles, CA, Fire Department, Earthquake Preparedness Handbook: http://lafd.org/eqbook.pdf.

Federal Emergency Management Agency, Earthquake Preparedness: www.fema.gov/areyouready/-earthquakes.shtm.

Federal Emergency Management Agency, Earthquake Publications and Resources: www.fema.gov/plan/prevent/earthquake/publications.shtm.

Institute for Business and Home Safety (IBHS), Is your home protected from earthquake disaster? A homeowner's guide to earthquake retrofit: www.ibhs.net/natural_disasters/downloads/earthquake.pdf.

Institute for Business and Home Safety (IBHS), Protect your home against earthquake damage: www.disastersafety.org/resource/resmgr/PDFs/earthquake10.pdf.

John A. Martin & Associates, Inc., Coping with children's reactions to earthquakes: www.johnmartin.com/earthquakes/EQCOPING/INDEX.HTM.

John A. Martin & Associates, Inc., Identification and reduction of nonstructural hazards in schools: www.johnmartin.com/earthquakes/eqschools/index.htm.

John A. Martin & Associates, Inc., Strengthening wood frame houses for earthquake safety: www.-johnmartin.com/earthquakes/Eqresid/index.htm.

Mid-America Earthquake Center: http://mae.ce.uiuc.edu/.

National Earthquake Hazards Reduction Program: www.nehrp.gov/.

National Geophysical Data Center (NGDC), Earthquake intensity database: www.ngdc.noaa.gov/hazard/earthqk.shtml.

Pacific Earthquake Engineering Research Center: http://peer.berkeley.edu/.

Southern California Earthquake Center, USC, Putting Down Roots in Earthquake Country and Related Resources: www.earthquakecountry.info/roots/index.php.

U.S. Geological Survey, American Red Cross, Asian Pacific Fund, California Earthquake Authority, Governor's Office of Emergency Services, New America Media, U.S. Department of Homeland Security, and the Federal Emergency Management Agency, Protecting Your Family From Earthquakes-The Seven Steps to Earthquake Safety (in English, Chinese, Vietnamese, and Korean): http://pubs.usgs.gov/gip/2007/42/.

U.S. Geological Survey, Earthquake Hazards Program: www.earthquake.usgs.gov/.

References

Better, O. S. (1997). History of the crush syndrome: from the earthquakes of Messina, Sicily 1909 to Spitak, Armenia 1988. *Am J Nephrol* 17(3–4):392–394.

Boyce, J. K. (2000). Let them eat risk? Wealth, rights and disaster vulnerability. *Disasters* 24:254–261.

Bryan, E. (1849). Letter describing the 1811 New Madrid, MO earthquake. In: Martin J, ed. *Lorenzo dow's journal*. Huntsville, Alabama: The Virtual Times; pp.344–346. www.hsv.com/genlintr/newmadrd/accnt1.htm.

Cooper, M. (2001). A nation challenged, The Trade Center: Guiliani declares that finding anyone still alive in the rubble would be "A miracle." *The New York Times*. September 25, 2001:B9.

Greely, A. W. (1906). *Relief operations conducted by the military authorities of the United States at San Francisco and other points, special report*. Washington, DC: U.S. Government Printing Office.

Hansen, G., & Condon, E. (1989). *Denial of disaster: the untold story and photographs of the San Francisco earthquake of 1906*. San Francisco, CA: Cameron and Co.; p.160.

Head, L. (1906). One boy's experience: a member of the Roosevelt Boys' Club writes of his experience during and after the Great Earthquake. *Roosevelt Boys' Club's Newsletter*. July 28, 1906. www.sfmuseum.net/1906/ew7.html.

Karmakar, S., Rathore, A. S., & Kadri, S. M., et al. (2008). Post-earthquake outbreak of rotavirus gastroenteritis in Kashmir (India): an epidemiological analysis. *Public Health* 122(10):981–989.

Leor, J., Poole, W. K., & Kloner, R. A. (1996). Sudden cardiac death triggered by an earthquake. *N Engl J Med* 334(7):413–419.

Mansoor, H. (2008). Disease hits freezing quake zone, Agence France-Presse. November 1, 2008. www.news.com.au/story/0,27574,24584905-23109,00.html.

Matsen, F. A., & Clawson, D. K. (1975). The deep posterior compartmental syndrome of the leg. *J Bone Joint Surg [Am]* 57:34–39.

McQuillan, W. M., & Nolan, B. (1968). Ischaemia complicating injury: a report of thirty-seven cases. *J Bone Joint Surg [Br]* 50:482–492.

Michaelson, M. (1992). Crush injury and crush syndrome. *World J Surg* 16(5):899–903.

Morrow, B. H. (1999). Identifying and mapping community vulnerability. *Disasters* 23:1–18.

Oda, J., Tanaka, H., & Yoshioka, T., et al. (1997). Analysis of 372 patients with Crush syndrome caused by the Hanshin-Awaji earthquake. *J Trauma* 42(3):470–476.

Olson, S. A., & Glasgow, R. R. (2005). Acute compartment syndrome in lower extremity musculoskeletal trauma. *J Am Acad Orthop Surg* 13(7):436–444.

Osaki, Y., & Minowa, M. (2001). Factors associated with earthquake deaths in the great Hanshin-Awaji earthquake, 1995. *Am J Epidemiol* 153:153–156. http://cmbi.bjmu.edu.cn/news/report/2008/eq/67.pdf.

Peek-Asa, C., Kraus, J. F., & Bourque, L. B., et al. (1998). Fatal and hospitalized injuries resulting from the 1994 Northridge earthquake. *Int J Epidemiol* 27(3):459–465. http://ije.oxfordjournals.org/cgi/reprint/27/3/459.pdf.

Peek-Asa, C., Ramirez, M., & Seligson, H., et al. (2003). Seismic, structural, and individual factors associated with earthquake related injury. *Inj Prev* 9:62–66.

Risse, G. B. (1992). A long pull, a strong pull, and all together: San Francisco and bubonic plague, 1907–1908. *Bull History Med.* 66:260–289.

Ron, D., Taitelman, U., & Michaelson, M., et al. (1984). Prevention of acute renal failure in traumatic rhabdomyolysis. *Arch Intern Med* 144:277–280.

Rorabeck, C. H. (1984). The treatment of compartment syndromes of the leg. *J Bone Joint Surg [Br]* 66:93–97.

Tanida, N. (1996). What happened to elderly people in the great Hanshin earthquake. *BMJ* 313(7065):1133. www.bmj.com/cgi/content/full/313/7065/1133.

Todd, F. M., ed. (1909). *Eradicating plague from San Francisco: report of the citizens' health committee and an account of its work*. San Francisco, CA: Murdock & Co.

University of Iowa, University Hygienic Laboratory Newsletter. (2008). Earthquake causes ripple effect at UHL. April 21, 2008. www.uhl.uiowa.edu/publications/tw@uhl/2008/04-21-08/04-21-08.pdf.

USA Today. (2005a). Many villages in Kashmir still seeking assistance. October 11, 2005. www.usatoday.com/news/world/2005-10-11-india-seeking-aid_x.htm.

USA Today. (2005b). Weather hampers Pakistani earthquake rescue and relief. October 16, 2005. www.usatoday.com/news/world/2005-10-16-earthquake-asia_x.htm.

Watkins, E. (1981). The 1906 San Francisco earthquake: a personal account. *California Geology* 34:12. www.johnmartin.com/earthquakes/eqpapers/00000047.

Floods

Objectives of This Chapter

- List the kinds of floods that can occur and distinguish how they are different.
- Recognize the warnings and watches associated with floods.
- Describe the human factors that contribute to flooding.
- List the meteorological and topographical factors that contribute to flooding.
- Describe the causes of flood-related morbidity and mortality.
- Explain the differences between flood-related illness and injury at different phases of a flood.
- Describe which demographic groups are at greatest risk during a flood and why.
- List individual actions that may be taken to reduce flood-associated risks.
- Explain what immediate actions may be taken to reduce risk during a flood.
- Describe the major risks associated with postflood recovery and how to reduce them.

Introduction—Common Issues of the 1988 Khartoum, Sudan Flood and the 2005 Post-Katrina Flooding of New Orleans

It is the capital city of Sudan and home to several million residents. The name of the city, Khartoum, means "elephant trunk" and is likely derived from the shape of the strip of land where the city is located at the confluence of the White and Blue Nile Rivers (Department of State, U.S. Embassy, About Sudan: the three towns: Khartoum, North Khartoum and Omdurman, http://sudan.usembassy.gov/3_towns.html). The location is also prone to flooding. Throughout the twentieth century, the city experienced a surge in population growth and urbanization. Weather pattern changes have increased the frequency of area rainstorms whereas changes in irrigation practices have reshaped the terrain (Walsh et al., 1994). Although these factors can be successfully managed by local officials to reduce the growing risk of flooding, there was little planning prior to 1988 to prepare the population for the coming disaster. It has been suggested that the primary condition for reducing future flood risks in Khartoum is overcoming complacency and taking sensible measures to control the risks (Davies and Walsh, 1997).

In early August 1988, heavy rains fell on Khartoum. The rainfall exceeded the total annual rainfall amounts for the region on a single day more than once during that time period. By mid-August, over 100,000 homes had been destroyed, displacing over a million people. Making matters worse, thousands of refugees from civil conflicts in the nation had settled in low-lying, flood prone areas around the city. The Sudanese Ministry of Health received assistance from the World Health Organization and a variety of public and private health organizations around the world to mobilize surveillance activities for monitoring the health status of the population. Gathering information from dozens of clinics and several large hospitals, they tracked illness trends. Diarrheal disease was observed in nearly one third of those seeking healthcare. Among more than 600 patients hospitalized for diarrhea, 68 died. The prevalence of malaria in the region doubled in the weeks following the flooding (Centers for Disease Control and Prevention, 1989). Of course, the full health impact of this disaster will never be known. Most of those displaced by the flooding were already refugees, and many illnesses and deaths among the refugee population were not investigated or reported.

Poor data collection in the region among refugee populations limits the availability of concise morbidity and mortality information from the Khartoum flood. However, there are still valuable lessons to be learned from it. Existing gaps in local infrastructure and current public health and social problems in a community will be dramatically amplified by a flood. There are seldom new diseases emerging from floods but rather increases in existing diseases that are not kept in check due to environmental and social decline that occur during and after the disaster. Though access to healthcare is diminished, displaced populations are exposed to more disease-carrying vectors, poorer sanitation, and have limited access to safe food and water. This trend is repeated with many historical floods, including the flooding of New Orleans that followed Hurricane Katrina in 2005.

Even prior to Hurricane Katrina, New Orleans had serious problems. The poverty rate of 23.2% is about twice the national average, and the vast majority of poor residents are African American (El Nasser, 2005). Much of the housing for poor residents was already dilapidated before the hurricane and subsequent flooding from the breached levees. The more expensive homes are built in areas that are typically safer with better flood protection. More affluent residents also have the means to evacuate whereas the poorer

FIGURE 6–1 New Orleans, LA, September 8, 2005—FEMA's US&R teams en route by helicopter to conduct a search in St. Bernard Parish, view the flooding in New Orleans. Photo: Michael Rieger/FEMA.

population and other vulnerable groups sometimes lack the means to protect themselves or evacuate. Following Hurricane Katrina, the death toll exceeded 1500 as evacuees continued to die in other states (Hunter, 2006). Elderly and ailing individuals that had been pulled from the floodwaters continued to succumb to their illnesses for weeks and months following their evacuation. According to local mortuary records, 64% of those who perished in New Orleans were of age 65 and older (State of Louisiana, Louisiana Family Assistance, Reports on deceased, www.dhh.louisiana.gov/offices/page.asp?ID=303&Detail=7047). Most of them died as a direct result of the flooding caused by the breached levees.

Like Khartoum, New Orleans had very clear flood risks but did little to carry out comprehensive planning prior to Hurricane Katrina and the associated flooding (See Figure 6–1). Both cities had their most vulnerable populations in high-risk flood areas and lacked effective evacuation and emergency sheltering preparations. They certainly had "plans" but plans are useless if they are not part of an ongoing, dynamic planning process that moves preparedness forward. Based on the response and outcomes, these processes were inadequate. In the end, the lack of preparedness had the greatest impact on the most susceptible populations of both cities.

Flood-Related Definitions

Acre-foot: The volume of water that can cover one acre of land (43,560 square feet) with one foot of water. One acre-foot is equal to 325,851 gallons.

Cubic feet per second (ft^3/s): A common unit of measurement used to describe water discharge. One cubic foot per second is equal to 448.8 gallons per minute.

Coastal flood warning: Issued by the National Weather Service when coastal flooding is occurring, imminent, or expected within 12 hours.

Coastal flood watch: Issued by the National Weather Service when coastal flooding is possible within 12–36 hours.

Discharge: The rate of flow—a volume of fluid passing a point per unit time, commonly expressed in cubic feet per second, million gallons per day, or gallons per minute.

Flash flood: A flood caused by heavy or excessive rainfall in a short period of time, generally less than 6 hours causing water to rise and fall rapidly. Flash floods are usually characterized by raging torrents after heavy rains that rip through river beds, urban streets, or mountain canyons. They can also occur even if no rain has fallen, for instance after a levee or dam has failed, or after a sudden release of water by a debris or ice jam.

Flash flood watch: Flash flooding is possible. Be prepared to move to higher ground; listen to NOAA Weather Radio, commercial radio, or television for information.

Flash flood warning: A flash flood is occurring; seek higher ground on foot immediately.

Flood: The excessive overflow of water onto normally dry land. The inundation of a normally dry area caused by rising water in an existing waterway, such as a river, stream, or drainage ditch. Ponding of water at or near the point where rain falls. Flooding is a longer-term event than flash flooding. It may last for weeks.

Flood frequency: Refers to a flood level that has a specified percent chance of being equaled or exceeded in any given year. For example, a 100-year flood occurs on average once every 100 years and thus has a 1% chance of occurring in a given year.

Flood stage: The stage at which overflow of the natural stream banks begins to cause damage within the area where the elevation is measured. Flood stages for each USGS gauging station are usually provided by the National Weather Service.

Precipitation: Rain, snow, hail, or sleet.

Surface runoff: That part of the runoff that travels over the soil surface to the nearest stream channel. It also is defined as that part of the runoff of a drainage basin that has not passed beneath the surface following precipitation.

Tsunami: A large ocean wave caused by an underwater earthquake or volcanic eruption; often causes extreme destruction when it strikes land. Tsunami waves can be up to 98 ft (30 m) high and reach speeds of 589 mph (950 km/hour). They are characterized by long wavelengths of up to 124 miles (200 km) and long periods, usually between 10 and 60 minutes.

Floods

Floods are nature's most widespread disaster. In the last century there were over eight million fatalities attributed to floods (EM-DAT: OFDA/CRED International Disaster Database, 2004). These disasters are growing in frequency worldwide due to a variety of human and environmental factors (See Table 6–1). The human factors that contribute to flooding are mostly associated with development and land use. Expanding populations and urbanization along shorelines and waterways place more people in flood hazard areas. Most major cities are built near waterways that have historically provided a means for transportation, drinking water, and other urban necessities. As the environment near the water is altered, the flood risks are changed. Although some flood scenarios may be avoided through flood walls, levees, and other engineered solutions, the development and habitation of areas that have historically flooded comes with risks.

In addition to the habitation of areas at risk, the alteration of the natural environment by human development contributes to new flood risks. When precipitation occurs in undeveloped areas, it saturates soil and is picked up by vegetation. When these processes meet their capacity, the moisture accumulates in lower areas that naturally become ponds, lakes, streams, and rivers. As people begin to inhabit an area, they remove much of the vegetation and cover the soil with buildings, roads, and other nonpermeable structures. This inevitably causes runoff challenges that are managed to some extent through storm drains, ditches, and culverts. However, the runoff from impermeable surfaces travels much faster than the runoff of natural terrains. This accumulated runoff speeds up even more when it enters a drainage system. This rapid buildup of fast moving water poses a risk for urban flooding.

Dam and Levee Failure

Low-lying habituated areas near rivers are often protected by levee systems. Though they may be fairly effective, levees can create other problems. Those across the river from a levee are flooded worse as the floodwaters that once spread over both sides of the river is pushed across to one side. Also, those inside the levee may develop a false sense of security. Prior to the levee, those in the flood plain would have originally had time to evacuate as the river slowly rose. With a levee system, there may be far less time to evacuate if a levee is breached. Levee breaks can create such destructive, fast rising waters, that those living in a flood plain may be unable to escape. This was the case with the levee failures in New Orleans following Hurricane Katrina. Even frequent inspections of levee systems cannot guarantee their safety. Seismic activity, extreme weather, and other factors can cause catastrophic levee damage and failure. Levee damage is not always apparent.

Table 6–1 Factors Contributing to Flood Risks

Meteorological Factors	Human Factors	Topographical Factors
Precipitation	Flood plain habitation	Topography (slope)
• Rainfall duration, amount and intensity	Poor drainage infrastructure	Impermeable cover (urban areas)
	Land use	Soil type and moisture
• Snow/snowmelt	• Urbanization (sealing surfaces)	Vegetation
Temperature	• Deforestation	Ground surface permeation
Wind	Overly efficient drainage upstream	Groundwater depth
Humidity	Climate change	Runoff patterns
Storm activity and direction	Failed dams or levees	Ponds lakes and other basins that catch runoff
Hurricanes/cyclones	Driving or wading in swift water	
Season		Bank and channel characteristics
		High tide drainage

There are several examples of other engineering feats that have achieved some success in effectively controlling flooding. The most impressive is the Delta Works, called the Deltawerken in Dutch. Nearly half of the Netherlands are below sea level and this network of flood protection systems is designed to guard those low-lying areas. It was built after a devastating flood in 1953 where 1836 people died as a direct result of the flooding (Gerritsen, 2005). Thousands of homes were destroyed and the salt water made the region's farmlands unusable for years. In response, the Delta Works began and has become the largest flood control effort in history. The project's scale is similar to the Great Wall of China (Infoplease Almanac, The seven wonders of the modern world, www.infoplease.com/ipa/A0923082.html). Even this engineering marvel will need shoring up in several decades if the oceans continue rising.

Although dams have excellent social benefits including flood control, power generation, water storage, irrigation, recreation, and more, they have not always been safely maintained. There are about 80,000 dams of various sizes, conditions, and vulnerabilities in the United States alone (Government Security, 2008). Unfortunately, many of these dams were not built according to any specific standards or have been poorly maintained. They are owned by federal, state, and local government, industry, and private citizens. Managing them safely becomes more difficult as they age and is particularly challenging with the growing threat of terrorist attacks against critical infrastructure in recent years.

The loss of life from dam failures depends upon how many people live in the dam failure flood plain, how severely it may flood if there is a dam collapse, and how much warning they receive that the dam failure is imminent (U.S. Department of Interior, 1999).

In a study of over 300 dam failures across several different continents, the primary reasons for collapse are improper design (40%), followed by flooding that exceeded the spillway capacity (35%), and foundation problems such as seepage and settlement (25%) (Biswas and Chatterjee, 1971). Most of the failures are avoidable through proper design and close monitoring (See Figure 6–2). Of course, the exceptions to this include extreme circumstances such as terrorist attacks or unforeseen natural disasters such as earthquakes that compromise a dam's structural integrity.

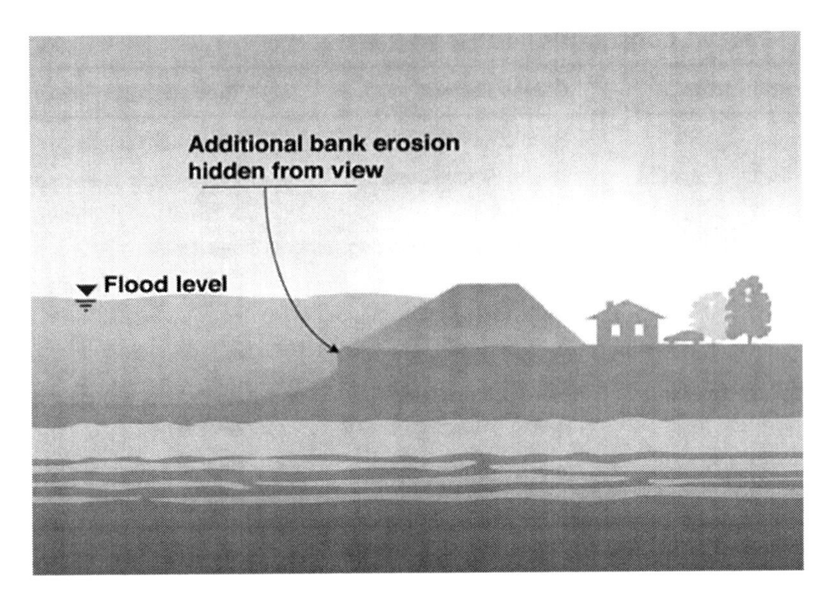

FIGURE 6–2 Hidden levee erosion. Image courtesy of the California Department of Water Resources, Office of Public Affairs. http://www.publicaffairs.water.ca.gov/newsreleases/2006/03-22-06levee.jpg.

The worst dam failure in history occurred in 1975. The Banqiao Dam in China was one of the largest in the world. It was completed in the late 1950s and spanned about one and a quarter mile in length (2 km). It held a reservoir the size of Lake Superior. As a large typhoon hit the region, the Banqiao Dam became the first of dozens of dams in the region to fail, killing tens of thousands and displacing millions. Following the failure of the dams, there were subsequent disease outbreaks and famine that killed hundreds of thousands more (U.S. Department of Interior, 1999).

Riverine Flooding

Riverine flooding includes a variety of flood scenarios and types. The nature of a riverine flood is shaped by the terrain and weather patterns. One type of riverine flooding is the common overbank flooding seen in many Midwestern U.S. flood disasters (Figure 6–3). In regions with relatively flat terrain, the water is often shallow and slow moving. It lingers for days or weeks. This is the type of flooding that sandbagging operations are sometimes used to control. It is also the classic scenario that most people think of when floods are mentioned.

Flash floods are another type of riverine flood. Narrow, steep valleys quickly collect floodwaters in a concentrated area making their floods faster moving and shorter in duration. Rainfall with sufficient intensity and duration on the right topography can cause flash floods in a very short period of time. These floods are distinguished by their quick rise and intense, destructive flow. It is not unusual for a flash flood to have a channel velocity of about 9 feet per second. At that speed, it can move a 90 pound rock downstream. In 1976, Colorado's Big Thompson Canyon experienced a flash flood that had a channel velocity

FIGURE 6–3 St. Genevieve, MO, July 9, 1993, clean-up commences at the waterworks following the devastating flood. A total of 534 counties in nine states were declared for federal disaster aid. Photo by Andrea Booher/FEMA Photo. http://www.fema.gov/photodata/original/2106.jpg.

of about 30 feet per second. At that speed, it moved 250-ton boulders (FEMA/EMI, Floodplain Management Course, Session 6). The density of water makes it exceptionally destructive. Several inches of swift moving water can easily pick up and move vehicles. Moving water can also quickly destroy a home. Water moving at 10 miles per hour exerts the same pressure on a building as a 270 mile per hour blast of wind (Shell, 2005).

Near Glenville, West Virginia, a flash flood hit on a Saturday evening in June of 1950. After only 2 inches of local rainfall, local residents did not realize that upstream in Leading Creek, about 8 inches of rain fell shortly after 11 p.m. creating a flash flood that would destroy homes across six counties and kill 31 people as the Little Kanawha River crested at over 31 feet (West Virginia Archives and History, 1950). The risk from this kind of scenario is significantly reduced today by the National Weather Service Alert System.

Lake Level Fluctuation Flooding

Lake level fluctuations can result in lake area flooding with a variety of effects. Shoreline erosion can threaten property and homes. Floodwater can intrude into drinking water sources. Septic systems can be compromised and cause contamination. The best way to minimize the risk of illness, injury, and property damage, is to build everything at a higher elevation than the lake's natural outlets. This includes placement of well heads and septic systems. Although that outlet may be considerably higher than recorded lake levels, it is the only way to ensure protection from flood risks.

FIGURE 6–4 This is a good illustration of why one should never drive into flooded roadways. Even water only six or eight inches deep may float and carry away some cars. http://www.fema.gov/photodata/original/2106.jpg.

Ground Failure and Flooding

Ground failures include a variety of circumstances including landslides, mudflows, and alluvial fans (Figure 6–5). The factors contributing to ground failure risks are topography, geology, and precipitation. A hillside that is comprised of certain types of rock and soil can become saturated by precipitation. When the moisture lubricates and loosens the soil and rock formations to the point where the slope can give way, a landslide can occur. A landslide sometimes blocks a stream or river and causes flooding upstream from the site. The accumulated debris may also give way and cause a flash flood downstream. Similarly, a mudslide, mudflow, or alluvial fan can focus or redirect debris that can reroute streams and rivers causing floods. Mudflows can be up to 50% solid material and can cause serious property damage. They can also cause long-term environmental damage and compromise drinking water sources. Alluvial fans are fan-shaped deposits of small debris that occur at the foot of a focused drainage area, usually at the base of a mountain range where it opens up to a plain. The unstable soil and excessive moisture lead to debris accumulation from repeated movement of debris by swift moving water. Alluvial fan floods are usually shallow and fast moving. They can carry a substantial amount of sediment and debris, making their flooding difficult to manage.

Coastal Flooding

Coastal flooding occurs when a powerful low-pressure system drives ocean water inland as a storm surge that exceeds the normal tidal surge. The tides are produced by the movement of the sun and moon and are very predictable. The natural cycle includes high and low tides with especially high tides in the spring and autumn. The tide levels also fluctuate with different weather patterns. Low-pressure systems can result from tropical or winter storms and cause storm surges. These storm surges happen several times each year

FIGURE 6–5 Alluvial fan. During the 1996 Buffalo Creek Fire, thousands of acres southwest of Denver, Colorado, were destroyed. The wildfires lowered the erosion threshold and as a consequence, a 100-year rainstorm in July 1996 caused erosion upstream and deposition of this alluvial fan at the mouth of a tributary to Buffalo Creek. Buffalo Creek is flowing to the right at the bottom of the photograph. Photo courtesy of the USGS, by R. H. Meade. http://www.brr.cr.usgs.gov/projects/Burned_Watersheds/AlvFan.jpg.

at any given coastal location but are usually not severe enough to cause serious flooding. But if a severe storm hits during high tide, the coastal flooding can be severe. Coastal flooding can also result from a tsunami.

Coastal flooding levels are defined as minor, moderate, and major. The same are used for the lakeshore alerts of the Great Lakes. These are qualitative classifications that provide a general description of anticipated flood results (New York City, Office of Emergency Management, NYC hazards: coastal flooding, www.nyc.gov/html/oem/html/hazards/storms_coastalflooding.shtml):

- Minor: nuisance coastal flooding of locations adjacent to the shore. Minor beach erosion can occur. Minor coastal flooding is not expected to close roads or do any major structural damage to homes and other buildings.
- Moderate: more substantial coastal flooding, threatening life and property. Some roads may become impassable due to flooding. Moderate beach erosion will occur along with damage to some homes, businesses, and other facilities.
- Major: a serious threat to both life and property. Numerous roads will likely become flooded. Many homes and businesses along the coast will receive major damage. People should review safety precautions and prepare to evacuate if necessary. Major beach erosion is also expected.

Another type of coastal flooding is a tsunami. Although some tsunamis may look similar to a storm surge, they are very different. Also called "tidal waves," these potentially massive waves are usually caused by seismic activity such as earthquakes and volcanoes beneath the ocean. They can also be caused by landslides or even an asteroid impact. Swelling to nearly 100 feet high, they travel at hundreds of miles per hour. Although some progress has been made in the development of warning systems, there is much work remaining to establish effective systems that provide sufficient warning for those at risk to evacuate to safety.

Effects on the Human Populations

The primary causes of flood-related morbidity and mortality are not as apparent and simple as they may seem. Many flood-related deaths are not attributed to drowning. Other causes of death include physical trauma, fire, electrocution, carbon monoxide poisoning, and heart attacks. There are different levels of risk associated with the type and phases of each flood disaster. A flash flood will have a very different impact on human health than a slowly rising, persistent riverine flood. There are also different risks at each phase of the disaster. These may be referred to as direct and indirect injuries or as immediate and delayed injuries. There are also many infectious disease risks associated with the post-flood environment. These risks include contaminated water, spoiled food, and hazardous mold.

Any time a weather disaster occurs in the United States that has at least 30 deaths or more than $100 million in damage, the National Weather Service (NWS) investigates and reports on the causes. The most deadly types of floods in the United States are flash floods (Figure 6–4). A review of NWS flash flood reports from 1969 to 1981 identified 1185 deaths from 32 flash flood events. Detailed information was obtained for 190 of the 1185 fatalities. It was observed that 93% of these deaths were due to drowning. Of those who drowned, 43% were in vehicles; another 43% were swept into the water when they attempted to cross on foot or they were swept into the water from a campsite or home (French et al., 1983). These trends suggest that many flood deaths involve risk taking behavior and are avoidable.

In an Australian flood fatality review, 2213 flood deaths that occurred over a 200-year time span (1788–1996) were analyzed. This included information from a variety of sources, including newspaper stories, government reports, and scientific studies. The majority of fatalities were male. Although the male:female ratio fluctuated over time from 1:1 to 10:1, the total ratio over the entire period of time was 4:1. More than 80% of the flood fatalities in Australia have been male. It is interesting to note that more than 38% of these fatalities were due to the risk-taking behavior of attempting to cross a flooded creek, road, or bridge (Coates, 1999). Based on the data from this and other flood fatality studies, there are clear opportunities to reduce deaths through enhanced risk communication. The majority of drowning deaths are risk-taking males who attempt to drive or wade through flooded areas (Figure 6–6). Appropriate awareness initiatives may be taken in flood prone areas to emphasize appropriate messages for those at risk.

Several interesting observations emerged through an analysis of 247 flood-related deaths from 13 different floods across the United States and Europe. The records from these floods were unique in that most of them included detailed circumstances of each victim's death. This provides more insight into lethal hazards than many previous flood fatality studies. Nearly 70% of flood deaths were due to drowning, while about 12% were due to traumatic injuries. The balance of resulting fatalities included an interesting assortment of nondrowning causes including heart attacks, fires, electrocution, and carbon monoxide poisonings. Here is a summary of the top 10 causes of death among the 247 fatalities (Jonkman and Kelman, 2005):

1. Drowning while in a vehicle = 81 (32.8%).
2. Drowning as a pedestrian = 62 (25.1%).
3. Drowning in a building = 15 (6.1%).

FIGURE 6–6 A car is submerged in floodwaters near Moss Bluff, Louisiana, on October 27, 2006. Photo courtesy of the National Weather Service. Available at: http://www.srh.noaa.gov/lch/jamb/flood1006-2.png.

4. Physical trauma in a vehicle = 14 (5.7%).
5. Heart attack = 14 (5.7%).
6. Fire = 9 (3.6%).
7. Physical trauma in a building = 8 (3.2%).
8. Electrocution = 7 (2.8%).
9. Drowning from a boat = 7 (2.8%).
10. Physical trauma as a pedestrian = 4 (1.6%).

The morbidity and mortality outcomes vary according to the type of flood but also change with each flood phase. The impact phase is when the majority of injuries and deaths occur. This is especially true with flash floods, dam and levee failures, and other rapid onset floods. However, there are a variety of other risks in slow rising, persistent floods and in the postimpact phase of all floods that contribute to a significant number of injuries and illnesses. During the U.S. Midwest Flood of 1993, surveillance activities included a standardized questionnaire to gather flood-related injury and illness information. From nearly 2 months during the height of the flooding, 524 flood-related conditions were reported and split almost evenly between injuries (250 {47.7%}) and illnesses (233 {44.5%}). The remaining reports were listed as "other" or "unknown." The primary injuries included sprains/strains (86 {34%}), lacerations (61 {24%}), and abrasions and contusions (27 {11%}). Common illnesses included gastrointestinal (40 {17%}), rashes/dermatitis (38 {16%}), and heat illness (31 {13%}) (Centers for Disease Control and Prevention, 1993).

The postdisaster risk of infectious disease outbreaks among displaced populations is a common concern after a flood. Historically, many people have died from postdisaster infectious disease conditions including malaria, respiratory infections, and diarrhea (Lignon, 2006). However, flood-related outbreaks are rare in more developed nations. Poorer nations with more endemic disease and inadequate public health infrastructure are far more likely to see postdisaster outbreaks. In developed nations, it is rare to observe increases in serious infectious conditions following a disaster (Shultz et al., 2005).

Based on disaster research findings, the nondrowning hazards that follow a flood include the risk of fire, electrocution, carbon monoxide poisoning, sprains, strains, lacerations, bruises and other injuries, and gastrointestinal illness, rashes, respiratory conditions, and other illnesses. These hazards can be dealt with by following recommendations offered by the Centers for Disease Control and Prevention. They include avoiding electrical hazards, hazardous materials, carbon monoxide, and dangerous wild or stray animals. The CDC also provides instructions for sanitation and safe food and drinking water after a flood. But the most problematic postflood hazard is mold (See Figure 6–7).

There are thousands of varieties of mold that are ubiquitous in the environment and can survive under harsh conditions for long periods of time. As a part of our natural environment, the majority of these molds are harmless to most people. The problem arises when a flood provides an ideal environment for molds to quickly grow inside buildings. Depending on the type of mold, it can cause severe problems for those cleaning up after a flood and for those who reoccupy flooded buildings. Some people are sensitive to molds and experience allergic reactions whereas others with preexisting respiratory conditions are at risk for more severe consequences. Those who are sensitive typically have allergic reactions including nasal congestion, wheezing, and eye or skin irritation. Conditions can be severe in those with overwhelming exposures or with preexisting conditions. For example, those with chronic obstructive pulmonary disease (COPD) may acquire a mold infection of the lungs. There has been a particular emphasis on "toxic molds." This is a poor definition because even the molds that are more prone to producing mycotoxins

FIGURE 6–7 Excessive mold growth on the ceiling of a water damaged residence. Photo courtesy of the Centers for Disease Control and Prevention. Available at: http://www.cdc.gov/nceh/publications/books/housing/Graphics/chapter_05/Figure5.01.jpg.

do not always produce them and there are so many different molds that are capable of producing the same toxin, that it is not an accurate or appropriate way to frame the problem (Institute of Medicine, Committee on Damp Indoor Spaces and Health, 2004).

Regardless of the suspected type of mold, those with respiratory conditions should avoid mold exposure. Following a significant flooding episode, the EPA recommends that even among those who are healthy, if the visible mold growth exceeds 10 square feet, that is an area of about 3 feet by 3 feet, you should consider hiring a professional remediation service (Environmental Protection Agency, A brief guide to mold, moisture, and your home, EPA Publication 402-K-02-003, www.epa.gov/iaq/molds/images/moldguide.pdf).

The Far Reaching Impact of Floods

In 1927, an unusually large rainfall amount began threatening the new levee system in place along the Mississippi River. Levees broke in 145 places from Missouri to Louisiana unleashing the worst disaster in U.S. history up to that point in time. Over 26,000 square miles of land was flooded disrupting the lives of nearly a million residents. More than 200 people died, thousands of buildings were destroyed, and crops and farm animals were lost (American National Red Cross, 1929). Relief shelters, referred to as "concentration camps" were established to house more than 300,000 displaced persons. Nearly 70% of those were African Americans. With a shortage of laborers to work on the relief efforts and levees, the local authorities of Greenville, Mississippi began recruiting African American laborers at gunpoint. The African American population comprised 75% of the area population and 95% of the labor for the regional farms and plantations. At one point, with little shelter or supplies, 13,000 African American workers were left on a levee that had been cut off by flooding. Under the direction of Will Percy, the head of the Flood Relief Committee, boats arrived to evacuate the trapped residents to safety but only evacuated white survivors. Will Percy later requested that the Red Cross make the town a regional distribution center for relief supplies. As the supplies were distributed, only white residents and African American workers with "Laborer" tags showing they were "volunteer" laborers were allowed access to rations. The black residents without tags were excluded from distribution (Public Broadcasting Service (PBS), American experience, fatal flood timeline, www.pbs.org/wgbh/amex/flood/timeline/timeline2.html).

The influence this disaster had on the United States was enormous. After the disaster, the 1928 Flood Control Act was passed and initiated the building of today's modern levee system on the Mississippi (Pearcy, 2002). It was a major factor in the migration of large numbers of African Americans to Northern cities. For those staying in the region, it contributed to shaping the musical style, the Blues. Charlie Patton recorded "High Water Everywhere" and Lonnie Johnson released the "Broken Levee Blues." It also influenced the political tide and began moving African American support from the Republican to the Democratic Party (Watkins, 1997). Many changes set in motion by the 1927 Mississippi flood continue to influence public health and emergency management in the United States today.

Prevention

Preparing for a flood requires the same initial preparation as any other disaster scenario. A family preparedness plan should be developed. This includes a minimum of several days of supplies and a communication plan that includes out of town contacts that will be notified by separated family members. You should also understand your level of flood risk. This can be established with the assistance of local government planning offices. There is also a searchable flood map on the Federal Emergency Management Agency Website for U.S. residents. Several other nations have similar online tools available. Based on the degree of risk, a decision should be made concerning flood insurance. In the United States, a congressionally mandated National Flood Insurance Program is available and is recommended for those in high-risk areas. It is also a good idea for those in moderate and even low-risk areas. It is a federally backed program available through private insurance companies.

Despite the availability of flood insurance, it is best not to build in flood-prone areas unless it is absolutely necessary and incorporates elevated and reinforced designs. The water heater, furnace, electrical panel, and other essential equipment should be elevated as much as possible above the potential flood level, and check valves should also be included in sewer traps to prevent floodwater backups. Basement areas should be sealed to minimize floodwater intrusion.

Ten Individual Preparedness Actions for Floods

1. Identify the types of flood risks you face where you live, work, go to school, or travel.
2. Avoid building in flood plains but if you do, include design features that elevate the HVAC and other essential equipment.
3. Store essential supplies in a home disaster kit.
4. Establish a communication plan with your family.
5. Identify a safe evacuation location outside the potential flood zone.
6. Identify vulnerable friends and neighbors that may need assistance.
7. Keep a smaller preparedness kit in your vehicle.
8. Plan transportation options if bridges or roads become impassable.
9. Learn how to use a fire extinguisher and maintain at least one in your home.
10. Get trained in first aid and be prepared to help others during a crisis.

We must constantly build dikes of courage to hold back the flood of fear.

Martin Luther King, Jr.

FIGURE 6–8 Midwest flood of '93, extensive flooding outside the Missouri capital of Jefferson City. Photo courtesy of the U.S. Geological Survey.

Immediate Actions

The initial actions taken during a flood should first be based upon public announcements and warnings of local officials. Any time a flood is possible due to weather or other factors, listen to the radio and television for specific hazard and evacuation information. Officials will communicate important information to the public using the emergency alert system. If water is slowly rising and there is time to prepare, secure your home by bringing in outdoor furniture and other items that can be carried away by floodwaters. Move important items and valuables to the highest level of the structure. Turn off utilities using the main switches and valves and unplug electrical appliances. Keep in mind that major causes of morbidity and mortality during and following a flood are from electrocution, fires, and other nondrowning causes. The exception to these suggestions is if you are aware of flash flooding risks, do not wait for local officials to tell you to evacuate. You should also not take the additional time to shut down utilities or move furniture and other items to higher levels of a structure. You should immediately move to a safe location. As you evacuate, keep the following in mind:

- Be aware of streams, channels, canyons, and other areas known to flood suddenly. Flash floods can occur in these areas with or without official warnings or signs such as rain clouds or heavy rain.
- Do not walk through moving water. As little as 6 inches of moving water can cause you to fall. If you must wade through water, use a stick to check for the depth of the water and the firmness of the ground in front of you.
- The leading cause of death in floods is drowning in a vehicle. Do not drive in flooded areas. You and your vehicle can be swept away. If floodwaters rise around your car, abandon your car and move to higher ground.

- Six inches of water will reach the bottom of most passenger cars. It can cause loss of control or possible stalling.
- A foot of water will float most vehicles.
- Two feet of rushing water can carry away most vehicles including sport utility vehicles (SUV's) and pick-ups.

An event occurred as a result of a slow moving thunderstorm along a canyon near Big Cove, North Carolina, in September 1990. Although the thunderstorm produced only a quarter of an inch of rainfall in the town of Cherokee, rainfall estimates upstream exceeded 6 inches in less than 2 hours. The heavy rainfall produced a wall of water rushing down a small stream. Water levels rose up to 25 feet in minutes in some areas. Because of the flash flood watch issued by the National Weather Service, the Emergency Manager for the Cherokee Indian Reservation reacted immediately and the 911 dispatcher sent fire/rescue units to begin an evacuation. As a result of the flash flood watch, subsequent flash warnings, and quick actions by the local authorities, no lives were lost. It is reasonable to assume in similar flash flood events occurring at night in canyons, over 20 people would have drowned. In addition, much property damage was averted (National Hydrologic Warning Council, 2002).

Response and Recovery Challenges

Once a flood is over, the immediate actions taken will be dictated by local officials and by the observed environmental impact. Buildings can be damaged and structurally weakened by floodwaters. Landslides or mudslides may also occur. People and livestock may die due to drowning and, in sufficient quantities, these deceased bodies may pose a public health risk.

Secondary effects of floods include the contamination of drinking water and general poor environmental health conditions in buildings that have been flooded (See Figure 6–9). Dead animals may be in private wells and other important sources of water that will need to be physically cleaned through removal of debris and then shocked with chlorine to restore healthy conditions. Floods can also cause food shortages because of direct crop damage.

Long-term effects of flooding includes economic adversity from clean-up and rebuilding costs, food shortages, declines in tourism, and other key businesses. These impacts are often carried more by the poor of the affected community. There will also be long-term psychological issues among those who have lost loved ones or experienced serious personal or economic setbacks from the disaster.

After a flood evacuation, you should listen closely to news reports that announce when it is safe to enter affected areas. Once floodwaters recede, some roads may have weakened to the point where they may collapse under the weight of a car and these weaknesses are not always apparent to those driving on weakened roads. If instructed that it is safe to return, continue to watch for other hazards such as downed power lines, and residual floodwaters. Avoid any moving flood water and any buildings surrounded by floodwaters. Standing floodwater may also be contaminated with sewage, petroleum

FIGURE 6–9 Wallace, NC, September 28, 1999—Nearly 750,000 turkeys were lost to flooding in Duplin Co. alone as well as 100,000 hogs, over 23,000 on this poultry farm in Wallace. The owner, Alan Reynor, lost at least $85,000 plus cleanup costs. He also lost 2,500 hogs and his corn crop. Photo by Dave Gatley/FEMA. Available at: http://www.fema.gov/photodata/original/97.jpg.

products, or other hazardous substances. The mud that may be coating everything in a flooded building may also contain hazardous substances. You must clean and disinfect or dispose of anything that has been wet. You will also need to have damaged septic and sewage systems serviced to avoid health hazards.

As the affected population begins to reenter the area, it is important to establish a visible support system across the community. Everyone must take care of themselves and look out for each other through these types of chronic disasters. Floods can take months or even years for full recovery and those most affected must be reminded to take care of themselves. This includes fundamental reminders to eat and rest well, keep a manageable schedule, and discuss concerns with friends, family, or professional counselors.

Basic postflood safety reminders are also important. This includes reminders to:

- Avoid walking or driving into floodwaters.
- Be careful about possible slip, trip, and fall hazards.
- Practice strict personal hygiene and sanitation including frequent handwashing.
- Have structural hazards assessed, including sagging, unstable ceilings, electrical hazards, and gas leaks, by qualified professionals.
- Avoid using open flames or smoking in a home until gas leak hazards have been assessed.

- Have mold cleaned by professionals if it exceeds 10 square feet.
- Watch for other natural hazards, including rodents, insects, and snakes. You should always wear sturdy boots and gloves when moving debris. If mosquitoes become a problem, cover skin with long sleeves and pants and use repellant.
- If pets or other animals have been found deceased, dispose of them according to local instructions.
- Be careful when retrieving food that has been contaminated by floodwater or may have spoiled during a power outage. Inspect stored food items carefully. If there are any doubts, it should be thrown out. The only food items that are usually salvageable after floodwater submersion are metal cans or sealed waterproof packets that can be sanitized by removing labels and placing them in boiling water or in a bleach solution (one tablespoon of chlorine bleach per gallon of potable water) for 15 minutes. A marker can then be used to relabel containers.
- Do not drink the water unless it has been declared safe. Otherwise, you will need to use bottled water or disinfect water for drinking and other purposes.
- Keep children and pets away from the area until clean-up is complete.

The best detailed reference for home flood recovery activities is from the Red Cross. The publication, "Repairing Your Flooded Home" is a 56-page detailed reference available through the Red Cross (www.redcross.org). It describes:

- How to safely reenter your home after a flood.
- How to protect your home and belongings from additional damage.
- How to record damage and request assistance.
- How to check for utility leaks and restore service.
- How to clean up various surfaces, appliances, furniture, and other items.

Summary

Floods are the most common natural disasters. They affect millions of people globally each year and are responsible for many disaster-related deaths. Although it is no surprise that the majority of these deaths are from drowning, there are other causes of death from flooding that are less apparent including electrocution, fires, traumatic injury, and carbon monoxide poisoning. There are also many preventable deaths associated with flooding disasters. Because the majority of fatalities are from risk-taking people crossing flooded areas on foot or in vehicles, basic risk communication efforts in flood-prone areas may reduce potential fatalities. Other nonfatal injuries and illnesses can be reduced through regular reminders for those in flooded areas on the range of risks they may encounter and how to reduce them.

There are a variety of flood types that pose different challenges at various stages of the disaster. Flash floods are often predictable and effective planning and warning systems can reduce community risks. There are unique hazards associated with flood walls, levees, and dams. Though many of these systems provide protection from slow rising

FIGURE 6–10 New Orleans, LA, September 1, 2005—Evacuees and hospital patients arrive at New Orleans airport where FEMA's D-MATs have set up operations. New Orleans is being evacuated as a result of flooding caused by Hurricane Katrina. Photo: Michael Rieger/FEMA.

riverine floods, they may also increase the likelihood of rapid, violent flooding resulting from structural failure of these systems.

The greatest public health challenges following flooding are associated with the environmental health impact of flooded infrastructure. Utilities are disrupted and can pose risks including explosions and fires from gas leaks, electrocutions from downed or submerged power lines, and infectious disease or chemical exposures from contaminated water or food supplies. The general sanitation of flooded buildings poses on-going postflood risks, and the rapid growth of dangerous molds is a problem that often requires professional remediation. Ensuring a safe environment during recovery is extremely difficult and requires preplanning.

Flooding disasters will continue to grow in frequency as urbanization near waterways and coastal areas continues. Healthcare and public health preparedness measures associated with floods must be specifically tailored for each region according to the types of potential floods that may occur and the populations that may be impacted. At the heart of an effective flood response and recovery plan is a thorough assessment of flood risks across each region and a tailored public education effort before, during, and after a flood.

Websites

American Red Cross, Flood preparedness (in English and Spanish): www.redcross.org/services/prepare/0,1082,0_240_,00.html.

American Red Cross, Repairing Your Flooded Home: What to do After a Flood or Flash Flood (In English and -Spanish): www.redcross.org/static/file_cont333_lang0_150.pdf.

British Columbia, Provincial Emergency Program, Personal Flood Preparedness and Prevention: www.pep.bc.ca/floods/preparedness.html.

Centers for Disease Control and Prevention, Flood Preparedness: www.bt.cdc.gov/disasters/floods/.

City of Houston Department of Health and Human Services, Floods and Flash Floods Fact Sheet: www.houstontx.gov/health/osphp/Natural%20Disasters/Floods.pdf.

City of New Orleans, Flash Floods and Flooding Fact Sheet: www.cityofno.com/portal.aspx? portal=46&tabid=5.

Federal Emergency Management Agency, Flood Preparedness: www.fema.gov/hazard/flood/index.shtm.

Federal Emergency Management Agency, National Flood Insurance Program (General Information): www.fema.gov/about/programs/nfip/index.shtm.

Floodsmart.gov, The Official Site of the National Flood Insurance Program: www.floodsmart.gov/-floodsmart/.

Harris County, Texas, Family Flood Preparedness: www.hcfcd.org/famfloodprepare.html.

National Weather Service, Floods and Flash Floods Brochure: www.nws.noaa.gov/om/brochures/ffbro.htm.

North Dakota State University, Flood Recovery Information: www.ag.ndsu.edu/disaster/flood.html.

U.S. Geological Survey, Flood Hazards: www.usgs.gov/hazards/floods/.

U.S. Geological Survey, Large Floods in the United States: Where They Happen and Why, U.S. Geological Survey Circular 1245: http://pubs.usgs.gov/circ/2003/circ1245/.

World Health Organization, Flood Preparedness and Response Kit: www.searo.who.int/en/Section1257/Section2263/info-kit/Flood-Information-Kit.htm.

References

American National Red Cross. (1929). *The Mississippi valley flood disaster of 1927: official report of relief operations of the American National Red Cross*. Washington, DC: American National Red Cross.

Biswas, A. K., & Chatterjee, S. (1971). Dam disasters—an assessment. *Eng J (Montreal)* 54(3):EIC-71-HYDEL3.

Centers for Disease Control and Prevention. (1989). International notes health assessment of the population affected by flood conditions—Khartoum, Sudan. *MMWR* 37(51–52):785–788. www.cdc.gov/mmwr/preview/mmwrhtml/00001323.htm.

Centers for Disease Control and Prevention. (1993). Morbidity surveillance following the midwest flood—Missouri, 1993. *MMWR* 42(41):797–798.

Coates, L. (1999). Flood fatalities in Australia, 1788–1996. *Aust Geogr* 30(3):391–408.

Davies, H. R. J., & Walsh, R. P. D. (1997). Historical changes in the flood hazard at Khartoum, Sudan: lessons and warnings for the future. *Singap J Trop Geogr* 18:123–140.

El Nasser, H. (2005). A New Orleans like the old one just won't do. USA Today. September 18, 2005. www.usatoday.com/news/nation/2005-09-18-new-orleans-rebuilding_x.htm.

EM-DAT: OFDA/CRED International Disaster Database. (2004). Brussels, Belgium: Université Catholique de Louvain. www.em-dat.net.

French, J., Ing, R., Von Allmen, S., & Wood, R. (1983). Mortality from flash floods: a review of National Weather Service Reports, 1969–81. *Public Health Rep* 98(6):584–588.

Gerritsen, H. (2005). What happened in 1953? The big flood in the Netherlands in retrospect. *Philos Trans R Soc A* 363:1271–1291. www.wldelft.nl/rnd/publ/docs/Ge_2005a.pdf.

Government Security. (2008). Report says U.S. dams require better security. September 3, 2008. http://govtsecurity.com/news/us-dam-security-0904/.

Hunter, M. (2006). Deaths of evacuees push toll to 1,577: out-of-state victims mostly elderly, infirm. *The Times-Picayune*. May 19, 2006. www.nola.com/news/t-p/frontpage/index.ssf?/base/news-5/1148020620117480.xml&coll=1.

Institute of Medicine, Committee on Damp Indoor Spaces and Health. (2004). *Damp Indoor Spaces and Health*. Washington, DC: The National Academies Press.

Jonkman, S. N., & Kelman, I. (2005). An analysis of the causes and circumstances of flood disaster deaths. *Disasters* 29(1):75–97.

Lignon, B. L. (2006). Infectious diseases that pose specific challenges after natural disasters: a review. *Semin Pediatr Infect Dis* 17:36–45.

National Hydrologic Warning Council. (2002). Use and benefits of the National Weather Service River and Flood Forecasts. May 2002. http://nws.noaa.gov/oh/ahps/AHPS%20Benefits.pdf.

Pearcy, M. T. (2002). After the flood: a history of the 1928 Flood Control Act. *J Ill State Hist Soc*. http://findarticles.com/p/articles/mi_qa3945/is_200207/ai_n9105154.

Shell, A. (2005). Floods can cause damage to structure. *USA Today*. September 9, 2005:3B. www.usatoday.com/money/perfi/housing/2005-09-08-wet-homes-usat_x.htm.

Shultz, J. M., Russell, J., & Espinel, Z. (2005). Epidemiology of tropical cyclones: the dynamics of disaster, disease, and development. *Epidemiol Rev* 27:21–35.

U.S. Department of Interior. (1999). A procedure for estimating the loss of life caused by dam failure, Bureau of Reclamation Report DSO-99-06. www.usbr.gov/ssle/dam_safety/Risk/Estimating%20life%20loss.pdf.

Walsh, R. P. D., Davies, H. R. J., & Musa, S. B. (1994). Flood frequency and impacts at Khartoum since the early nineteenth century. *Geogr J* 160(3):266–280.

Watkins, T. H. (1997). Boiling over: the 1927 Mississippi flood was a caldron of racism and greed. *New York Times*. April 13, 1997. www.nytimes.com/books/97/04/13/reviews/970413.13watkint.html.

West Virginia Archives and History. (1950) Worst flash flood in history sweeps troy and other communities: thirty-one known dead in 6-county area; Little Kanawha river hits crest of 31.1 at Glenville. *Glenville Pathfinder*. June 30, 1950. www.wvculture.org/history/disasters/flood195001.html.

7 Heat Waves

Objectives of This Chapter

- Define types of heat injuries and illnesses.
- Explain how the body manages excessive heat.
- Describe the risk factors for those who are most vulnerable to excessive heat.
- List environmental factors that contribute to the severity of heat waves.
- Describe how a community can prepare for and reduce the impact of severe heat conditions.
- Explain the steps in an effective local government response to a heat wave.
- Identify the triggers used by government officials to activate heat wave response activities.
- Describe the mistakes historically made by government leaders in managing heat waves.
- Explain the content and delivery methods for heat warnings shared with vulnerable populations.
- List the practical steps everyone should take to reduce the risk of heat injury.

FIGURE 7–1 Photo courtesy of the National Oceanic and Atmospheric Administration. http://www.srh.noaa.gov/lch/prep/heat.jpg.

Introduction (Case Study: European Heat Wave, 2003)

A long, scorching European summer took a dreadful turn for the worst in August 2003. By the end of the month, it was evident that this would be the worst weather-related health disaster to hit Western Europe in centuries. Higher than normal temperatures began in May and by the end of August, the average temperatures were about 30% higher than normal for the season. This was accompanied by above normal days of sunshine and below normal rain and humidity in regions where rain and clouds are the norm during the summer months (Rebetez et al., 2006). High temperatures of 95–104 °F (35–40 °C) were regularly recorded in July and August. These conditions resulted in more than 25,000 fires across Western Europe that destroyed over 1.6 million acres (>647,000 hectares) and cost billions of euros (United Nations Environment Programme, 2004). In Portugal and Spain, wildfires raged out of control for weeks. Thousands of firefighters from across Europe deployed to bring the fires under control and equipment was on loan from as far away as Russia (CNN.com, 2003).

Although those in the rural areas of Europe struggled with the drought, heat, and fires, the human toll in the cities was mounting. One of the hardest hit cities was Paris. It is impossible to know when the first heat wave fatalities began that summer. Those at greatest risk have other conditions and complications that are exacerbated by the heat. The cause of death among initial casualties in Paris was likely attributed to preexisting conditions and not the heat. Many casualties were elderly, immobile, and lived alone. Sometimes several days had passed before they were found in their homes after succumbing to the heat. The normally mild summer temperatures of Paris usually top out in July and August at about 75 °F (24 °C). As a result, air conditioners are not commonplace. When the temperatures in August 2003 exceeded 100 °F (38 °C) for nearly 2 weeks straight, the human toll quickly mounted (Figure 7–2). A daily surge of over 150 patients was experienced by Parisian healthcare facilities. A 130% mortality increase was observed above what is normally expected in the month of August across the city (Hemon and Jougla, 2003). Over one-third (35%) of these fatalities occurred in homes, 42% occurred in hospitals, and the remaining 19% occurred in long-term care facilities, most of which did not have air conditioning (Dhainaut et al., 2004).

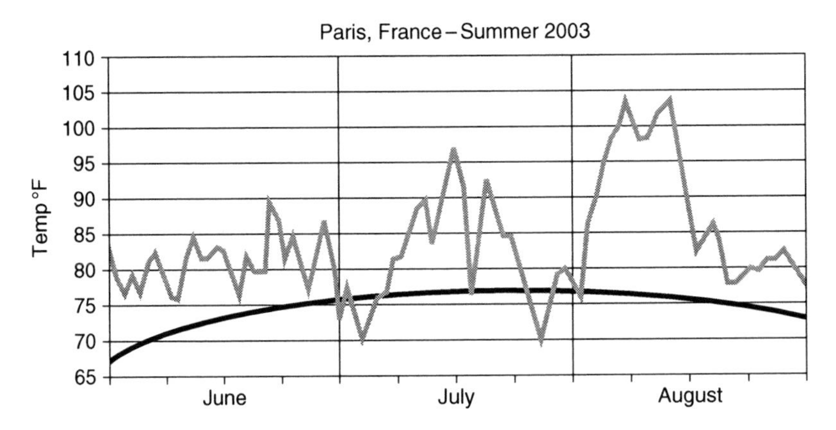

FIGURE 7–2 Actual versus average daily maximum temperatures in Paris, France, 2003.
Source: EPA Excessive Heat Events Guidebook.

A variety of controversies emerged from the tragic events in Paris. Questions were raised concerning the response of government officials. A full week had passed during the peak of the heat wave before officials acknowledged the problem and declared an emergency. Socialist Party officials blamed the administration of Prime Minister Jean-Pierre Raffarin and asked for the resignation of his Health Minister, Jean-Francois Mattei. The Prime Minister's office responded with charges that the establishment of a 35-hour work week by the Socialist Party was responsible for reducing the number of healthcare workers available to manage the crisis. In the end, Dr. Lucien Abenhaim, who served as the Director General for Health, submitted his resignation. This would be comparable to the resignation of the U.S. Surgeon General. Following the resignation of Dr. Abenhaim, Prime Minister Raffarin blamed the heat wave's impact on the larger community state of affairs. He pointed out in interviews with the French press that over half of the fatalities occurred outside healthcare facilities and many who were brought to local emergency departments were in such poor condition by the time they arrived that little could be done to save them. He shifted the blame instead to global warming and the general public. August is a popular time for family vacations. Many friends and relatives of elderly victims who would normally check in on them were out of town during the heat wave (Tagliabue, 2003).

In reality, *all* of the issues raised by government leaders, party officials, and the French media contributed to the tragic outcomes experienced by the French. Most buildings in Paris do not have air conditioning. The city is simply not designed to protect the population from excessive temperatures. Public health officials failed to anticipate the impact of the heat wave and institute community measures and public warnings. The healthcare system was not adequately prepared and staffed to handle such an emergency. It was vacation season and many elderly residents did not have the social support they would normally have checking on them. These and other factors contributed to a death toll that was reported to exceed 30,000. Nearly half of the fatalities were in France. The others were throughout Germany, Spain, Italy, UK, Netherlands, Portugal, and Belgium.

Heat-Related Definitions

Heat wave: A heat wave is an extended period of excessive heat and humidity resulting in health threats to the community. There is not a precise quantitative formula to define a heat wave because there are so many variables from one community or geographical region to another. A qualitative definition is simply a period of abnormal heat that generates a public response. This response may be due to the impact on human or animal health, crops, utilities, or commerce. However, for the purposes of this section, we are referring primarily to the human health threat.

Heat index: The heat index is a number in degrees (Fahrenheit or Celsius) that tells how hot it really feels when relative humidity is added to the actual air temperature. As humidity increases, the evaporation of perspiration from a person is slowed and more heat is retained. The heat index accounts for that and is a more accurate representation of the temperature related to human health risks (Figure 7–5).

Heat cramps: Heat cramps are muscular pains and spasms resulting from physical exertion in a hot environment. Insufficient fluid intake may increase their severity. Although heat cramps are the least severe type of heat injury, they are an early indication that the body is having trouble with the heat.

Heat exhaustion: Heat exhaustion occurs when people physically exert themselves in a hot, humid environment. As body fluids are lost through heavy sweating, blood flow to the skin begins to increase. This results in decreased blood flow to the vital organs leading to a form of mild shock. If the victim is not quickly cooled and rehydrated, a heat stroke may result.

Heat stroke: Heat stroke, also called sunstroke, is a life-threatening condition. If heat exhaustion is left unchecked, the body's temperature control system may stop working. This includes the loss of ability to sweat. The body temperature may increase so high that brain damage and death may result if the body is not cooled quickly.

Sun stroke: Another name for a heat stroke.

Heat Waves

Heat waves are one of the most subtle yet most lethal disasters (See Figure 7–3). Local infrastructure shows no damage, commerce and travel continue with little noticeable impact. Hospitalizations and deaths slowly increase among vulnerable populations, while causes often remain unattributed to the heat or even unreported. Although heat emergencies do not exhibit the drama of many other disasters, they produce greater mortality than hurricanes, tornadoes, and earthquakes combined. This lack of drama often results in a lack of media attention, public outcry, and political will for focusing adequate attention and resources toward improving heat emergency preparedness and making it a higher priority. The worst toll of extreme heat is on the elderly, children, the poor, and the homeless. In addition, persons with chronic cardiovascular, pulmonary, or other health conditions and those who are mentally impaired are at an increased risk. Future predictions of heat wave occurrences and severity look harsh. Global warming and other environmental changes are driving the likelihood of extreme heat events higher at a time when we are seeing continued urbanization and an aging population. The nexus between these environmental and social factors places us on a certain path toward severe and lethal heat waves in the future. It is important to understand effective measures to prepare for, predict, and manage heat waves.

Effects on the Human Body

The human body maintains an internal temperature within a very narrow range. A normal temperature is 98.6 °F (37 °C) but can fluctuate in a healthy person by about 1 °F (0.6 °C) over the course of a day. These normal fluctuations can result from varying activity levels and in women they can result from regular hormonal changes. As heat builds in the body, the blood vessels close to the skin dilate to bring the blood closer to the surface and release heat. This vasodilation places additional stress on organs such as the heart, brain, and kidneys by reducing the available supply of blood needed for normal function. It can also reduce blood pressure when the heart's output cannot compensate to normalize blood pressure. Sweat glands also become active, releasing moisture on the skin. As sweat evaporates, more heat is dissipated through evaporation (Simon, 1993). Heat is also released through the lungs during respiration. During everyday activities and moderate temperatures, the skin and lungs can efficiently regulate internal body temperature. However, when environmental conditions or physical activity generate more heat than the body can regulate, there is increasing risk for heat-related illnesses. The heat burden of the body significantly increases when engaged in physical activity. During times of heavy exertion, additional energy is expended and nearly 75% of that energy is directly converted to heat rather than energy (Mellion and Shelton, 1997).

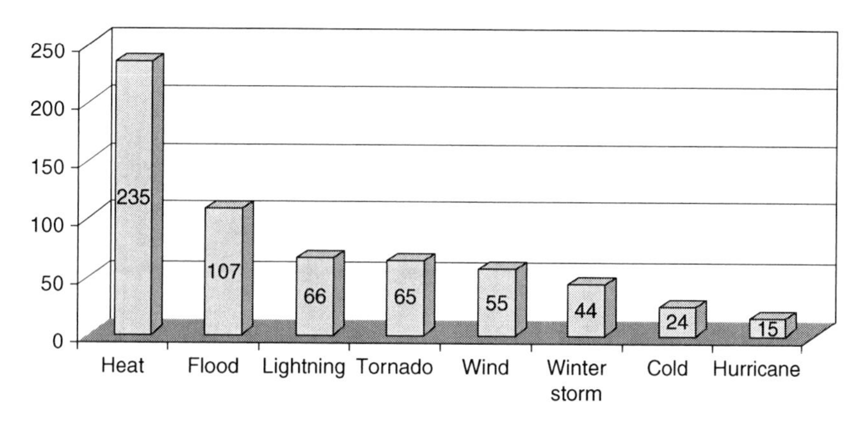

FIGURE 7–3 U.S. National average weather fatalities (1975–2004).

Source: U.S. National Oceanic and Atmospheric Administration (NOAA), National Weather Service Heat Wave Health Threats. http://www.srh.noaa.gov/shv/Heat_Awareness.htm.

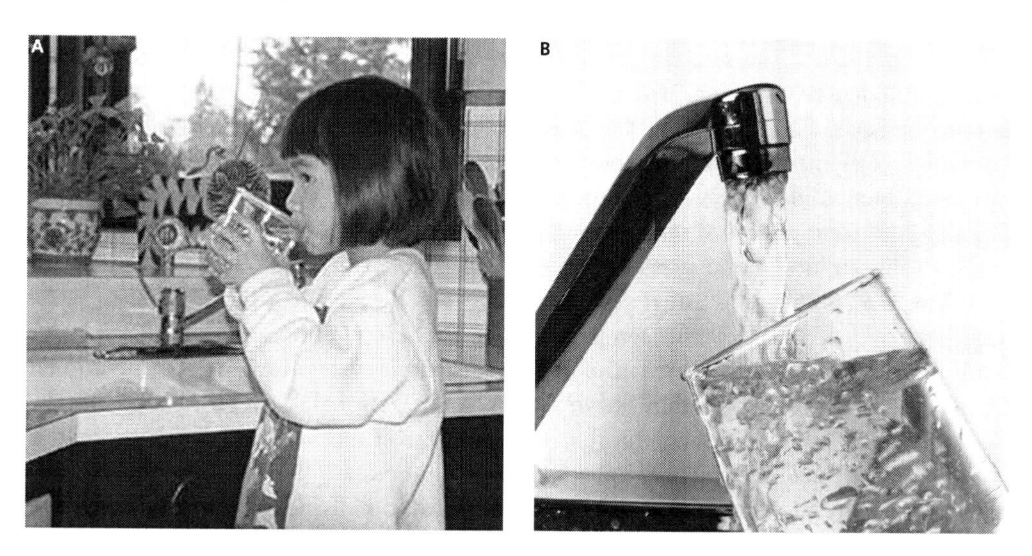

FIGURE 7–4 (A, B) Hydration is the key to reducing heat stress.

Source: U.S. Geological Survey. http://pubs.usgs.gov/fs/fs-027-01/ (a). South Carolina Department of Health and Environmental Control. http://www.scdhec.gov/environment/water/dwoutreach.htm (b).

Risk Factors for Heat Illness and Injury

Physical Factors

- Elderly (age > 65)
- Infants (age < 1)
- Physically disabled
- Mentally impaired
- Engaged in outdoor exercise or work
- Users of alcohol or illegal drugs
- Users of certain prescription medications (e.g., medications for high blood pressure, motion sickness, gastrointestinal medications containing atropine, sleeping pills, antidepressants, antihistamines, and Parkinson's disease)

Social Factors

- Low income
- Homeless
- Socially isolated
- Urban residence
- Poor access to healthcare or cooling shelters

Dehydration is also a critical risk factor for heat illness. If fluid loss exceeds only 3% of normal levels in the body, dehydration has begun and the risk of heat illness quickly increases (Aerospace Medical Association, 2005). Fluid loss can occur due to inadequate water consumption during physical activity or simply with extended exposure to low humidity. This includes long trips on airlines where humidity levels are typically below 20% (Allen et al., 1998). Other injuries or illnesses may contribute to fluid loss. This includes shock, trauma, burns (including sunburn), gastrointestinal tract disease causing diarrhea or vomiting, malnutrition, and hyperglycemia from chronic conditions such as diabetes. Fluid loss is further complicated by consumption of diuretics such as alcohol- or caffeine-containing beverages. Some medications cause individuals to sweat or urinate more and may contribute to dehydration as well. These drugs include motion sickness and gastrointestinal medications containing atropine, sleeping pills, antidepressants, antihistamines, blood pressure medications, and Parkinson's disease medications.

The U.S. National Weather Service (NWS) has developed a Heat Index to more accurately reflect the apparent temperature or the temperature experienced by your body when humidity is factored into the conditions (See Figure 7–5) (U.S. National Oceanic and Atmospheric Administration (NOAA), National Weather Service, 1994). The increasing humidity reduces the ability for sweat to evaporate and provide cooling. If you are in direct sunlight, the heat index increases by about 15 °F. Conditions that include high temperatures, low humidity, and strong winds can rapidly dehydrate your body. The NWS Heat Index Chart does not fully account for the effect of direct sunlight or dry, windy conditions. The heat index chart calculations are based on a variety of assumptions. This includes assuming environmental parameters to include shady, slightly windy conditions. These assumptions reduce the precision of the heat index as a predictive tool. However, the chart does provide a simple measure of risk, particularly among higher risk groups. This is a basic tool that provides us with a trigger for instituting community response measures during a heat wave.

Category	Heat Index	Possible Heat Illnesses for Exposed Vulnerable Populations
Extreme danger	130 °F or higher (54 °C or higher)	Heat stroke likely with prolonged exposure
Danger	105–129 °F (41–54 °C)	Heat exhaustion/muscle cramps likely, heat stroke possible
Extreme caution	90–104 °F (32–40 °C)	Heat exhaustion/muscle cramps possible
Caution	80–89 °F (27–32 °C)	Heat fatigue possible

HEAT INDEX °F (°C)													
	RELATIVE HUMIDITY (%)												
Temp.	**40**	**45**	**50**	**55**	**60**	**65**	**70**	**75**	**80**	**85**	**90**	**95**	**100**
110 (47)	138 (58)												
108 (43)	130 (54)	137 (58)											
106 (41)	124 (51)	130 (54)	137 (58)										
104 (40)	119 (48)	124 (51)	131 (55)	137 (58)									
102 (39)	114 (46)	119 (48)	124 (51)	130 (54)	137 (58)								
100 (38)	109 (43)	114 (46)	118 (48)	124 (51)	129 (54)	136 (58)							
98 (37)	105 (41)	109 (43)	113 (45)	117 (47)	123 (51)	128 (53)	134 (57)						
96 (36)	101 (38)	104 (40)	108 (42)	112 (44)	116 (47)	121 (49)	126 (52)	132 (56)					
94 (34)	97 (36)	100 (38)	103 (39)	106 (41)	110 (43)	114 (46)	119 (48)	124 (51)	129 (54)	135 (57)			
92 (33)	94 (34)	96 (36)	99 (37)	101 (38)	105 (41)	108 (42)	112 (44)	116 (47)	121 (49)	126 (52)	131 (55)		
90 (32)	91 (33)	93 (34)	95 (35)	97 (36)	100 (38)	103 (39)	106 (41)	109 (43)	113 (45)	117 (47)	122 (50)	127 (53)	132 (56)
88 (31)	88 (31)	89 (32)	91 (33)	93 (34)	95 (35)	98 (37)	100 (38)	103 (39)	106 (41)	110 (43)	113 (45)	117 (47)	121 (49)
86 (30)	85 (29)	87 (31)	88 (31)	89 (32)	91 (33)	93 (34)	95 (35)	97 (36)	100 (38)	102 (39)	105 (41)	108 (42)	112 (44)
84 (29)	83 (28)	84 (29)	85 (29)	86 (30)	88 (31)	89 (32)	90 (32)	92 (33)	94 (34)	96 (36)	98 (37)	100 (38)	103 (39)
82 (28)	81 (27)	82 (28)	83 (28)	84 (29)	84 (29)	85 (29)	86 (30)	88 (31)	89 (32)	90 (32)	91 (33)	93 (34)	95 (35)
80 (27)	80 (27)	80 (27)	81 (27)	81 (27)	82 (28)	82 (28)	83 (28)	84 (29)	84 (29)	85 (29)	86 (30)	86 (30)	87 (31)

FIGURE 7–5 U.S. National Weather Service Heat Index Chart.
Source: U.S. National Weather Service. http://www.crh.noaa.gov/jkl/?n=heat_index_calculator.

The effects of extreme heat on the body worsen as humidity rises slowing the evaporation of moisture from the skin, challenging the body to maintain a normal internal temperature. The resulting heat illness can range from heat cramps or heat exhaustion to heat stroke. These typically occur when you exert yourself beyond your physical capacity based on age, condition, and other risk factors. This is particularly challenging in urban areas. Brick, concrete, asphalt, and other surfaces ubiquitous in an urban environment tend to hold heat and release it throughout the evening and nighttime hours, therefore the temperatures remain elevated for a longer period of time. Exposure to heat from these warmed surfaces is through both conductive and radiant heat exchange. Conductive heat transfer is through direct contact with the heated surfaces and radiant heat transfer is heat given off surfaces in waves. These radiant heat waves are similar to waves of light but have a much greater wavelength. The greater number of heated surfaces in an urban environment increases the accumulation of heat and the risk to the urban population.

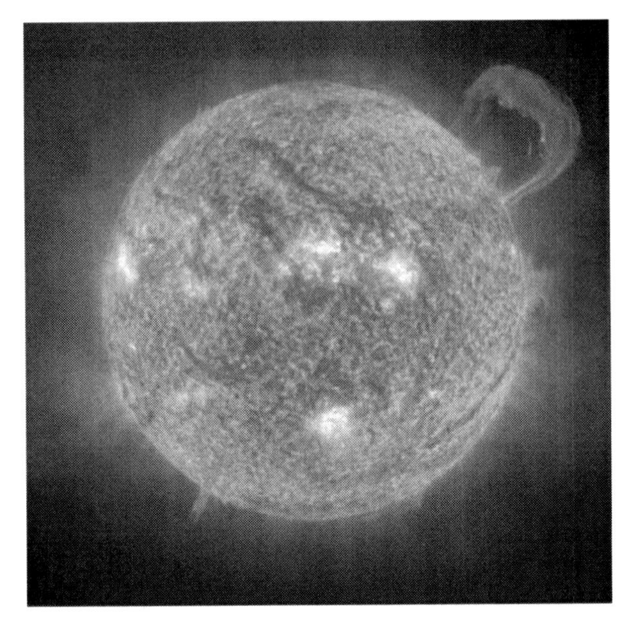

FIGURE 7–6 Photo courtesy of the National Aeronautics and Space Administration. http://discovery.nasa.gov/images/sun.jpe.

Heat Illness—Environmental Factors

- Air temperature
- Humidity
- Air movement
- Radiant heat
- Conductive heat
- Activity level
- Clothing

Prevention

Unlike many other types of disasters, all deaths from heat waves are preventable. Heat waves can be predicted and the measures needed to successfully manage their impact are well understood and inexpensive. Before the emergency begins, it is important that the community engage in preparedness activities specifically for a heat emergency. As decisions are made concerning infrastructure, particularly in urban areas, the use of cooler pavement and roofing surfaces can mitigate some of the effects of a heat wave. In addition, employers with workers potentially exposed to heat must plan ahead for possible work delays and institute policies that will protect those at risk.

Urban Heat Island Mitigation

The risk of heat emergencies increases as ongoing industrialization and urbanization drive a larger proportion of the human population to settle in or near cities. A major factor in this increasing heat risk is the phenomenon of the "urban heat island effect" (See Figure 7–7). This refers to the alteration of the environment resulting from urban development combined with the added heat generated by population density. The reduction in vegetation across urban areas reduces the shading of buildings and surfaces. These surfaces absorb a tremendous amount of heat and release it slowly. As a result, nighttime urban temperatures drop more gradually than in rural areas. This keeps nighttime temperatures higher and prolongs the daily heat burden on the urban population. Another factor in increasing urban temperatures associated with vegetation is the reduction in evapotranspiration. Just as sweat evaporating from the surface of your skin provides a cooling effect; plants transpire or emit moisture through their leaves. As the water evaporates, it provides a cooling effect to the surrounding area. A reduction in vegetation also reduces these natural cooling processes.

If you fly over any urban area, you will notice a patchwork of typically dark colored roofs and pavements. These darker surfaces are less reflective and absorb more heat. The widespread use of dark surface materials in urban areas creates a setting where large amounts of solar energy are absorbed and retained driving the daytime temperatures higher and sustaining higher temperatures throughout the night. The taller structures in metropolitan areas also block air movement and further reduce the rate at which these surfaces can cool. When these factors are combined with the heat generated by urban industrial and transportation activities, the accumulating heat can drive urban and suburban temperatures up by up to 10 °F (6 °C) above what is seen in surrounding rural areas (U.S. Environmental Protecton Agency (EPA), Heat island reduction initiative, Heat island effect. n.d, www.epa.gov/hiri/index.html).

Mitigation measures to reduce urban heat islands are simple and offer benefits far beyond the reduction in heat. A study conducted by the U.S. Forestry Service in Brooklyn, New York estimated that the borough's trees, which cover just over 10% of the Brooklyn

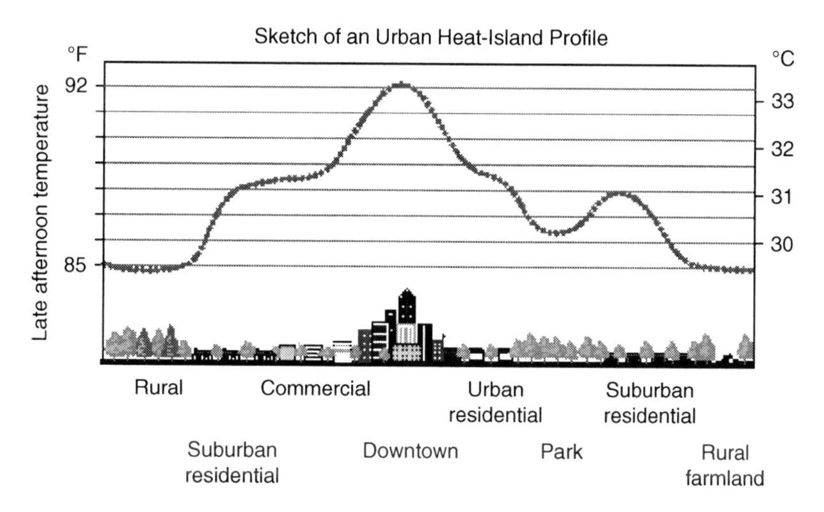

FIGURE 7–7 Urban heat island profile.

Source: Lawrence Berkeley National Laboratory, Heat Island Group. http://eetd.lbl.gov/HeatIsland/HighTemps/.

area, remove over 250 metric tons of air pollution each year (Nowak et al., 2002). This has a dual effect on the health of the public. There is a reduction in pollutants, decreasing stress on the respiratory systems of vulnerable populations, and a reduction in temperatures through the shading and transpiration of the vegetation. This vegetation also increases the overall quality of life for urban residents by positively influencing perceptions of land use and emotions (Sheets and Manzer, 1991).

In addition to the introduction of extra vegetation, the use of cooler roof and pavement surfaces has far reaching effects. Older roofing materials, particularly those that are dark colored, absorb heat and have low thermal emittance or ability to release heat. "Cool roof" materials are light colored to reflect heat and have the ability to cool much more quickly than older roofing materials (See Figure 7–8). This reduces the urban heat island contribution of the roof by absorbing less heat and quickly releasing absorbed heat at night. Cool roof surfaces on flat roofs are usually white to maximize reflection. This is the same principle as wearing light colored clothing when spending time in the sun. It simply keeps you cooler by reflecting the sunlight. Not only do cool roofs reduce urban heat, they also reduce air conditioning bills for building owners and residents and even reduce air pollution and smog formation. The same principles apply to pavement surfaces. Lighter colored pavements (light brown, light gray, etc.) reflect more solar energy and absorb less heat. The porosity of the paved surface also contributes to the emittance of absorbed heat.

Preseason Community Preparedness Activities

The most important steps a community can take to reduce morbidity and mortality associated with extreme heat is to identify vulnerable populations in advance and determine the best methods to get information into their hands and, when necessary, move them to a safer environment. To successfully accomplish this requires planning. Unfortunately,

FIGURE 7–8 Workers apply cool roof coating on a row house in Baltimore, MD.
Source: EPA. http://www.epa.gov/hiri/images/Baltimore.jpg.

many cities have poor heat emergency plans or none at all. Some cities focus few or no resources toward heat emergency planning because it is a rare or unlikely event in their region. That was the case in Paris, France prior to the 2003 heat wave. It was such an improbable event for Paris that more pressing issues demanded the time of those responsible until it was too late for preparedness measures. Regions where the occurrence of heat emergencies is less likely are also more likely to have higher mortality during a heat wave (Kalkstein and Davis, 1989). For example, if the same extreme hot weather conditions existed in Minneapolis, Minnesota and Houston, Texas, mortality will likely be higher in Minneapolis. Some of this is due to infrastructure being based upon the lower temperature norms. In areas where air conditioners are rarely needed, many people simply do not purchase them. The population is also less acclimated to the heat. This increases the risk of serious health outcomes from heat waves in cities that are typically cooler. The population and infrastructure are not prepared to cope with it.

Community planning for heat emergencies is inconsistent from city to city in spite of the fact that on average, heat waves kill more people than any other type of natural disaster. No specific standards exist that prescribe appropriate preparedness and response actions. Some cities have no plans that are specific to heat emergencies. Even if an "all-hazards" planning approach is taken, there must be sufficient details that are specific to each type of potential emergency. A recent review of 18 city heat emergency plans found that six of the 18 cities had no heat emergency planning at all and many who did gave the topic inadequate attention (Bernard and McGeehin, 2004). Effective plans must have a phased approach that addresses the measures taken by each agency leading up to the summer season. The role of each local agency must be defined and specific measures leading up to a heat emergency must be clear. A standard warning system must be instituted and communicated in a consistent manner. Communication efforts must emphasize and target vulnerable populations. These measures may include both print and broadcast public services announcements, as well as establishment of hotlines that can provide instructions for personal preparedness and locations of cooling shelters. Surveillance of heat casualties must also be instituted (CDC, 1995a; Kizer, 2000). This is especially challenging since a variety of chronic and acute conditions may be triggered by extreme heat but often not attributed to the heat conditions. An effective plan will reduce the impact of a heat wave on the health of the affected population (Naughton et al., 2002; Weisskopf et al., 2002).

A good example of heat wave planning can be found in the city of Milwaukee, Wisconsin. Their plan assigns the primary responsibility for development of messages and coordination of an excessive Heat Task Force to the Milwaukee City Health Department. Their primary partners include public and private agencies throughout the region that serve vulnerable populations such as the elderly, disabled, homeless, and mentally ill. Their metropolitan medical society is asked to encourage healthcare providers to prepare for heat emergencies. The meteorologists in the area are encouraged to cover stories on heat illness prevention and maintain contact with public health authorities to update health notices. The Red Cross maintains a 24-hour heat health tip hotline. Behavioral health organizations prepare in-patient clientele with information on heat and medications as well as support in developing emergency plans with family members. Two to four days prior to the arrival of possible extreme heat conditions, a "Heat Health Outlook" statement is issued to task force members asking them to begin anticipating a heat advisory. By adding a preemptive "outlook" warning, preparations may begin prior to the arrival of the heat emergency and an enhanced preparedness posture is achieved. From that point, increasing heat alert levels and associated actions are ramped up based upon meteorological or epidemiological thresholds as described in Table 7–1 (City of Milwaukee Health Department, Heat health information, www.city.milwaukee.gov/HeatHealth Informatio2689.htm).

The meteorological threshold for Milwaukee and many other cities is the heat index. Because the index assumes shady conditions with a light wind, it does not compensate for other important weather variables such as wind speed and cloud cover. It also does not take into account the differences in human acclimatization in various regions and at different times of the year. There is also an escalating impact that heat has on human health over the course of multiple days of excessive heat that is also not captured by the heat index. One of the few systems that capture a broad range of meteorological and epidemiological predictive data started as the Philadelphia, Pennsylvania, Hot Weather Health Watch Warning System (PWWS) (See Figure 7–9). The system was developed in part as a result of an historical heat wave in Philadelphia in 1993 and is now becoming the new standard for triggering heat wave response activities. The PWWS uses a Temporal Synoptic Index (TSI) derived from an algorithm using complex statistical analysis of historical meteorological and epidemiological data (Kalkstein et al., 1987). An approaching air mass can be evaluated 48 hours in advance of arrival for the potential impact it could have on the health of a metropolitan population. Data from similar historical weather occurrences in the same city at the same time of year are factored in along with the epidemiological data

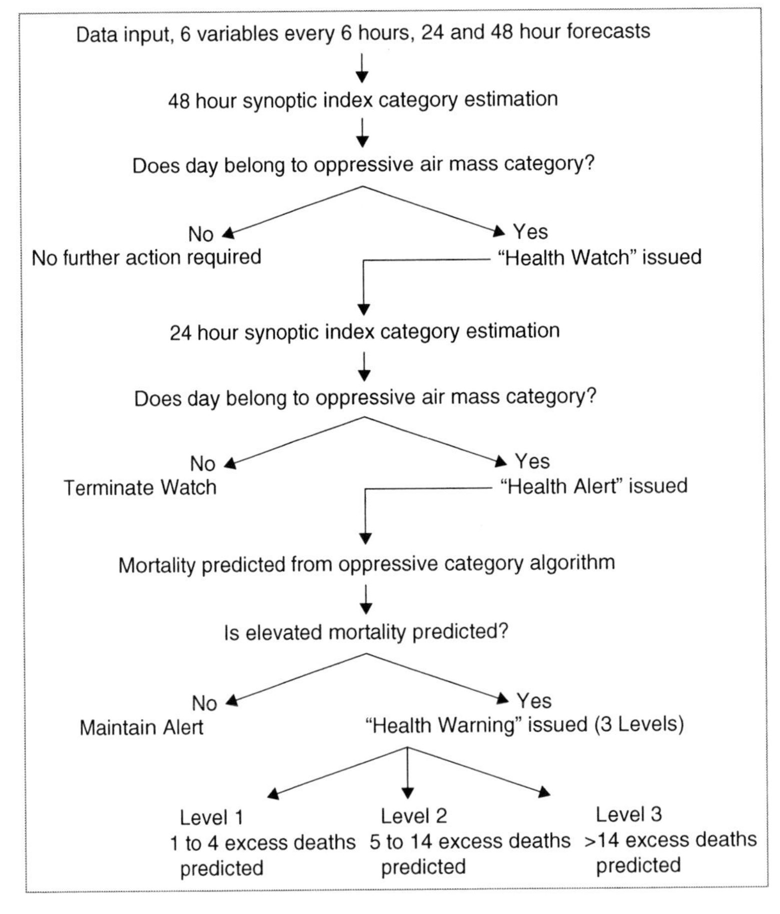

FIGURE 7–9 Configuration of the Philadelphia hot weather health watch/warning system (Kalkstein et al., 1996).

Table 7–1 Milwaukee Heat Alert Table

Level of Alert	Meteorological Threshold	Epidemiological Threshold	Heat Alert Level	General Actions
Level 1	Heat Advisory or excessive heat warning conditions anticipated within the next 2 to 4 days	None	Heat Health Outlook	• Send alert to Task Force to review Heat Plan and anticipate implementation.
Level 2	Issued 6–24 hours in advance of any 24-hour period in which daytime heat indices are expected to be 105–110 for 3 or more hours with nighttime heat indices ≥75.	A case of heat-related illness, heat stroke, or heat-related death from a residential setting.	Heat Health Advisory	• Begin active surveillance for heat injury. • Initiate 24-hour response for heat-related calls to the local public health agency. • Alert Task Force to prepare for Heat Watch or Warning within 24 hours. • Open cooling sites as demand warrants. • Initiate daily checks on high-risk populations by Task Force member organizations. • Alert state agencies of possible resource needs. • Consider multiagency command post activation.
Level 3	Excessive heat warning conditions expected in 24–48 hours.	None	Excessive Heat Health Watch	• Continue activities as described earlier.
Level 4	Issued 6–24 hours in advance of any 24-hour period in which daytime heat indices are expected to exceed 110 for 3 hours or more and nighttime heat indices ≥80.	Multiple cases of heat related illness, heat stroke, or heat-related deaths from a residential setting.	Excessive Heat Warning	• Continue activities as described earlier.

that resulted. The resulting warning provides an estimate of how many heat casualties may be expected based on the history of that city. This kind of system is much more rigorous than the heat index. The Director of the National Weather Service, retired U.S. Air Force Brigadier General David Johnson, has stated that the PWWS is a model program for the rest of the nation (U.S. National Oceanic and Atmospheric Administration [NOAA], 2005). The proven efficacy of the PWWS is now resulting in widespread implementation and continued improvements of a Heat/Health Watch Warning System (HHWW) by weather service offices serving metro areas throughout the nation. This system is already saving lives by providing more advanced and enhanced warnings.

Workplace Heat Stress Prevention

Another important risk factor for heat illness or injury is occupational exposure to heat. The Occupational Safety and Health Administration (OSHA) does not have a specific standard for heat exposure but addresses it in several general industry standards and also considers it to fall under the Occupational Safety Act, Section 5(a)(1), General Duty Clause. This simply states that employers must "furnish to each of his employees employment and a place of employment which are free from recognized hazards that are causing or are likely to cause death or serious physical harm." OSHA interprets this with three key points (Occupational Safety and Health Administration [OSHA], 2001).

1. Allow workers to drink water at liberty;
2. Establish provisions for a work/rest regimen so that exposure time to high temperatures and/or the work rate is decreased;
3. Develop a heat stress program which incorporates the following
   a. A training program informing employees about the effects of heat stress, and how to recognize heat-related illness symptoms and prevent heat-induced illnesses;
   b. A screening program to identify health conditions aggravated by elevated environmental temperatures;
   c. An acclimation program for new employees or employees returning to work from absences of 3 or more days;
   d. Specific procedures to be followed for heat-related emergency situations;
   e. Provisions that first aid to be administered immediately to employees displaying symptoms of heat-related illness.

If you work in a hot environment, you must be oriented before regular heat exposure. This includes being taught about the signs and symptoms of heat stress and appropriate measures to take when they are recognized. A screening process should be in place to identify and quantify your risk factors. A time must be established for acclimatization of your body to the hot environment through incremental exposures. The time required to become acclimated will vary from one person to another. If you have certain risk factors, you may never be able to comfortably acclimate to extreme conditions. Under ideal conditions, a young, healthy individual with no risk factors should be able to fully acclimate to a hot environment in about 2 weeks. This rate of acclimatization requires at least 2 hours of exposure each day combined with gradual increases in cardiovascular exercise. Placing this incremental stress on the body will result in biological adjustments that reduce physiological strain, protect the body, and increase the comfort level. These changes include a faster physiological response to heat, such as earlier and greater sweating and blood flow to the skin. However, the changes can be lost at nearly the same rate that they are gained. If you are acclimated but do not continue periodic heat exposures, it will begin to diminish after several days and you will lose about 75% of your acclimatization in 3 weeks (U.S. Army Center for Health Promotion and Preventive Medicine [CHPPM], 2003). Acclimatization will also impact the amount of water you need to consume. Because an acclimated individual will sweat more quickly and profusely, more fluid intake is required. The quantity of fluids needed to sustain a healthy, highly active, acclimated person may exceed a quart (or liter) per hour.

Case Study: Chicago Heat Wave, 1995

One of the deadliest heat waves in the history of the United States occurred in July 1995. The city of Chicago, Illinois, recorded its highest sustained temperatures in decades. A thermal inversion aggravated the situation by trapping pollution close to the ground. The heat was so intense that roads buckled, power failed, and train rails twisted. The heat index during this heat wave set a new record on July 13 at 119 °F (48.3 °C). The Cook County Medical Examiner's Office established a criteria for heat-related mortality including:

- Core body temperature >105 °F (40.6 °C) at the time of death.
- Clear evidence of heat as a contributing factor (e.g., discovered in a home without air conditioning, etc.)
- Victim discovered in decomposed condition with no other apparent causes of death and evidence that the victim was alive during the heat wave.

Identification of heat casualties is a difficult task. During the week of July 4 (July 4 through 10), there was none reported. The period immediately following, July 11 through 27, there were 465 heat-related deaths reported (See Figure 7–10) (CDC, 1995b). The dramatic increase led to skepticism among local leaders and

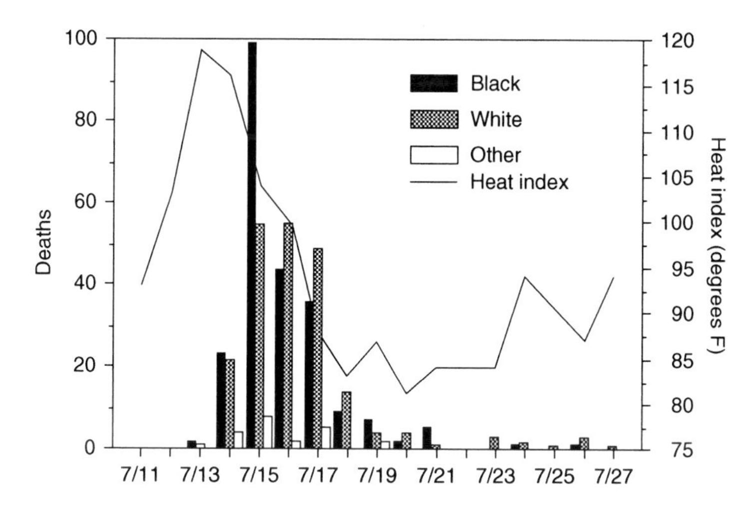

FIGURE 7–10 Number of heat-related deaths, (n = 465) by date of occurrence and race (The Cook County Medical Examiner's Office categorizes race of decedents as black, white, or other) of decedent, and heat index, by date, Chicago, July 11–27, 1995.

the media. Mayor Daley was criticized for belated activation of Chicago's "Heat Plan." At the time, the written plan consisted of less than two pages. The Mayor quickly blamed utilities for the power outages and accused the Medical Examiner of exaggerating the heat casualty count. Because the primary causes of death are often preexisting conditions among vulnerable populations, determining how much heat conditions contribute to specific deaths is very difficult. The Chicago Medical Examiner used the best available information in tracking heat victims.

Some difficult lessons were learned through this event. Perhaps, the most significant lesson is the importance of good communication with the public emphasizing delivery of warnings and survival tips to those who are poor or elderly. This was clearly a challenge for Chicago. This lack of heat wave preparedness knowledge among local leaders was most apparent in public statements made by the Commissioner of Human Services, Daniel Alvarez. He suggested the victims were to blame when he said, "We are talking about people who died because they neglected themselves ... people who don't read the newspaper, who don't watch TV. We did everything possible" (Van Biema, 1995). He later apologized for those comments. The key take-home message from this incident is that heat wave deaths are preventable and the key is communication with vulnerable populations before and during the occurrence.

Immediate Actions

The basic steps needed for an effective response to a heat wave begin with community planning and personal preparedness. Vulnerable populations must be identified prior to an emergency. Effective strategies should begin with the formation of a local or regional Heat Wave Task Force that includes representatives from organizations representing vulnerable populations so specific approaches, unique to each community, may be planned with appropriate triggers for actions such as establishing cooling centers and checking on the elderly. Building a heat wave resilient community includes measures that reduce the possibility of failure of critical infrastructures such as power grids. This includes advising the public on basic energy conservation measures during peak hours of heat. Reducing the health impact of heat on vulnerable populations is best achieved through communicating basic personal preparedness measures and emphasizing actions taken to cool yourself down and connect with the help you may need.

Personal Actions

As many environmental variables of a heat wave cannot be controlled, you need to consider all the personal variables that may influence your risk and plan ahead for the warm season. If you typically spend time outside in direct sunlight, clothing selection is particularly important. Lighter colored clothes reflect light and will keep you much cooler than dark-colored clothing. Loose-fitting, light weight clothing will allow air to circulate and speed evaporation of sweat, cooling your body more quickly. Plan activities that keep you in the shade or prepare to wear a hat and use sunscreen with a high SPF to reduce the risk of sunburn which increases the risk of all heat illnesses by making it more difficult for your skin to help regulate body temperature. It is also important to simply slow down.

Those with risk factors should stay in the coolest place available and avoid strenuous activities in the heat. If you engage in outdoor work or exercise, it is important to avoid excessive exertion during peak temperatures of the day and consider rescheduling more strenuous tasks for cooler times of the day. It is also important to avoid heavy meals. Many foods that are typically considered healthy, such as foods high in protein, are not a good choice during extreme heat. These foods increase the production of metabolic heat and also contribute to water loss. As some medications affect your sweat production or blood circulation, it is important to evaluate prescription medications such as blood pressure and psychiatric drugs for their impact on heat stress. Over-the-counter medications like antihistamines and decongestants can also affect your body's cooling system. The use of illegal drugs, alcohol, and caffeinated products are also factors that increase risk. Those who live alone should be identified and checked on regularly. The majority of heat wave deaths are among individuals who are alone.

If you are at risk for heat injury or illness, particularly if you have risk factors, there are simple things you can do to reduce the likelihood of experiencing serious health problems. If you do not have air conditioning, or if your power goes out during peak hours of heat, identify nearby air conditioned safe havens. These include community centers, libraries, and other public buildings. There are also malls and other facilities that offer cool environments. You should also have important phone numbers of friends and family members, as well as hotline numbers readily available. A home preparedness plan and kit should be prepared in advance (as described in Chapter 2). Before the hot season hits, it is a good idea to insulate spaces around window air conditioners, check weather stripping on doors and windows, and make sure your home is well insulated. A good inspection of your air conditioner, ducts, and vents is advisable each year. Shades, blinds, or drapes can substantially reduce the heat entering your home by direct sunlight and attic fans are very effective in reducing the build-up of heat in attic spaces.

Community Actions

There is no single plan or program that fits every country or even every region within a country. The central focus of community heat wave preparedness is identifying and communicating with vulnerable populations. The most effective government actions are focused regionally and locally working through a wide variety of organizations that represent vulnerable populations. Each public and private organization involved must clearly understand its function before the emergency and be prepared to quickly carry out its assigned roles. Nonprofit and faith-based organizations play a central role in reaching out to those at risk. Meal delivery aides, case workers for the elderly and disabled, and bus operators should be trained to watch for signs of heat illness among those they encounter. Much of this effort can be accomplished well in advance of a heat emergency. Readiness preparations include these ten essential steps:

1. Create heat mitigation public education materials.
2. Identify media and public resources to distribute heat wave information to at-risk populations.
3. Identify vulnerable populations in the community for future campaign focus.
4. Identify and prepare potential cooling centers and transportation resources for vulnerable individuals.
5. Coordinate community resources for a heat wave response.

6. Create clinical alerts providing guidance on diagnosis and treatment of heat illness and share them with providers.
7. Initiate a heat-related public awareness campaign.
8. Request that local agencies serving vulnerable populations contact clients to ensure their safety and provide heat preparedness information.
9. Encourage the public to check on high-risk persons that might be affected by the heat.
10. Initiate heat-related death and illness data collection by Local Public Health Agencies.

The media can assist in reaching a large portion of these individuals. Local newspapers are often willing to publish a special section with emergency information such as locations of cooling centers and other emergency services. Many more may be reached through school-based initiatives that provide information to children on the importance of checking on those at high risk, including neighbors and family members. The key messages include:

- Stay indoors. Limit your exposure to direct sunlight and if air conditioning is not available, stay on the lowest floor of your home out of direct sunlight. If temperatures there are high, go to a local cooling center or public facility with air conditioning.
- Avoid direct sunlight. Sunburn compromises your skin's ability to cool itself. The sun will also raise your internal core temperature leading to dehydration. Use a sunscreen lotion with a high sun protection factor (SPF) rating if you need to be outside.
- Drink plenty of fluids. The leading cause of heat-related death is dehydration. It can happen quickly and not be noticed until it is too late. The symptoms are frequently confused with other causes. Individuals with existing health issues or who have problems with fluid retention (individuals with epilepsy, heart, kidney, or liver disease) should consult a doctor before increasing liquid intake.
- Wear light-colored, lightweight, loose-fitting clothing. Cover as much skin as possible with light-colored, loose-fitting clothing to avoid sunburn and over-heating effects of sunlight on your body. Light-colored, lightweight, loose-fitting clothing reflects heat and helps maintain normal body temperature.
- Protect face and head. A hat will keep direct sunlight off your head and face. Sunlight can burn your skin and also increase your core temperature.
- Take frequent breaks. If you are working outside, take frequent breaks in a cool area. Also drink fluids so you can stay hydrated and tolerate heat better.
- Avoid strenuous activity during peak hours of heat. Reduce, eliminate, or reschedule strenuous activities. High-risk individuals should stay in cool places during the hottest part of the day. Get plenty of rest to allow your natural "cooling system" to work. If you must do strenuous activity, do it early in the morning or late in the evening during the coolest parts of the day.
- NEVER leave children or pets in vehicles. Temperatures inside vehicles can reach deadly levels within minutes. Exposure to such high temperatures can quickly kill any occupants of the vehicle. Never leave a person or pet in a vehicle.
- Limit use of alcohol or caffeinated beverages. These beverages can make you feel better for a brief period of time but increase your vulnerability to heat illnesses.

- Eat light, regular meals. Eat smaller meals more often. Avoid large, heavy meals that can generate more internal body heat during digestion. Also, avoid foods that are high in protein. These foods also generate more metabolic body heat. Avoid using excess salt or salt tablets unless directed to do so by a physician.
- Avoid excessive changes in temperature. A cool shower immediately after being in excessive heat may sound refreshing but the drastic temperature shift can result hypothermia, particularly for vulnerable populations such as the elderly and very young.
- Check on family, friends, and neighbors who do not have air conditioning and/or who spend time alone.
- High-risk individuals should consider spending the peak hours of heat in public buildings such as libraries, schools, shopping malls, and community facilities.
- High-risk individuals should also use a buddy system. Partners can keep an eye on each other and can assist each other when needed. Sometimes exposure to heat can cloud judgment. Chances are if you work alone, you may not notice this.

Summary

Heat waves are one of the most lethal and yet subtle disasters. The worst impact is felt among vulnerable populations and many casualties of heat waves die alone. Their deaths are often attributed to other causes or conditions even though the excessive heat serves as the tipping point for the fatal outcome. Unlike many other disasters where infrastructures experience catastrophic failure resulting in unavoidable fatalities, all heat wave losses are avoidable. The measures that need to be taken by those at risk are simple, and the interventions required of local and state government officials are clear. The keys to success are political will to make heat wave preparedness and response a priority and effective communication with vulnerable populations. There is historical evidence suggesting that public outcry and political will are essential in gaining the necessary awareness and resources to effectively prepare and respond. The subtle nature of heat waves has proven to be the principal barrier. Global warming and other environmental changes combined with the growth of urban environments and increasing numbers of susceptible populations are all contributing to the increasing likelihood of future catastrophic heat waves such as those experienced in Chicago in 1995 and Paris in 2003.

Websites

CDC Extreme Heat Health Information: http://www.bt.cdc.gov/disasters/extremeheat/.

FEMA Heat Wave Information: http://www.fema.gov/hazard/heat/index.shtm.

National Weather Service: http://www.nws.noaa.gov/om/brochures/heat_wave.shtml.

NIOSH Publication No. 86-112: Working in Hot Environments: www.cdc.gov/niosh/hotenvt.html.

NIOSH Safety and Health Topic: Heat Stress: www.cdc.gov/niosh/topics/heatstress/.

OSHA Safety and Health Topics: Heat Stress: www.osha.gov/SLTC/heatstress/index.html.

OSHA Standard Interpretation, Acceptable methods to reduce heat stress hazards in the workplace.

Red Cross Information: http://www.redcross.org/services/prepare/0,1082,0_243_,00.html.

San Francisco State University, Department of GeoSciences: Safety Recommendations Fact Sheet: Never Leave Your Child Alone in the Car! http://ggweather.com/heat/fact_sheet_2006.pdf.

U.S. Army Center for Health Promotion and Preventive Medicine. Heat Injury Prevention: http://chppm-www.apgea.army.mil/heat/;SunProtection: Questions and Facts for Soldiers; http://usachppm.apgea.army.mil/HealthTipOfTheWeek/sunProtection/default.asp.

U.S. Army Center for Health Promotion and Preventive Medicine: Sun Protection: Questions and Facts for Soldiers: http://usachppm.apgea.army.mil/HealthTipOfTheWeek/sunProtection/default.asp.

U.S. Environmental Protection Agency, Excessive Heat Events Guidebook: www.epa.gov/hiri/about/pdf/EHEguide_final.pdf.

U.S. Environmental Protection Agency, SunWise Program: www.epa.gov/sunwise/.

References

Aerospace Medical Association. (2005). Useful tips for airline travel. Retrieved January 19, 2008, from www.asma.org/pdf/publications/Tips_For_Travelers2001.pdf.

Allen, R. G., Pereira, L. S., Raes, D., & Smith, M. (1998). Crop evapotranspiration—guidelines for computing crop water requirements, FAO irrigation and drainage paper 56. Food and Agriculture Organization (FAO) of the United Nations. Retrieved January 13, 2008, from www.fao.org/docrep/X0490E/x0490e00.HTM.

Bernard, S. M., & McGeehin, M. A. (2004). Municipal heat wave response plans. *Am J Public Health* 94(9): 1520–1522.

CDC. (1995a). Heat-related illnesses and deaths—United States, 1994–1995. *MMWR* 44:465–468.

CDC. (1995b). Heat-related mortality—Chicago, July 1995. *MMWR* 44:577–579.

CNN.com. (2003). Portugal fires 'mostly controlled'. Retrieved January 19, 2008, from http://edition.cnn.com/2003/WORLD/europe/08/05/portugal.fires/index.html.

Dhainaut, J., Claessens, Y., Ginsburg, C., & Riou, B. (2004). Unprecedented heat-related deaths during the 2003 heat wave in Paris: consequences on emergency departments. *Crit Care* 8:1–2.

Hemon, D., & Jougla, E. (2003). Surmortalité liée à la canicule d'août 2003-Rapport d'étape (Translation: Mortality related to the heat wave of August 2003-Report/ratio of stage 1). Retrieved January 13, 2008, from www.snphar.com/news/stock/caniculeINSERM.pdf.

Kalkstein, L. S., & Davis, R. E. (1989). Weather and human mortality: an evaluation of demographic and interregional responses in the United States. *Ann Assoc Am Geogr* 79(1):44–64.

Kalkstein, L. S., Jamason, P. F., Greene, J. S., Libby, J., & Robinson, L. (1996). The Philadelphia hot weather-health watch/warning system: development and application, Summer 1995. *Bull Am Meteorol Soc* 77(7): 1519–1528.

Kalkstein, L. S., Tan, G., & Skindlov, J. A. (1987). An evaluation of three clustering procedures for use in synoptic climatological classification. *J Climate Appl Meteor* 26:717–730.

Kizer, K. W. (2000). Lessons learned in public health emergency management: personal reflections. *Prehospital Disaster Med* 15:209–214.

Mellion, M. B., & Shelton, G. L. (1997). Safe exercise in the heat and heat injuries. In: Mellion, M. B, Walsh, W. M, & Shelton, G. L, eds. *The team physician's handbook*. 2nd ed. Philadelphia, PA: Hanley & Belfus; pp.151–165.

Naughton, M. P., Henderson, A., Mirabelli, M. C., et al. (2002). Heat-related mortality during a 1999 heat wave in Chicago. *Am J Prev Med* 22:221–227.

Nowak, D. J., Crane, D. E., Stevens, J. C., & Ibarra, M. (2002). Brooklyn's urban forest. *Gen Tech Rep NE-290*.

Occupational Safety and Health Administration (OSHA). (2001). Standard interpretations: 10/17/2001, acceptable methods to reduce heat stress hazards in the workplace. Retrieved February 20, 2008 from www.osha.gov/pls/oshaweb/owadisp.show_document?p_table=INTERPRETATIONS&p_id=24008.

Rebetez, M., Mayer, H., Dupont, O., et al. (2006). Heat and drought 2003 in Europe: a climate synthesis. *Ann For Sci* 63:569–577.

Sheets, V. L., & Manzer, C. D. (1991). Affect, cognition, and urban vegetation: some effects of adding trees along city streets. *Environ Behav* 23:285–304.

Simon, H. B. (1993). Hyperthermia. *N Engl J Med* 329:483–487.

Tagliabue, J. (2003). French official quits over toll in heat wave. *New York Times*. August 19, 2003.

United Nations Environment Programme. (2004). Environmental alert bulletin: impacts of summer 2003 heat wave in Europe. Retrieved January 12, 2008, from www.grid.unep.ch/product/publication/download/ew_heat_wave.en.pdf.

U.S. Army Center for Health Promotion and Preventive Medicine (CHPPM). (2003). Ranger and airborne school students heat acclimatization guide. http://chppm-www.apgea.army.mil/doem/pgm34/HIPP/Heat%20Acclimatization%20Guide_6.pdf.

U.S. National Oceanic and Atmospheric Administration (NOAA), National Weather Service. (1994). *Regional Operations Manual Letter*. Eastern Region National Weather Service. p.3.

U.S. National Oceanic and Atmospheric Administration (NOAA). (2005). News Release, NOAA heat/health watch warning system improving forecasts and warnings for excessive heat. NOAA Magazine. Retrieved February 22, 2008, from www.noaanews.noaa.gov/stories2005/s2366.htm.

Van Biema, D. (1995). More heat than light. *Time* 31:37.

Weisskopf, M. G, Anderson, H. A, Foldy, S., et al. (2002). Heat wave morbidity and mortality, Milwaukee, Wis, 1999 vs 1995: an improved response? *Am J Public Health* 92:830–833.

Hurricanes

Objectives of This Chapter

- Explain how tropical storms develop.
- List the factors that dictate how hurricane intensity is determined.
- Describe the variables that influence hurricane-associated morbidity and mortality patterns.
- Explain how storm surges occur and what influences their severity.
- Describe where the greatest impact of a storm is seen in relation to the eyewall and why.
- List the primary and secondary causes of hurricane-associated deaths.
- Explain why recent years have seen dramatic decreases in hurricane-associated morbidity and mortality.
- Describe what can be done to mitigate the impact of a storm on a home or other structure.
- Besides evacuation, list things that an individual can do to reduce the risk of injury during a hurricane.
- Describe the hazards associated with hurricane recovery operations.

Introduction—2005 Hurricane Katrina

Tony was at home in Biloxi with his two daughters when Hurricane Katrina made landfall. They lived in a small apartment on the first floor of a two-story building. They huddled in their home as the wind howled on Monday morning. The intensity of the windswept rains grew steadily. Not far away, at the beach, the storm surge started off small like the tide coming in but continued to grow. Each wave was a little larger than the last until the water reached beyond the beach and into the city. The murky water surged little by little across Biloxi and grew deeper as the winds swirled and the rain blew. As the flood surrounded buildings, the pulsating rush of water began to weaken and collapse one structure after another.

At their apartment, the growing waves reached the building and ruptured the bottom panel of the front door. They raced to the back bedroom and gathered some of their things. Tony tried to reassure his daughters, 3-year-old Brooke and 8-year-old Shania that everything would be alright. He tried to hide his own fear as he considered the possibility that they may not make it. They waded through the rising waters in their front room and opened a window. Tony, with a daughter in each arm, climbed out of the apartment into the rising waters and waded to a set of outside stairs leading to the second floor. Although they found refuge above the rising waters on the second floor, Tony could also see buildings around them beginning to collapse. Rather than risking the possibility of their building collapsing, he decided to pick up the girls and try to get to higher ground. A house just behind their apartment building was likely to be a safer place. Tony carried the girls through water that was now neck deep. They made their way behind their apartment building and waded out of the water onto a hill. They knocked on the front door of their neighbor's home that stood above the rising waters and were invited in. They rode out the rest of the storm with their neighbor. Although they lost everything in the storm, Tony and the girls came through it. After spending some time staying at a friend's home,

FIGURE 8–1 In the top image, taken in 1998, notice the pier, pier house, and the antebellum house in Biloxi, MS. The bottom image shows the same location on August 31, 2005, 2 days after Hurricane Katrina made landfall. This photo shows the complete destruction of these landmarks.

Source: U.S. Geological Survey. Available at: http://coastal.er.usgs.gov/hurricanes/katrina/photo-comparisons/images/katrina_biloxi_pair2-lg.jpg.

they began rebuilding their lives (United Nations Children's Fund [UNICEF], Katrina: one family's survival story, www.unicef.org/infobycountry/usa_28275.html).

Hurricane Katrina was by far the largest disaster in the history of the United States (Figure 8–1). Even though there have been other hurricanes recorded that were more powerful, the timing and location of landfall made Katrina more devastating than any storm in U.S. history. The measure of a storm's intensity is based on how low the barometric pressure goes. Barometric pressure is the weight of a column of air from the ground, or water surface, to the top of the atmosphere. It is expressed as inches of mercury or millibars.

Lower pressure means worse weather, whereas high pressure usually means clearer weather. There have only been two U.S. hurricanes more intense than Katrina. In 1935, an unnamed Category 5 hurricane hit the Florida Keys on Labor Day with a barometric pressure of 26.35 inches of mercury and winds of 160 miles per hour (mph). In 1969, Hurricane Camille, also a Category 5 storm, carried a pressure of 26.84 and winds of 190 mph. Hurricane Katrina had a pressure reading of 27.11 and winds of 140 mph, making it lower than Hurricane Andrew at 27.23 (See Table 8–1) (Rice, 2005; Jarrell et al., The deadliest, costliest, and most intense United States hurricanes from 1900 to 2000, NOAA Technical Memorandum NWS TPC-1, www.aoml.noaa.gov/hrd/Landsea/deadly/index.html). Even though Hurricane Andrew had stronger winds of 165 mph compared to Hurricane Katrina's 140 mph winds, Katrina's damage and loss of life were more severe.

The statistics of Katrina are staggering. The storm surge directly impacted about 93,000 square miles. That is about the size of the entire nation of Great Britain. It included multiple states and over 130 parishes or counties. It killed more than 1300 people and completely destroyed about 300,000 homes. More than a million people were evacuated and damage estimates have exceeded $96 billion (White House, The federal response to Hurricane Katrina: lessons learned, www.whitehouse.gov/reports/katrina-lessons-learned.pdf). Although there have been more intense hurricanes and others with a higher death toll, there has not been another storm in the history of the United States that brought devastation over such a large region or one that has been more cloaked in controversy.

Table 8–1 Most Intense Hurricanes in United States History

Hurricane	Year	Landfall Location	Wind Speed (mph/km/hour)	Barometric Pressure (Inches/Millibars)	Category
Unnamed	1935	FL Keys	160/258	26.35/892	5
Camille	1969	MS, LA, VA	190/306	26.84/909	5
Katrina	2005	FL, LA, MS, AL	140/225	27.11/918	4
Andrew	1992	FL, LA	165/266	27.23/922	5
Unnamed	1919	FL Keys, TX	150/240	27.37/927	4

Source: National Weather Service (Jarrell et al., The deadliest, costliest, and most intense United States hurricanes from 1900 to 2000, NOAA Technical Memorandum NWS TPC-1, www.aoml.noaa.gov/hrd/Landsea/deadly/index.html). Available at: http://www.aoml.noaa.gov/hrd/Landsea/deadly/Table4.htm

Hurricane-Related Definitions

Center: Generally refers to the vertical axis of a tropical storm defined by the location of minimum wind or minimum pressure.

Cyclone: A storm circulation in the atmosphere that rotates counter-clockwise in the Northern Hemisphere and clockwise in the Southern Hemisphere.

Direct hit: A close approach of a hurricane or cyclone to a specific location. For locations on the left-hand side of a hurricane or cyclone's track (looking in the direction of motion), a direct hit occurs when the cyclone passes to within a distance equal to the cyclone's radius of maximum wind. For locations on the right-hand side of the track, a direct hit occurs when the cyclone passes to within a distance equal to twice the radius of maximum wind.

Gale warning: A warning that 1-minute sustained surface winds are in the range of 34 knots (39 mph or 63 km/hour) to 47 knots (54 mph or 87 km/hour). This may or may not be associated with a hurricane or tropical cyclone.

High wind warning: A warning that 1-minute average surface winds are 35 knots (40 mph or 64 km/hour) or greater lasting for 1 hour or longer, or winds gusting to 50 knots (58 mph or 93 km/hour) or greater regardless of duration that are either expected or observed over land.

Hurricane (Typhoon): A tropical cyclone with maximum sustained surface wind (using the U.S. 1-minute average) is 64 knots (74 mph or 119 km/hour) or more. It is normally used for storms in the Atlantic Basin and the Pacific Ocean east of the International Date Line. The term typhoon is used for Pacific tropical cyclones north of the Equator and west of the International Dateline.

Hurricane eye: The low-pressure center of a hurricane or cyclone. Winds are normally calm as the eye passes over and sometimes the sky is clear. It is surrounded by the eyewall cloud.

Hurricane eyewall: A ring of cumulonimbus clouds and thunderstorms that surround the hurricane eye. The heaviest rain and strongest winds are normally in the eyewall.

Hurricane local statement: A release prepared by local National Weather Service offices that are in or near an area at risk for a hurricane. It provides details for each county/parish warning area on weather conditions, evacuation decisions, and other precautions necessary to protect life and property.

Hurricane season: The time of year with a relatively high incidence of hurricanes. The hurricane season in the Atlantic, Caribbean, and Gulf of Mexico is from June 1 to November 30. The hurricane season in the Central Pacific Basin is the same but the season in the Eastern Pacific Basin starts a couple weeks earlier and is from May 15 to November 30.

Hurricane warning: A warning means sustained winds of 64 knots (74 mph) or higher associated with a hurricane are expected in a specified coastal area in 24 hours or less. A hurricane warning can remain in effect when dangerously high water or a combination of dangerously high water and exceptionally high waves continues, even though winds may be less than hurricane force. If told to move to a shelter or evacuate the area, do so immediately.

Hurricane watch: A hurricane watch means a hurricane is possible in your area, generally within 36 hours. Keep listening to NOAA weather radio, local radio, or local television for updated information. Hurricanes can change direction and speed, and they can gain strength very quickly. It is important to keep listening for updated information several times a day.

Indirect hit: Generally refers to an area near landfall, that does not experience a direct hit from a hurricane or cyclone but does experience high winds (sustained or gusts) and/or tides of at least 4 feet above normal.

Landfall: The location of intersection of the center of the storm with a coastline.

Major hurricane: A hurricane classified as Category 3 or higher.

Radius of maximum winds: The distance from the center of a hurricane or cyclone to the location of the cyclone's maximum winds. In a well-developed hurricane, the radius of maximum winds is generally at the inner edge of the eyewall.

Storm surge: An abnormal rise in the sea level accompanying a hurricane or other intense storm. It is usually estimated by subtracting the normal or astronomic high tide from the observed storm tide.

Storm tide: The actual level of sea water resulting from the astronomic tide plus the storm surge.

Tropical depression: A tropical cyclone with closed wind circulation around a center with wind speeds (using the U.S. 1-minute average) of 20–33 knots (38 mph or 62 km/hour) or less.

Tropical storm: Maximum sustained surface wind speed (using the U.S. 1-minute average) ranges from 34 knots (39 mph or 63 km/hour) to 63 knots (73 mph or 118 km/hour). The storm is named once it reaches this tropical storm strength.

Tropical storm warning: A warning that a storm with sustained winds of 34–63 knots (39–73 mph or 63–118 km/hour) is expected in a specified coastal area within 24 hours or less.

Tropical storm watch: An announcement for a specific coastal area that a tropical storm is possible within 36 hours.

Hurricanes

Hurricanes are severe tropical storms. They form over warm water and draw energy from the evaporation of water as they pass over the ocean surface. The storms begin simply as an area of low pressure and thunderstorms called a tropical disturbance. As the warm air causes water to evaporate, it creates lift in the low pressure area. Cold air replaces the warm air as it rises, setting in motion a process that can erupt into a violent storm encompassing hundreds

Table 8–2 The Saffir-Simpson Hurricane Scale

Category	1, Minimal	2, Moderate	3, Extensive	4, Extreme	5, Catastrophic
Winds	74–95 mph	96–110 mph	111–130 mph	131–155 mph	>155 mph
	64–82 knots	83–95 knots	96–113 knots	114–135 knots	>135 knots
	119–153 km/hour	154–177 km/hour	178–209 km/hour	210–249 km/hour	> 249 km/hour
Storm Surge	4–5 Feet	6–8 Feet	9–12 Feet	13–18 Feet	>18 Feet
Pressure	>980 mb	965–979 mb	945–964 mb	920–944 mb	<920 mb
	>28.94 in	28.50–28.91 in	27.91–28.47 in	27.17–27.88 in	<27.17 in
Anticipated Wind Damage	No significant damage to building structures. Damage primarily to unanchored mobile homes, shrubbery, and trees. Some damage to poorly constructed signs.	Some roofing material, door, and window damage. Considerable damage to shrubbery and trees. Considerable damage to mobile homes, poorly constructed signs, and piers.	Some structural damage to small residences and utility buildings with a minor amount of curtain wall (building facade) failures. Damage to shrubbery and trees with foliage blown off trees and large trees blown down. Mobile homes and poorly constructed signs destroyed.	More extensive curtain wall (building facade) failures with some complete roof failures on small residences. Shrubs, trees, and signs blown down. Complete destruction of mobile homes. Extensive damage to doors and windows.	Complete roof failure on many buildings. Some complete building failures with small utility buildings blown over or away. All shrubs, trees, and signs down. Complete destruction of mobile homes. Severe and extensive window and door damage.

(Continued)

Table 8–2 *(Continued)*

Category	1, Minimal	2, Moderate	3, Extensive	4, Extreme	5, Catastrophic
Anticipated Storm Surge Damage	Some coastal road flooding and minor pier damage.	Coastal and low-lying escape routes flood 2–4 hours before arrival of the hurricane center. Small craft in unprotected anchorages break moorings.	Low-lying escape routes may be cut off by rising water 3–5 hours before hurricane center arrival. Flooding near coast destroys small structures. Large structures damaged by battering from floating debris. Terrain lower than 5 feet above mean sea level may be flooded inland 8 miles (13 km) or more. Evacuation of low-lying residences within blocks of the shoreline may be required.	Low-lying escape routes may be cut by rising water 3–5 hours before arrival of the center of the hurricane. Major damage to lower floors of structures near the shore. Terrain lower than 10 feet above sea level may be flooded requiring massive evacuation of residential areas as far inland as 6 miles (10 km).	Low-lying escape routes cut off by rising water 3–5 hours before hurricane center arrival. Major damage to lower floors of all structures less than 15 feet above sea level and within 500 yards of the shoreline. Massive evacuation of residential areas on low ground within 5–10 miles (8–16 km) of the shoreline may be required.

Adapted from the National Weather Service, National Hurricane Center, Description of the Saffir–Simpson Hurricane Scale. Available at: http://www.nhc.noaa.gov/aboutsshs.shtml.

of square miles. As the winds begin to rotate over the warm water, energy builds. When the winds reach 20–33 knots, or just under 38 mph (62 km/hour), it is referred to as a tropical depression. Once the surface wind speeds grow to between 34 and 62 knots (39–73 mph or 62–118 km/hour), it is called a tropical storm and receives a name. When the winds reach 64 knots (74 mph or 119 km/hour) (Weather.com, The Weather Channel, Hurricanes and tropical storms, www.weather.com/encyclopedia/tropical/forecast.html), it is called a hurricane. In the Pacific, west of the International Dateline, these storms are called typhoons.

It was not until the 1970s that hurricanes were ranked in categories 1 through 5 based on the intensity. An engineer in Florida, Herbert Saffir, worked with the Director of the National Hurricane Center, Robert Simpson to develop a scale that is now referred to as the Saffir–Simpson Hurricane Scale (Table 8–2). This simple rating is used to estimate the anticipated damages and threats to coastal residents posed by a hurricane. There are geographical differences in each coastal region that can have a strong influence on the interpretation of this ranking. A lower ranked storm can produce more damage in one area than a higher ranked storm in another area depending upon where it makes landfall. However, the ranking system has served as an effective tool in predicting the hurricane damage.

The naming of tropical storms was started years ago to make it easier to identify and associate warnings with specific storms. It was an arbitrary practice that started in the early twentieth century. Originally, the storms were named by when or where they hit (e.g., Galveston Hurricane, Labor Day Hurricane). During World War II, military meteorologists needed to be able to quickly communicate the status of multiple storms that could impact military missions throughout the Pacific. They began naming them. After the war, the meteorology community saw the benefits of storm naming and made it a formal process. In 1953, the U.S. National Hurricane Center developed a list of female names used to label tropical storms in the Atlantic. In 1979, they added alternating male names as well. Over the years, this practice was extended to every region of the world and the lists for the next 6 years are maintained by the World Meteorology Organization. Once a list is used, it is retired for 6 years and then reused. The exceptions to this process are the names of storms that become major hurricanes. Those names are replaced. For example, the name "Katrina" is retired and will not be used for another storm but other storm names from 2005 will be used again in 2011. Each year the alphabetized list denotes the order of the arriving storms. The first tropical storm of the season is a name beginning with the letter "A," the second with the letter "B," and so on. The lists of names are recognizable for those at risk. For example, the 2012 North Atlantic storms that threaten the United States will include 21 names beginning with Alberto, Berl, Chris, and Debby. Twenty-one letters of the alphabet are used; five of them (Q, U, X, Y, and Z) are not used. In other regions of the world, the names are selected from a list of names contributed by each nation at risk in the region. For example, in the Northern Pacific and South China Sea region the names on an upcoming list include Sarika (Cambodia), Haima (China), Meari (Korea), etc (World Meteorological Organization, Tropical storm naming, www.wmo.int/pages/prog/www/tcp/Storm-naming.html). The alphabetical order used is based on the name of the nation that contributed the tropical storm name.

The hazards associated with hurricanes are different with each one. Although they consistently include storm surge, flooding, high winds, and tornadoes, the predominant hazard for a given hurricane is not always predictable. This is most apparent in the morbidity and mortality recorded for each storm. Though one will have deaths primarily attributed to drowning, another will have few drowning deaths, a large number of traumatic injuries are due to high winds. Others will have a balance of both.

The primary factor in the loss of life from most hurricanes is the storm surge. Simply put, it is water that is pushed ashore by the winds of a hurricane. It is worst near the eye of the storm. A dome of water ranging from several feet to as high as 25 feet builds and is carried ashore as the hurricane makes landfall. The severity can be influenced by a number of factors. The intensity and speed of the storm determine the size of the water dome. The slope of the sea floor and the angle the storm hits the shoreline influence how much of the dome is carried ashore. In the Northern Hemisphere, the storm rotates counter-clockwise making the right front quadrant of the storm at landfall, the area with the greatest surge. The areas left of the eye at landfall have winds blowing toward the ocean, reducing the storm surge. Also, if the sea floor has a gradual slope, the surge is likely to be far more severe than areas where the sea floor has an abrupt drop that can help block the surge. The astronomical tide is also an important factor. If the hurricane makes landfall during high tide, the storm surge will be that much deeper. Because water weighs about 1700 pounds per cubic yard, the rush of a storm surge is capable of leveling nearly any structure (National Oceanic and Atmospheric Administration, National

FIGURE 8–2 NOAA satellite image of Hurricane Katrina, taken a day before the storm made landfall on the U.S. Gulf Coast. While in the Gulf of Mexico, Katrina's winds peaked near 175 miles per hour. Available at: http://apod.nasa.gov/apod/image/0508/katrina_goes12.jpg.

Hurricane Center, Storm surge, www.nhc.noaa.gov/HAW2/english/storm_surge.shtml). Only specially designed buildings can withstand a major surge.

The deepest recorded storm surge to ever hit the United States was in 1995 with Hurricane Opal. The storm hit the Florida panhandle with surges between 10 and 20 feet deep, and a maximum storm tide of 24 feet near Fort Walton Beach causing nearly $3 billion in damages. It is interesting to note that the most deadly storm surge was about half as deep and it was the Galveston, Texas, Hurricane of 1900. Between 8 and 15 feet of water covered the entire island and about 8000 people died, mostly from the storm surge (National Oceanic and Atmospheric Administration, National Hurricane Center, Storm surge, www.nhc.noaa.gov/HAW2/english/storm_surge.shtml). The difference today is warning and preparedness. The National Hurricane Center does an outstanding job predicting where hurricanes will make landfall, and the coastal areas have established excellent systems to warn and evacuate the population.

As storm surge predictions have improved, the drowning deaths from hurricane storm surges have dropped dramatically. However, a closely related hazard continues to take many lives further inland during and after the landfall. The intense rainfall from a hurricane or tropical storm is a significant threat to areas farther away from the shoreline. Although weaker, slow-moving storms may have less of a storm surge and spare the

FIGURE 8–3 The Galveston, Texas, Hurricane of 1900. The high winds and storm surge took over 8000 lives. It remains the most deadly disaster in U.S. history.

Source: National Oceanic and Atmospheric Administration. Available at: http://celebrating200years.noaa.gov/magazine/devast_hurricane/image3_full.jpg.

coastal areas, they can stall even hundreds of miles inland and pour a massive amount of rain. In recent history, this has become the primary cause of hurricane-related U.S. fatalities.

The wind speeds of a hurricane are the primary ranking factor on the Saffir–Simpson Scale. The winds of a hurricane will be most severe on the right-side of the eye. These winds are coming directly off the water verses the left-side winds that have slowed some as they pass over land. Although hurricane force winds can reach inland for a hundred or more miles, they usually slow dramatically within 12 hours of landfall (National Oceanic and Atmospheric Administration, National Hurricane Center, High winds, www.nhc.noaa.gov/HAW2/english/high_winds.shtml). The risks associated with hurricane force winds include the destruction of mobile homes and poorly built residences, or the blowing out of windows, especially in high-rise structures. Falling trees are often responsible for fatalities and property damage. Power lines can be damaged posing electrocution hazards and cutting off vital utilities.

In addition to the high winds generated by a hurricane, some storms also produce tornadoes. Although they are most likely to occur in the right front quadrant of a hurricane, they can be found anyplace in the rain bands. These tornadoes are different than those occurring with nonhurricane storms. They are typically smaller and do not have the associated lightning and hail seen with other tornadoes. Although they are more likely during the day, they can occur at any time for several days following landfall. Though some hurricanes will produce no tornadoes, some produce many. In 1967, Hurricane Beulah is believed to have generated 141 tornadoes (National Oceanic and Atmospheric Administration, National Hurricane Center. Tornadoes. http://www.nhc.noaa.gov/HAW2/english/tornadoes.shtml). These added hazards complicate an already difficult disaster scenario.

Top 5 Most Lethal U.S. Hurricanes

Rank	Year	Name/Location	Fatalities	Category
(1)	1900	Unnamed/Galveston, TX	8,000	4
(2)	1928	Unnamed/SE Florida	1,836	4
(3)	2005	Katrina—FL, LA, MS, AL	1,300	4
(4)	1919	Unnamed/FL Keys/S Texas	600	4
(5)	1938	Unnamed/New England	600	3

Source: National Weather Service and White House Reports (Jarrell et al., The deadliest, costliest, and most intense United States hurricanes from 1900 to 2000, NOAA Technical Memorandum NWS TPC-1, www.aoml.noaa.gov/hrd/Landsea/deadly/index.html); (White House, The federal response to Hurricane Katrina: lessons learned, www.whitehouse.gov/reports/katrina-lessons-learned.pdf).

Effects on Human Health and Safety

Unlike some disasters where the causes of morbidity and mortality are relatively consistent and predictable, hurricanes are not. Though some hurricanes cause more injury and loss of life from trauma due to high winds and building collapse, others have few injuries related to wind but have large numbers of drowning victims (Figure 8–4). Accurately accounting for hurricane-related morbidity and mortality is difficult. For example, death certificates have limitations on the quantity and quality of information recorded. This kind of information is important to learn from each disaster and understand what may be improved to limit the loss of life in the future. Even the estimated hour of death can be an important piece of

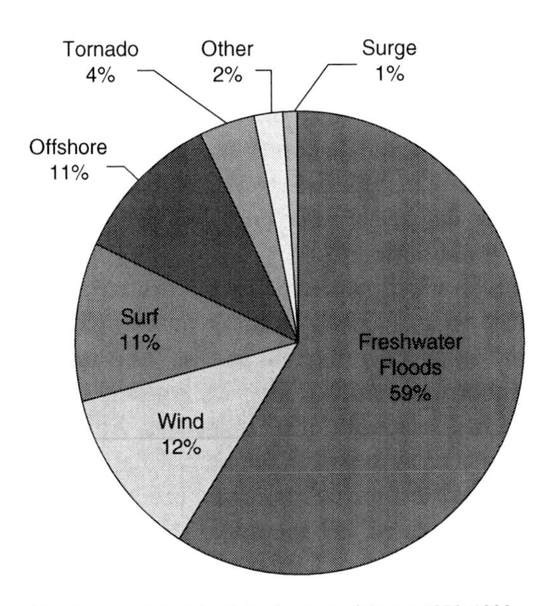

FIGURE 8–4 Leading causes of hurricane-related death in the United States 1970–1999.
Source: National Hurricane Center. Available at: http://www.nhc.noaa.gov/HAW2/english/surge/cyclone_deaths.gif.

information. If a person is lost due to a collapsed structure while winds are high, there may be a building code issue that should be considered. If it is several hours later, when winds have slowed, there may be a problem with response or communication of postdisaster risks.

In 1992, Hurricane Andrew made landfall in Florida. It was one of only three Category 5 storms to hit the United States in recorded history. It caused massive damage and killed 44 people. These deaths all occurred in only 2 of the 11 counties most affected. Of the 44 deaths recorded, 15 (34%) were directly related to the hurricane and 29 (66%) were indirectly related. Among those who died as a direct result of the storm, only two were drowning victims and the other 13 (87%) were trauma and asphyxiation victims from collapsing structures. The indirect deaths included 29 lives. Of those, 12 were stress-induced cardiovascular events, 8 were traumatic injuries from falls (5), auto collisions (1), and a plane crash (2). The remaining fatalities were from a wide range of causes including premature natural causes, a fire, electrocution, lightning strike, and asphyxiation (Combs et al., 1996). There are no consistent trends for morbidity and mortality related to Hurricane Andrew aside from a trend toward trauma-related deaths and a few drowning deaths.

Seven years later, another hurricane hit the East Coast. This time landfall was in North Carolina. Hurricane Floyd made landfall as a Category 3 storm, although it covered a much larger area than most Category 3 storms. Despite the Category 3 ranking, this storm killed more people than the Category 5, Hurricane Andrew. Hurricane Floyd took 52 lives. The leading cause of death, including 36 (69%) was drowning. The second was motor vehicle accidents that killed 7 (13%) excluding those that drowned in vehicles lost in floodwaters (Centers for Disease Control and Prevention, 2000). The deaths attributed to Hurricane Floyd were 69% drowning victims compared to less than 5% drowning victim fatalities with Hurricane Andrew (Compare Figures 8–5 and 8–6).

The characteristics of the storm, the topography of the area of landfall, and the demographics and behavior of those at risk are the primary factors that will determine morbidity and mortality from hurricanes and typhoons. Although the United States has

FIGURE 8–5 Aerial view of Hurricane Andrew devastation.
Source: Department of Defense.

FIGURE 8–6 Aerial view of Hurricane Floyd inland flooding.
Source: U.S. Army Corps of Engineers, J. Jordan. Available at: http://www.nws.noaa.gov/oh/hurricane/4cFloyd99.jpg.

seen a drastic reduction in the number of fatalities due to the storm surge of a hurricane, it remains the primary risk in other parts of the world. As populations continue to explode along coastal areas in places like Bangladesh, the poor construction codes and difficulties in conducting evacuations will continue to allow the storm surges of typhoons to be a leading cause of death (Chowdhury et al., 1992; Diacon, 1992).

The morbidity and mortality of a specific hurricane is difficult to predict. In general, the top three types of hurricane- or typhoon-related injuries are blunt trauma, lacerations, and puncture wounds. About 80% of those injuries are to the feet and lower extremities (Noji, 1993). There are also infectious disease concerns following these storms. As critical utilities, such as water, sewer, and electricity, as well as public health infrastructure are affected, the risk of an outbreak increases. This is further aggravated by the displacement and sheltering of populations (United Nations Development Programme, 2004; Western, 1982).

Fortunately, outbreaks associated with hurricanes are rare in developed nations and usually limited to minor noncommunicable gastrointestinal or respiratory illness (Centers for Disease Control and Prevention, 1993, 2000; Toole, 1997). The key is that these storms make the spread of existing infectious diseases easier because of the compromised infrastructure and conditions. It is rare that any new disease is introduced. In developing nations, infectious disease outbreaks associated with hurricanes or typhoons are more likely because those infectious conditions are already there and given the opportunity, will spread.

Prevention

There are many things that governments and individuals can do to reduce the morbidity and mortality associated with hurricanes or typhoons. The most important challenge is public awareness In the United States and other nations around the world, the coastal

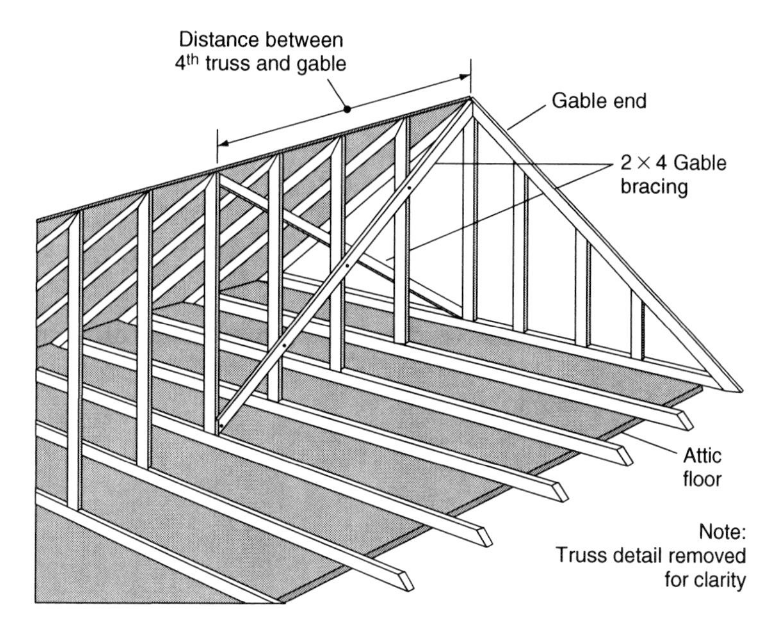

FIGURE 8–7 Bracing the gable end of the roof framing to prevent roof collapse.
Source: Federal Emergency Management Agency.

regions are becoming more populated with residents and visitors. This makes storm surge an ongoing threat to health and safety. Advances in storm forecasting have reduced the populations at risk through evacuations whereas the growing number of residents who choose to live on the coast continues to make evacuation more challenging. Human development continues to encroach on the natural barriers that can protect human populations from hurricanes and typhoons. Restoration of coastal and estuarine habitats such as marshes, wetlands, oyster reefs, and tidal streams is an important step toward making populations at risk safer from the impact of storms (National Oceanic and Atmospheric Administration, National Ocean Service, Natural Resource Restoration, http://oceanservice.noaa.gov/topics/coasts/restoration/welcome.html). The U.S. National Ocean Service is engaged in a variety of initiatives that restore natural resources by providing tools and resources to enhance the habitats that provide natural barriers to hurricanes and other threats.

Local communities can mitigate the impact of hurricanes through the enforcement of building codes that strengthen roofing, water barriers, and roof to wall connections of buildings. The roof should be reinforced with straps or additional clips to secure the roof to the frame. Simple measures like adding addition bracing to the gable ends of the roof can prevent collapse during high winds (Figure 8–7). This is one of the weak points of most structures. When high winds hit the side of a home, roof collapse often results starting with the collapse of the gable (side wall in the attic). Details on this and other mitigation approaches are available through local planners and through the Federal Emergency Management Agency.

Windows are another weak point of a structure. Although permanent storm shutters offer the best window protection, 5/8 inch marine plywood can also be cut to fit and installed. The windows are likely to be blown out during high winds and a barrier is needed to prevent blowing rain from flooding the inside of the structure. Taping the windows offers little protection. Double entry doors are another potential point of weakness that should be reinforced with additional hardware. It is also important to bring in yard ornaments and trim trees or large shrubs. During a storm, the trees can collapse and loose items in the yard, as well as tree branches, can become damaging "missiles." Simple measures, such as cleaning out the gutters and downspouts, can reduce the risk of water intrusion. All of these measures should be taken to protect the safety of residents and property.

Top Ten Hurricane Hazard Mitigation Actions

Remove trees and other potential windborne missiles.
Cover windows with storm shutters of plywood.
Reinforce roof anchoring to the frame of the structure.
Brace the gable end of roof framing.
Secure metal siding and metal roofing.
Reinforce double entry doors and garage doors.
Install sewage backflow valves.
Build with or retrofit with flood proof materials.
Flood-proof utilities such as private wells, electrical equipment, and HVAC equipment.
Maintain a family preparedness plan and emergency supply kit.

FIGURE 8–8 Hurricane evacuation routes are typically well marked in high risk areas.
Source: National Weather Service.

There is an important distinction between the generic use of the term "shelter" and the formal criteria for a shelter or a safe room. Some buildings are designed to provide protection against natural hazards like hurricane force winds and some are not able to provide protection. Taking shelter in a building that appears sturdy does not constitute an approved shelter. Many buildings cannot withstand high winds. However, those that do not provide adequate protection can be fitted with an internal safe room. There are specific design criteria for shelters and safe rooms available through FEMA.

Immediate Actions

When a hurricane is forecasted, those in the path should follow the instructions of local authorities. If an evacuation order is issued, it is foolish to remain in danger. For those who cannot evacuate, they should seek refuge in a sturdy shelter on high ground. This is particularly true for individuals living in mobile homes or temporary structures but also applies to those in high-rise buildings. The winds are stronger as you go higher and individuals should evacuate to lower floors. Anyone living in low-lying areas along the coast, near inland waterways, or near rivers and streams should evacuate to higher ground to avoid the storm surge.

If there is not a designated shelter available and the building does not have a safe room, you should stay indoors and away from windows. Close blinds and curtains to help deflect glass if a window is blown out. Close all the doors in the building and brace the external doors. Take shelter in a small inner room or space such as a bathroom, closet, or hallway. Lie on the floor under a sturdy piece of furniture, if possible. Many survivors of major hurricanes have had to change their plans as the building they were in began to flood or structurally fail. There should always be a plan on where the next nearby safe location may be to evacuate to in the event the original shelter location is compromised.

Recovery

If you are in an area that has experienced significant damage from a hurricane, there are a myriad of concerns that must be addressed. The first priority is assisting the injured. You should provide whatever level of care you are trained in and can competently carry out. As a minimum, you may simply identify where there are injured or trapped individuals and help direct responders in getting help to them more quickly.

The posthurricane environment can be treacherous. A variety of health and safety threats result from wind damage and flooding. Power lines are often down, and building wiring damaged by the storm pose electrocution risks. Gas leaks can cause explosions and fire. Broken glass and sharp wood and metal debris pose injury risks. The area may be littered with tripping hazards that can result in additional injuries (Figure 8–9). Industrial and household hazardous materials are likely to be spilled during the storm or during clean-up. The roads may be flooded or washed out causing additional driving hazards. Over time, these risks will change. Mold growth is a major posthurricane problem and is discussed in more detail in the "Floods" chapter of this text.

FIGURE 8–9 The Lakes by the Bay subdivision took a direct hit from Hurricane Andrew's eyewall resulting in massive destruction.

Source: National Weather Service. http://www.srh.noaa.gov/mfl/events/andrew/1992andrew2.gif.

Summary

Hurricanes (typhoons) carry a variety of public health risks. However, unlike many disaster scenarios where the morbidity and mortality are relatively predictable, hurricanes have too many variables to predict health outcomes with any degree of accuracy. Although some storms generate a larger number of traumatic injuries and few drowning deaths, others have fewer direct injuries and many more drownings. The severity of storm surge depends upon the coastal terrain and the amount of human development along the shoreline. These factors also influence the amount of inland flooding but not as much as the speed and trajectory of the storm. If the weather system moves slowly or stalls, there may be massive quantities of rain causing flooding like Hurricane Floyd. If it moves quickly, like Hurricane Andrew, there may be little flooding but severe structural damage, and subsequent traumatic injuries from high winds.

Aside from the levee failure in New Orleans following Hurricane Katrina, the reduction observed in recent years of hurricane-associated morbidity and mortality has been impressive. It is a testament to the hard work of meteorologists, emergency management, and local planners. The continued reduction of injuries from these storms can only be achieved by enhancement and enforcement of building codes and the education of the public at risk. In developing nations, there is still much work to be done in protecting the growing populations of coastal residents. It will require international effort to achieve.

Websites

American Red Cross, Hurricanes: www.redcross.org/services/disaster/0,1082,0_587_,00.html.

Centers for Disease Control and Prevention, Hurricanes: www.emergency.cdc.gov/disasters/hurricanes/index.asp.

Centers for Disease Control and Prevention, Hurricane Response and Cleanup: www.emergency.cdc.gov/disasters/hurricanes/workers.asp.

Environmental Protection Agency, Hurricanes: www.epa.gov/naturalevents/hurricanes/.

Federal Emergency Management Agency, Hurricanes: www.fema.gov/hazard/hurricane/index.shtm.

Federal Emergency Management Agency, Hurricane Mitigation Best Practices: www.fema.gov/mitigationbp/sstoryFind.do.

Federal Emergency Management Agency, Protecting your property from disasters: www.fema.gov/plan/prevent/howto/index.shtm#4.

National Institute for Occupational Safety and Health: www.cdc.gov/niosh/topics/flood/.

National Oceanic and Atmospheric Administration, National Climatic Data Center: http://lwf.ncdc.noaa.gov/oa/climate/severeweather/hurricanes.html.

National Oceanic and Atmospheric Administration, National Weather Service, Hurricane Awareness: www.nws.noaa.gov/om/hurricane/index.shtml.

National Oceanic and Atmospheric Administration, National Weather Service, National Hurricane Center: www.nhc.noaa.gov/.

National Oceanic and Atmospheric Administration, National Weather Service, National Hurricane Center, Hurricane Preparedness Week: www.nhc.noaa.gov/HAW2/english/intro.shtml.

References

Centers for Disease Control and Prevention (1993). Injuries and illnesses related to Hurricane Andrew—Louisiana, 1992. *Morb Mortal Wkly Rep* 42:242–243, 249–251.

Centers for Disease Control and Prevention (2000). Morbidity and mortality associated with Hurricane Floyd—North Carolina, September–October 1999. *Morb Mortal Wkly Rep* 49(17):369–372.

Centers for Disease Control and Prevention (2002). Tropical Storm Allison rapid needs assessment—Houston, Texas, June 2001. *Morb Mortal Wkly Rep* 51:365–369.

Chowdhury, M., Choudhury, Y., Bhuiya, A., et al. (1992). Cyclone aftermath: research and directions for the future. In: Hossain, H., Dodge, C. P., Abed, F. H., eds. *From crisis to development: coping with disasters in Bangladesh*. Dhaka, Bangladesh: University Press Ltd; pp.101–133.

Combs, D. L., Parrish, R. G., McNabb, S. J., et al. (1996). Deaths related to Hurricane Andrew in Florida and Louisiana, 1992. *Int J Epidemiol* 25:537–544.

Diacon, D. (1992). Typhoon resistant housing in the Philippines: the core shelter project. *Disasters* 16:266–271.

Noji, E. K. (1993). Analysis of medical needs during disasters caused by tropical cyclones: anticipated injury patterns. *J Trop Med Hyg* 96:370–376.

Rice, D. (2005). Hurricane Katrina stronger than Andrew at landfall. *USA Today.* August 31, 2005. www.usatoday.com/weather/stormcenter/2005-08-31-Katrina-intensity_x.htm.

Toole, M. J. (1997). Communicable disease and disease control. In: Noji, E. K., ed. *The public health consequences of disasters*. New York: Oxford University Press: pp.79–100.

United Nations Development Programme (2004). *Reducing disaster risk: a challenge for development*. New York: John S. Swift Company.

Western, K. (1982). *Epidemiologic surveillance after natural disaster*. Washington, DC: Pan American Health Organization. PAHO Scientific Publication no. 420.

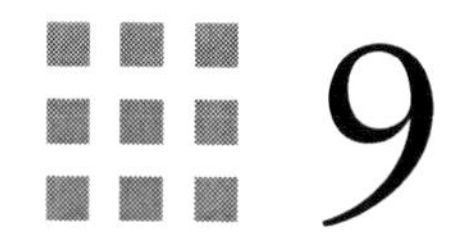

Nuclear and Radiological Disasters

Objectives of This Chapter

- Explain the difference between ionizing and nonionizing energy.
- Recognize the primary units of measurement for radiation.
- Describe the differences in risk among types of radiation.
- Explain the various stages and effects of acute radiation syndrome.
- List the effects of a nuclear detonation.
- Describe the difference between an improvised nuclear device and a dirty bomb.
- List the factors that determine the morbidity and mortality of a dirty bomb.
- Explain what can be done to reduce the threat of nuclear and radiological weapons.
- Describe the three most important protective factors when dealing with radiation.
- List the key issues that hospitals must consider when planning for radiological disasters.

Introduction (Case Study: Chernobyl Nuclear Accident)

In April 1986, a small city in Ukraine became the focus of international concern and the site of the largest technological disaster in history. The Chernobyl Nuclear Power Station was constructed in the 1970s and consisted of four reactors. The first reactor, Reactor Number 4, started generating power in 1983. Others were in various stages of construction in 1986 when a terrible catastrophe occurred. A routine test was under way to determine whether Reactor Number 4 could keep cooling pumps operating during a power outage until the backup generators could trigger. A power surge during the test led to a massive explosion in the reactor that sets a meltdown in motion (See Figure 9–1) (United States Nuclear Regulatory Commission, 2006). The fuel rods and the graphite reactor cover melted through the floor and a huge cloud of radiation was released. The amount released has been described as 30–40 times the amount of radiation released from the detonation of the Hiroshima and Nagasaki atomic bombs (Library of Congress, 1996).

The groups at risk from Chernobyl radiation exposures have been studied by dividing them into several groups including "liquidators" or those who responded and did on-site response and clean-up, evacuees from the "highly contaminated" zone, residents of the "strict-control" zones, and residents of other contaminated areas beyond the control zones. These areas span across Ukraine, Belarus, and Russia. However, measurable contamination from Chernobyl was detected in nearly every nation in the Northern Hemisphere (United Nations, 1988b).

The first populations exposed were plant personnel and first responders. They received whole body gamma irradiation, substantial skin exposure to beta radiation, and inhalation of a variety of nuclides including radioiodine and cesium. Just over 200 workers were involved in the initial response and over half of them suffered from Acute Radiation Syndrome (ARS). Over the next 3 weeks 19 died (United Nations, 1988a). In the 4 months following the accident, a total of 28 facility workers and responders died from their exposures (United States Nuclear Regulatory Commission, 2006).

Hundreds of thousands were recruited to carry out clean-up operations. To limit the exposure of individual workers, each one would work a shift of several minutes and quickly move out as the next crew came in to work several minutes. As individuals reached their limit for "allowable" exposures, more workers would be brought in to replace them. Besides the initial workers and responders who were present during and immediately after the explosion, these "liquidators" received the most significant radiation exposures. Most were 20–45 years old at the time of the accident. There are ongoing debates over the health impact among these workers. While some claim that tens of thousands have died from illnesses related to their exposures, the International Atomic Energy Agency claims that the morbidity and mortality studies carried out among the liquidators show no correlation between their exposures at Chernobyl and their cancer and death rates (International Atomic Energy Agency, 2005).

The other groups that have been observed for radiological health effects include evacuees from the area immediately surrounding the facility and those who reside in potentially contaminated areas. Besides the direct exposure to the initial radioactive plume that was emitted from the facility, area residents have received additional exposure from contaminated food, water, and air over the years following the disaster. Eating food grown in contaminated soil and drinking milk from cows grazing on contaminated grass have increased the exposure of those living throughout the region but not sig-

(a)

(b)

FIGURE 9–1 (a) Chernobyl Nuclear Power Station following the April 1986 explosion and (b) the "sarcophagus" built to entomb Reactor Number 4 and the radioactive remains.

Source: New York State Education Department and Nuclear Regulatory Commission. Image of damage on left available at: http://www.emsc.nysed.gov/ciai/images/chernobyl.jpg. Image of sarcophagus on right available at: http://www.nrc.gov/about-nrc/emerg-preparedness/images/chernobyl.jpg.

nificantly (See Table 9–1). While there are reports of increased thyroid cancers among children exposed, others claim that the increases are the result of enhanced surveillance activities and do not represent a true increase. The United Nations Scientific Committee on the Effects of Atomic Radiation comprises a staff of 15, a committee of 146, and 21 national delegations (Jaworowski, 2000–2001). It is considered by many to be the most knowledgeable and objective collection of nuclear experts in the world. With the possible exception of increased thyroid cancers among exposed children, they have not seen compelling evidence of any serious health problems resulting from Chernobyl. However, there are still many who disagree .

Chronic doses of radiation are cumulative over each person's lifetime and are known to cause thyroid cancer, leukemia, solid cancers, circulatory disease, cataracts, and birth defects. There are several possible reasons why these problems are not being observed in the exposed population as dramatically as many public health experts predicted. First, there still may be changes in health status among the exposed that has yet to occur. Several of these conditions develop over many years and the differences between those exposed and those not exposed may not become apparent for several more years. Second, there is ongoing mistrust of authorities and some doubt if results showing substantial increases in latent radiological morbidity and mortality will be shared with the international community. Finally, there may not be the long-term morbidity and mortality that many predicted. If this is the case, there will still be many conspiracy theorists and others steeped in their own dogma that will never believe it.

Table 9–1 Average Accumulated Dose of Chernobyl-Affected Populations Compared to Natural and Medical Radiation Sources (The Chernobyl forum, 2005; Health Physics Society, Health Physics Fact Sheet, Radiation exposure from medical diagnostic imaging procedures, http://www.hps.org/documents/meddiagimaging.pdf)

Population	Number Exposed	Exposure Timeframe	Average Dose (mSv)
Recovery workers ("Liquidators")	600,000	1986–1989	~100
Evacuees from contaminated zone	116,000	1986	>33
Residents of strict control zones	270,000	1986–2005	>50
Residents of other contaminated areas	5,000,000	1986–2005	10–20
Naturally occurring background levels	All	1986–2005	48
Medical—chest X-ray	N/A	Per procedure	0.08
Medical—mammogram	N/A	Per procedure	0.13
Medical—head CT scan	N/A	Per procedure	2
Medical—abdomen CT scan	N/A	Per procedure	10
Medical—angioplasty (heart study)	N/A	Per procedure	7–57

CT, computed tomography.

The psychological impact of Chernobyl may carry the most significant health impact of all. The nature of radiation exposure is frightening. It cannot be seen or detected with any human senses and the damage it does may take decades to emerge. When it appears, it could be in the form of a terminal cancer that could cause an extended time of suffering. It can also impact future generations and those exposed will worry for themselves as well as for their children. When this scenario is coupled with the propensity for much of regional political leadership to misinform the public and deny problems, it is a recipe for long-term stress that will undoubtedly contribute to the overall morbidity and mortality of Chernobyl.

Although Chernobyl was the first catastrophic reactor failure, there were others that preceded it. In 1957, Windscale Number 1, the first large reactor built in the United Kingdom experienced a fire but no explosion. Radiation was released across England. In 1961, an experimental reactor in Idaho experienced an explosion that killed three servicemen and contaminated the facility and surrounding area. In both cases, the reactors were closed down and abandoned. Walter Patterson, a nuclear expert and an author, has made some interesting comparisons of these incidents to Chernobyl (Patterson, 1986).

- Equipment failed or processes did not work as planned.
- Human error made the situation worse.
- The public seldom received accurate information in the immediate aftermath.

Nuclear and Radiological Disaster Definitions

Absorbed dose: The amount of energy deposited by ionizing radiation in a mass of tissue expressed in units of joule per kilogram (J/kg) and called "gray" (Gy).

Acute exposure: Exposure to radiation that occurs in a matter of minutes.

Alpha radiation (alpha particle): A positively charged particle consisting of two neutrons and two protons. It is the least penetrating but most ionizing of the three common forms of radiation. It can be stopped by a sheet of paper but can cause significant long-term damage if inhaled or ingested. It carries more energy than beta or gamma radiation.

ARS: An often fatal illness caused by receiving a high dose of penetrating radiation to the body in a short time (usually minutes).

Background radiation: Radiation from the natural environment originating primarily from the natural elements in rock and soil of the earth and from the cosmic rays.

Becquerel (Bq): Amount of a radioactive material that will undergo one decay (disintegration) per second.

Beta radiation (beta particle, beta ray): An electron of either positive or negative charge has been ejected by an atom in the process of a transformation. Beta particles are more penetrating than alpha radiation but less than gamma. They can cause serious skin burns with high exposures.

Cumulative dose: The total dose that accumulates from repeated or continuous exposures of the same part of the body, or of the whole body, to ionizing radiation.

Curie (Ci): The traditional measure of radiation that was based on the observed decay rate of 1 g of radium.

Cutaneous Radiation Syndrome: A complex syndrome resulting from excessive ionizing radiation exposure to the skin. The immediate effects can be reddening and painful swelling of the exposed area; large doses can result in permanent hair loss, scarring, altered skin color, deterioration of the affected body part, and death of the affected tissue (requiring surgery).

Decay (radioactive): The change of one radioactive nuclide into a different nuclide by the spontaneous emission of alpha, beta, or gamma rays or by electron capture. The end product is a less energetic, more stable nucleus.

Decontamination: The removal of radioactive contaminants by cleaning and washing.

Dirty bomb: A device designed to spread radioactive material by the explosion of a conventional device. It is relatively simple to make and kills or injures people through the initial blast of the conventional explosive and also spreads radioactive contamination.

Fission: The splitting of a heavy nucleus into two roughly equal parts accompanied by the release of a relatively large amount of energy in the form of neutrons and gamma rays. The three primary fissile materials are uranium-233, uranium-235, and plutonium-239.

Gamma: A highly penetrating type of nuclear radiation, similar to X-radiation except that it comes from the nucleus of an atom. Gamma rays penetrate tissue farther than beta or alpha particles but leave a low concentration of ions in their path to damage cells.

Geiger counter: A Geiger–Müller detector and measuring instrument contains a gas-filled tube that discharges electrically when ionizing radiation passes through it and a device that records the events. They are the most commonly used portable radiation detection instruments.

Gray (Gy): A unit of measurement for the absorbed dose of radiation. The unit Gy can be used for any type of radiation, but it does not describe the biological effects of the different radiations.

Ionization: The process of adding one or more electrons to, or removing them from atoms or molecules, thereby creating ions.

Ionizing radiation: Radiation that is capable of displacing electrons from atoms, thereby producing ions. High doses of ionizing radiation may produce severe skin or tissue damage.

Isotope: Isotopes of an element have the same atomic number but different atomic weights (different number of neutrons in their nuclei). Uranium-238 and uranium-235 are isotopes of uranium.

Protective Action Guide (PAG): A guide that tells state and local authorities at what projected dose they should take action to protect people from radiation exposures in the environment.

Radiation: Energy moving in the form of particles or waves. Familiar radiations are heat, light, radio waves, and microwaves. Ionizing radiation is a high-energy form of electromagnetic radiation.

Radiation absorbed dose (rad): The basic unit of absorbed radiation dose. It is a measure of the amount of energy absorbed by the body. The rad is the traditional unit of absorbed dose. It is being replaced by the unit Gy, which is equivalent to 100 rad.

Radioactivity: The process of spontaneous transformation of the nucleus, generally with the emission of alpha or beta particles often accompanied by gamma rays. This process is referred to as decay or as the disintegration of an atom.

Roentgen: A unit of radiation exposure defined as the amount of X- or gamma-radiation that produces 1 electrostatic unit of charge in 1 cm^3 of dry air under standard conditions.

Roentgen equivalent, man (rem): A unit of equivalent dose. Not all radiations have the same biological effect, even for the same amount of absorbed dose. Rem relates the absorbed dose in human tissue to the effective biological damage of the radiation. Although it is the traditional unit of equivalent dose, it is being replaced by the sievert (Sv), which is equal to 100 rem.

Sievert (Sv): The international standard unit for the amount of ionizing radiation required to produce the same biological effect as one rad of high-penetration X-rays, equivalent to a gray for X-rays. (100 rem or 8.38 roentgens). This relates to the absorbed dose in human tissue that varies by the type of radiation.

Basic Facts about Nuclear and Radiological Threats

For decades, U.S. military personnel were trained in nuclear, biological, and chemical (NBC) or chemical, biological, and radiological (CBR) defense. In recent years, a distinction has been drawn between nuclear and radiological events leading to a new acronym. Chemical, biological, radiological, and nuclear (CBRN) has become the common term and acronym for weapons of mass destruction among preparedness and response professionals to delineate differences between nuclear and radiological accidents or acts of terrorism.

A nuclear event is distinguished by a nuclear detonation or fission. This includes nuclear bombs and smaller improvised nuclear devices (INDs) that release blinding light, a blast wave with intense thermal heat, and radiation. A radiological event does not involve a fission reaction or nuclear explosion. This includes accidents at nuclear facilities or with radiological materials in transport. It can also refer to two radiological terrorism scenarios. A hidden radioactive source can be placed where people are exposed to dangerous levels of radiation, or if an explosive device is added to the source, it can be used as a radiological dispersion device (RDD) or dirty bomb. The distinction between these events is important to planning and preparedness. Responding to a nuclear event is different from responding to a radiological event.

Radiation is simply a form of electromagnetic energy. Energy with different wave lengths has different applications and effects, and the entire spectrum is typically divided into seven subsections including radio, microwave, infrared, visible, ultraviolet, X-ray, and gamma ray (See Figure 9–2). The most important difference between these forms of energy is the ionizing potential. At lower frequencies, from radio through ultraviolet energy, the energy is nonionizing. As the frequency increases, the energy becomes ionizing. That means it has the ability to remove an electron from an atom or a molecule. When this happens to water molecules inside the body, it creates highly oxidizing free radicals. These free radicals damage various parts of living cells that result in cell death, inhibition of cell division, or the production of abnormal or malignant cells.

The terms used to express radiological units of measure have been adopted internationally (gray, centigray, sievert, centisievert, and becquerel), but many references are still made to the old units of measure (rad, rem, roentgen, and curie). This will continue to cause confusion across the scientific community until the international values are

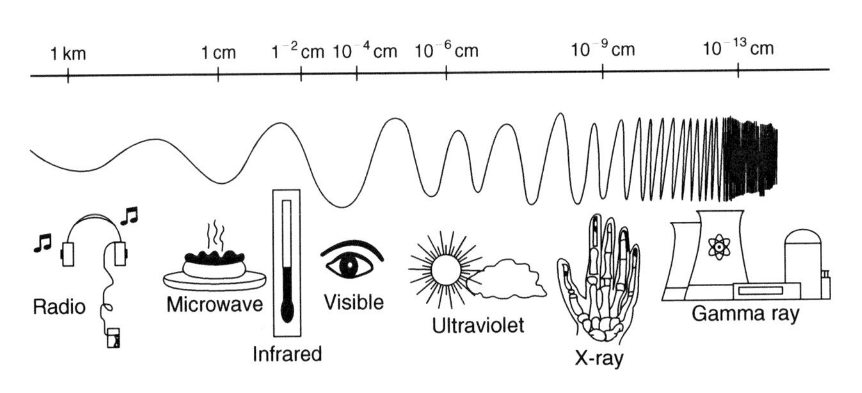

FIGURE 9–2 The electromagnetic spectrum.
Source: NASA. Available at: http://www.centennialofflight.gov/essay/Dictionary/ELECTROSPECTRUM/DI159G1.jpg.

completely adopted. The relationships between the older units of measurement and the international values are as follows (Figley, 1980):

- 1 Gy = 1 J/kg = 100 rad
- 1 centigray = 1 rad
- 1 Sv = 100 rem
- 1 centisievert = 1 rem
- 2.58×10^{-4} C/kg = 1 roentgen
- 1 Bq = 1 nuclear transformation per second = 2.7×10^{-11} Ci

Ionizing radiation exposures are measured by the amount of absorbed energy per unit of mass. The unit of absorbed dose is Gy. It represents a joule per kilogram (J/kg). If the total absorbed dose in a person exceeds 1 Gy, it can result in ARS. Since there are so many different levels of susceptibility among the organs of the body, another measurement is used to describe the effective dose. This is called a sievert. It is a measurement that takes into account the type of radiation, susceptibility of the most sensitive organs, internal versus external exposures, and uniform versus nonuniform exposures. One sievert is considered to be a very large exposure, and so millisieverts (mSv), or thousandths of a Sv, are used to describe typical exposures. For example, in Table 9–1, the 20-year background radiation exposure is based on the United Nations Scientific Committee on the Effects of Atomic Radiation estimates that the average annual exposure of a human to natural background radiation is 2.4 mSv (The Chernobyl forum, 2005).

There are four primary types of radiation: alpha, beta, gamma, and X-ray. Each one has a different origin or poses unique health threats to those exposed (Figure 9–3). During radioactive decay, these can all be present.

- Alpha radiation is a large particle compared with other particulate radiological threats. Examples of alpha emitters include radon and uranium. It is not able to penetrate human skin and poses no external hazard. However, as a larger particle, it poses a significant internal threat. If the particles are inhaled, ingested,

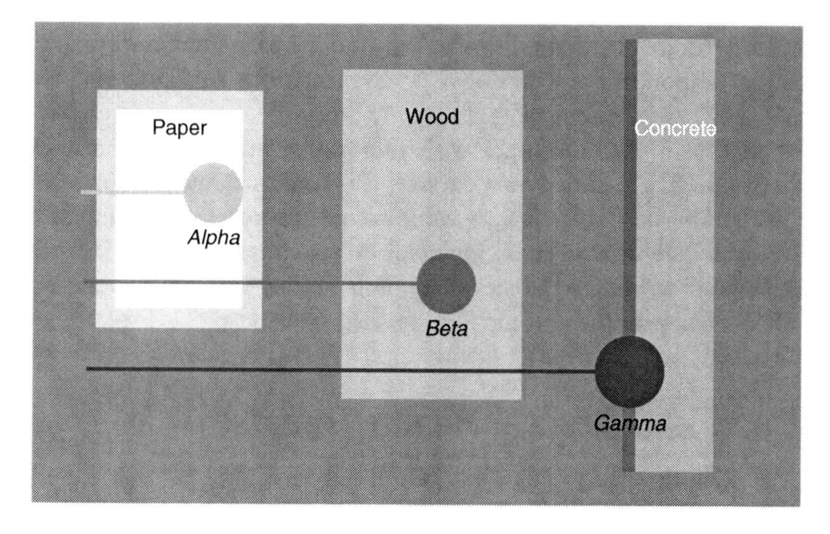

FIGURE 9–3 Relative penetrating power of radiation types.
Source: Environmental Protection Agency. Available at: http://www.epa.gov/radiation/images/rad-penetration.gif.

or absorbed through an open wound, they can irradiate surrounding living cells and cause chronic health risks. Alpha particles can only travel several inches in the air. There are a variety of detection devices available but they must be within a couple inches of the alpha source with no interferents, including heavy dust or water.

- Beta radiation is also a particle. Examples of beta-emitters include carbon-14 and strontium-90. These particles consist of ejected electrons and can travel several feet. They are much smaller than alpha particles and can penetrate several layers of skin cells resulting in "beta burns." Beta can also pose an internal hazard. They cannot be stopped by a sheet of paper like alpha particles but can be stopped by a sheet of plywood. Clothing provides some protection, though not complete. Beta radiation can be detected using specific beta survey instruments, but some beta emitters produce such low energy that they are difficult to detect with current instrumentation.
- Gamma radiation and X-rays are highly penetrating radiation. They consist of electromagnetic energy that can penetrate most surfaces. In many ways, they are like radio waves or ultraviolet light, only with much more energy. They can travel many feet in the air and through several inches of human tissue. Exposure to gamma radiation can result in acute or chronic injuries. Examples of gamma emitters include cesium-137 and cobalt-60. There are a variety of detection devices available that can detect gamma radiation.

With radiation taking on two primary forms, particulate and energy, there are two primary types of exposures possible: irradiation and contamination. During different types of nuclear and radiological accidents or attacks, one or both of these types of exposures may occur.

Irradiation occurs when gamma or X-ray radioactive energy passes directly through the body. This causes rapid cellular damage and can make a person sick almost immediately with higher doses causing a more rapid onset of symptoms. The acute illness that results is called ARS. Irradiation can also lead to chronic health problems. Damage to an exposed person's DNA can also lead to chronic illnesses such as cancer and birth defects. Many of the survivors of the atomic bomb attacks during World War II and the initial responders at Chernobyl experienced both ARS and long-term health concerns.

Contamination is direct contact with radioactive material. When a radioactive source, solid or liquid, is retained on a surface, it is contaminated. If that surface comes in contact with other surfaces, it can continue to spread the contamination. If the radioactive contamination is inhaled, ingested, or penetrates the skin, it is transported throughout the body where it irradiates surrounding cells. Although contamination does not cause ARS, it can produce chronic illnesses such as cancer.

Nuclear and Radiological Health Threats

ARS is one of the most challenging aspects of a public health and medical response to a nuclear or radiological incident. This condition is the result of a large exposure to a penetrating external radiation source over a short period of time. It includes four stages.

A prodromal stage includes gastrointestinal symptoms such as nausea, vomiting, and diarrhea. It can begin within minutes or days of the exposure and last up to several days. A latent stage follows where the patient will feel fine for a period of time ranging from hours to weeks. This is followed by a manifest illness stage that includes one or more of three classic syndromes (Centers for Disease Control and Prevention, 2005a):

- Bone Marrow Syndrome often leading to death from the destruction of bone marrow leading to infections and hemorrhage
- Gastrointestinal Syndrome likely leading to death from serious GI tract damage causing infections, dehydration, and electrolyte imbalance
- Cardiovascular/Central Nervous System Syndrome leading to death within a few days from circulatory system collapse and increased cranial pressure from edema, vasculitis, and meningitis

The final stage is either recovery or death. The entire process can take from a few weeks to a couple years. Table 9–2 provides additional details on the progression of each syndrome.

ARS usually involves such a large external radiation exposure that it includes cutaneous radiation injury (CRI) as well. CRI can also occur with no association to ARS. In fact, it is most often found in people who have come in direct contact with an industrial radiation source. There is usually a delay between the exposure and the onset of symptoms and the progression is in stages, similar to ARS. There is a prodromal, latent, and manifest illness stage, sometimes followed by a third wave before recovery (Centers for Disease Control and Prevention, 2005b). Skin problems at the CRI site can reemerge in months and years following recovery and may include skin cancer.

There are also special challenges in managing pregnant patients exposed to radiation. Although the fetus or embryo receives a lower exposure than the mother due to the protection of the mother's body, they are also far more vulnerable to the consequences of a radiation exposure. Since cells are dividing quickly in the growing fetus, the DNA and cellular damage that radiation exposure can cause can lead to birth defects, growth retardation, impaired brain function, and cancer (Centers for Disease Control and Prevention, 2005c). Managing these exposures requires a team of specialists that includes a health physicist to assist in determining the dose experienced.

Table 9–2 Acute Radiation Syndrome Doses and Stages

	Bone Marrow Syndrome	**Gastrointestinal Syndrome**	**Cardiovascular/Central Nervous System Syndrome**
Threshold dose when symptoms may appear	0.3 Gy (30 rad)	6 Gy (600 rad)	20 Gy (2000 rad)
Dose*	>0.7 Gy (>70 rad)	>10 Gy (>1000 rad)	>50 Gy (5000 rad)

(Continued)

Table 9–2 *(Continued)*

	Bone Marrow Syndrome	Gastrointestinal Syndrome	Cardiovascular/Central Nervous System Syndrome
Prodromal stage	Symptoms include anorexia, nausea, and vomiting. Onset in 1 hour to 2 days after exposure. Stage lasts from minutes to days.	Symptoms are anorexia, severe nausea, vomiting, cramps, and diarrhea. Onset occurs within a few hours after exposure. Stage lasts about 2 days.	Symptoms are extreme nervousness and confusion; severe nausea, vomiting, and watery diarrhea; loss of consciousness; and burning sensations of the skin. Onset occurs within minutes of exposure. Stage lasts for minutes to hours.
Latent stage	Stem cells in bone marrow are dying, although patient may appear and feel well. Stage lasts 1–6 weeks.	Stem cells in bone marrow and cells lining GI tract are dying, although patient may appear and feel well. Stage lasts less than 1 week.	Patient may return to partial functionality. Stage may last for hours but usually is less.
Manifest illness stage	Symptoms are anorexia, fever, and malaise. Drop in all blood cell counts occurs for several weeks. Primary cause of death is infection and hemorrhage. Most deaths occur within a few months.	Symptoms are malaise, anorexia, severe diarrhea, fever, dehydration, and electrolyte imbalance. Death is due to infection, dehydration, and electrolyte imbalance. Death occurs within 2 weeks of exposure.	Symptoms are return of watery diarrhea, convulsions, and coma. Onset occurs 5–6 hours after exposure. Death occurs within 3 days of exposure.
Recovery stage	In most cases, bone marrow cells will begin to repopulate the marrow. There should be full recovery for a large percentage of individuals from a few weeks up to 2 years after exposure. Death may occur in some individuals at 1.2 Gy (120 rad). The LD50/60† is about 2.5–5 Gy (250–500 rad).	LD100‡ is about 10 Gy (1000 rad).	No recovery expected.

*All absorbed doses listed are "gamma equivalent" values.

†The LD50/60 is the dose necessary to kill 50% of the exposed population in 60 days.

‡The LD100 is the dose necessary to kill 100% of the exposed population.

Adapted from the Centers for Disease Control and Prevention, Acute Radiation Syndrome: A Fact Sheet of Physicians. Available at: http://www.bt.cdc.gov/radiation/arsphysicianfactsheet.asp.

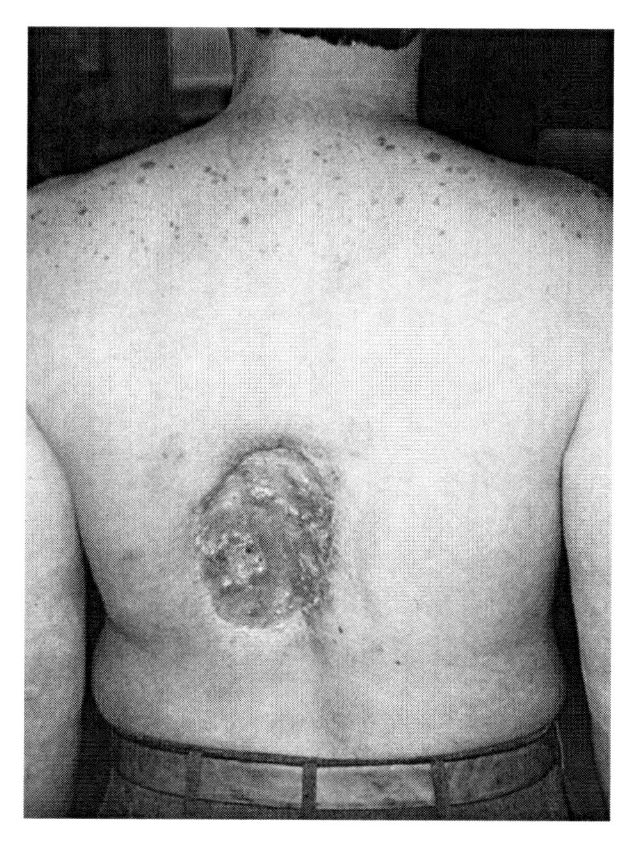

FIGURE 9–4 A cutaneous radiation injury from a fluoroscopy nearly 6 months after an exposure estimated at about 20 Gy.

Source: Food and Drug Administration. Available at: http://www.fda.gov/cdrh/figure2c.jpeg.

Nuclear and Radiological Accidents versus Terrorism

The U.S. Department of Homeland Security (DHS), 15 National Planning Scenarios discussed in Chapter 1 include one nuclear scenario and one radiological scenario. Both are terrorism related. Scenario number 1 is the detonation of a 10-kiloton IND, and scenario number 11 is an attack using the RDD. These scenarios are very distinct in terms of the challenges that each will carry. They are also fundamentally different from accident scenarios. In fact, there are five distinct differences (Department of Homeland Security, 2006):

- Most radiological emergency planning is focused on nuclear power facility accidents that are much smaller events than an IND and would not include the same blast and heat hazards.
- Due to the alarm systems in place at modern nuclear facilities, most radiological accident scenarios will have several hours of warning before a release begins. IND or RDD attacks are more likely to be unannounced.
- The IND or RDD incident is more likely to happen in a densely populated area and most nuclear reactors are located in rural areas. Fewer people and less infrastructure will be affected by an accident than an attack.
- Nuclear facilities have detailed preparedness plans and hold regular exercises. Tremendous resources are expended on their facility preparedness. The management

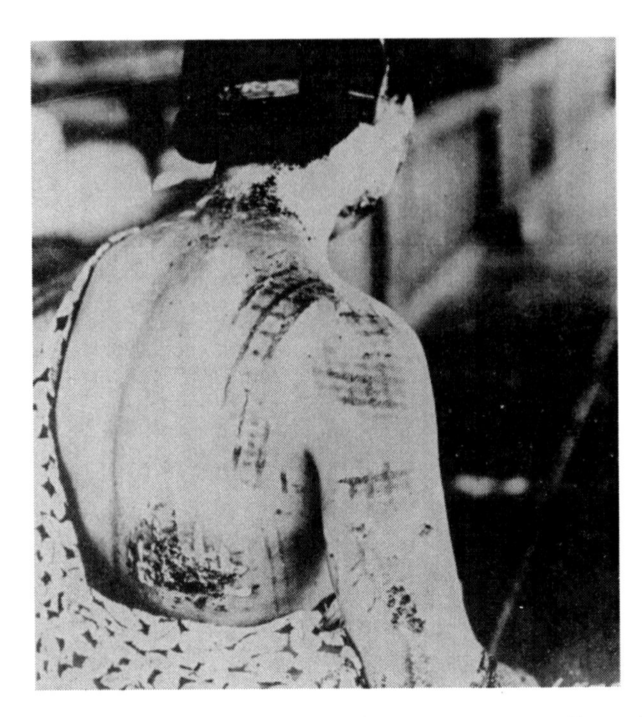

FIGURE 9–5 The victim of an atomic bomb with her skin burned in a pattern corresponding to the dark portions of a kimono worn at the time of the explosion.

Source: National Archives. Available at: http://www.archives.gov/research/ww2/photos/images/ww2-164.jpg.

of an emergency there is likely to be much better coordinated than it will be after an attack.

- The type of radiological material involved in a nuclear facility accident is likely to be different from what would be seen with the IND or RDD attack.

A nuclear detonation releases energy in three forms: blast, thermal, and ionizing radiation. Fifty percent of the energy is released in the blast as a crushing overpressure and hurricane-like wind. Thirty-five percent is released as intense heat. The remaining 15% is released as ionizing radiation with about 5% being released instantly as gamma and the remaining 10% in the fallout (Figure 9–6). What that means is the initial detonation will instantly release so much energy that anyone looking toward it will likely be blinded. While light surfaces will reflect the flash, darker surfaces will absorb heat from it as shown in Figure 9–5 where the pattern of her clothing was burned on her skin when the lighter fabric reflected the flash and the darker material absorbed it.

After the initial flash, a mach stem, or crushing change in atmospheric pressure, will precede the blast winds. As quickly as the overpressure pounds everything in its path, hurricane-like winds with searing heat will follow quickly behind. At the same time, a large amount of radiation is released that can result in many ARS cases. The fallout cloud will then carry radiation many miles downwind of the detonation site posing a chronic risk to many more people for an extended period of time.

The types of injuries resulting from a nuclear detonation include primary blast injuries from the overpressure, secondary traumatic injuries from the debris propelled by the blast, tertiary injuries from people being thrown and from collapsing structures, and quaternary injuries such as burns. The flash of the fireball will result in retinal burns that will vary by distance from the explosion. There will also be direct and indirect burn

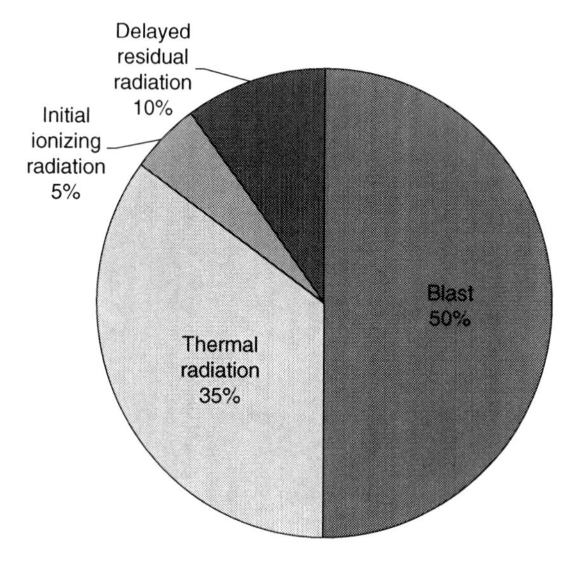

FIGURE 9–6 Nuclear weapon detonation effects.
Source: U.S. Armed Forces Radiobiology Research Institute.

injuries. Radiation will have both immediate effects for individuals with acute radiation illness and long-term effects over a much larger area for those in the path of fallout who inhale or ingest particles posing long-term health hazards.

The RDD poses a very different scenario. Although there is an explosion associated with some RDDs, they have no nuclear reaction or fission and have an explosion thousands of times smaller than a nuclear detonation. A "dirty bomb" is one type of RDD that simply consists of an explosive device paired with radiological material. The most immediate health impact of a RDD is blast injury. As first responders arrive at the scene of a mysterious explosion, they will use standard detection equipment to identify the radiological threat. The level of risk to human health that extends beyond the initial blast injuries will be determined by the quantity and type of radiological material used as well as a variety of other environmental and response variables.

Dirty Bombs: Top 10 Morbidity and Mortality Impact Factors

1. Type of radioactive material (alpha, beta, or gamma emitter)
2. Amount of radiological material
3. Size of explosion
4. Type of individual exposures (internal or external)
5. Time of individual exposure to radioactive material
6. Distance of individuals from explosion
7. Shielding of individuals at risk (indoors or outdoors)
8. Weather and wind patterns
9. Decontamination, evacuation, and sheltering effectiveness
10. Availability and efficacy of chelating agents and other drugs

FIGURE 9–7 Testing of a 21-kiloton nuclear device dropped from an aircraft on November 1, 1951.
Source: Nevada Division of Environmental Protection. Available at: http://ndep.nv.gov/boff/atomic.jpg.

Prevention

Prevention of nuclear or radiological incidents, both accidental and intentional, depends on control of radiological material. Careful security and control is needed for all existing radiological sources that could be used in INDs or dirty bombs. It is also important that the international community continues to press all nuclear capable nations to abide by the Nuclear Non-Proliferation Treaty (NPT). The treaty has been signed by 189 nations over the past 30 years and expresses support for nonproliferation of nuclear weapons programs, disarmament, and peaceful use of nuclear technologies for energy and other peaceful applications (United Nations Office for Disarmament Affairs, 1968). Although the five declared nuclear states (Russia, the United States, China, France, and United Kingdom) have signed the NPT, three known nuclear powers (India, Pakistan, and Israel) have refused to sign it. In addition, nations pursuing nuclear programs include North Korea, Syria, and Iran (See Table 9–3). Those nations not standing with the international community to limit the proliferation of nuclear programs will continue to enhance the risk for nuclear and radiological accidents and attacks.

Each nuclear-weapon State Party to the Treaty undertakes not to transfer to any recipient whatsoever nuclear weapons or other nuclear explosive devices or control over such weapons or explosive devices directly, or indirectly; and not in any way to assist, encourage, or induce any non-nuclear-weapon State to manufacture or otherwise acquire nuclear weapons or other nuclear explosive devices, or control over such weapons or explosive devices.

Nuclear Non-Proliferation Treaty, Article I (United Nations Office for Disarmament Affairs, 1968)

In addition to nations with nuclear weapons programs, dozens of other nations have weapons-usable material that could be used for the development of INDs or RDDs. Since the most essential radiological materials to build nuclear weapons (plutonium and highly enriched uranium or HEU) do not occur in nature, they must be manufactured. The problem is that these materials are also manufactured for civilian uses in nuclear energy. In addition to closely monitoring the thousands of nuclear weapons around the planet, there is an ongoing need to ensure proper security of tons of these materials across dozens of nations that have nuclear power initiatives (Figure 9–8). It is a challenging task but is absolutely essential to ensuring the nuclear safety of the international community.

The last barrier against the illicit use of radiological material is physical security and controls. In recent years, nations around the world have enhanced security measures to include detection of radiological materials, especially in ports where large quantities could potentially be transferred. There are also human portals at nuclear facilities that are able to detect workers with radioactive materials.

In recent years, there have been a variety of new developments in stand-off detectors. Detectors are positioned at major events with national or international security interests such as the Olympic games or major political events. They are sensitive enough to detect individuals walking by who are undergoing diagnosis or treatment with radiopharmaceutical drugs and could certainly detect the presence of an individual carrying a nuclear or radiological device. The future of this emerging technology looks promising.

Table 9–3 Nations with Nuclear Weapons, Treaty Status, and Number of Weapons (Arms Control Association, 2007; Norris and Kristensen, 2006; Norris and Kristensen, 2007a,b)

Nation	Nuclear Status	Weapons	Worldwide %*
Russia	Declared—signed NPT	15,000	58
United States	Declared—signed NPT	10,000	39
France	Declared—signed NPT	350	1
China	Declared—signed NPT	200	0.8
United Kingdom	Declared—signed NPT	200	0.8
Israel (estimated)	Undeclared—Non-NPT	60–80	0.5
India (estimated)	Undeclared—Non-NPT	50–60	0.3
Pakistan (estimated)	Undeclared—Non-NPT	40–50	0.2
North Korea	Undeclared—Non-NPT	10	0.04
Total		25,950	100

NPT, Nuclear Non-Proliferation Treaty.

*Totals slightly exceed 100% due to rounding error.

FIGURE 9–8 A truck passes through a radiation portal monitor at the port of Newark, New Jersey.
Source: White House, Office of Management and Budget. Available at: http://www.whitehouse.gov/omb/budget/fy2006/images/dhs-6.jpg.

Researchers at Purdue University are working with the State of Indiana to establish a network of cell phones with radiation sensors. The sensors are inexpensive and small enough to be added to laptops or cell phones. They have been tested with very small quantities of radioactive material and could detect it up to 15 feet away (Purdue University, 2008). Since cell phones already have the ability to be tracked by location, a sensor can be added to detect radioactive sources and locations. Since the signal gets weaker as the cell phone moves away from the source, the software being developed could triangulate the radioactive source by using the detection signal strength from multiple phones to track the movement of a possible dirty bomb or IND in real time.

Immediate Actions

The actions taken in response to a nuclear or radiological incident will depend on the specific circumstances. There are four broad scenarios that should be considered during planning so the proper actions are incorporated into preparedness activities.

1. A transportation accident involving radiological materials. Consider the possibility of dozens of injured or exposed people on the scene and determine who will make the final decisions and what guidelines will be used to inform that decision-making process.

2. An accident at a nuclear facility (Figure 9–9) with dozens initially injured and potentially exposed to excessive radiation that could lead to ARS, and thousands more potentially exposed in the immediate area and downwind. Messages should be preplanned in what needs to be communicated under various circumstances and plans should be in place to manage mass prophylaxis or evacuations. Everyone at risk must be registered in a database for appropriate follow-up.
3. The RDD or dirty bomb with the explosion of a package or a vehicle carrying radioactive materials. Consider the management of the scene where there may be dozens with traumatic injuries. Although the risk of ARS is unlikely, consider how the long-term risks will be managed in the exposed population. The scene must be characterized and secured. The exposed must be decontaminated and registered for follow-up.
4. A nuclear detonation could destroy an entire city but a small, IND could appear more like the September 11, 2001, attacks on the United States with a block or several blocks of buildings as "Ground Zero." The challenges will include large amounts of radioactive debris, many traumatic injuries complicated by many more cases of ARS, and chronic exposures for many miles downwind of the detonation.

FIGURE 9–9 Three Mile Island Nuclear Facility near Harrisburg, Pennsylvania.

Source: U.S. Department of Energy. Available at: http://www.fda.gov/centennial/this_week/images/13_06/3_mile_island.jpg.

The good news is that responding to nuclear and radiological incidents is more predictable than many other attack and accident scenarios. Some biological and chemical incident scenarios have such an array of unknown variables that the ability to rapidly detect and characterize the threat is difficult and time consuming. With radiological events, the principles are well understood and there are fewer unknowns. The detection instrumentation is ubiquitous, dependable, and quick to use. The ability to predict those at risk and offer recommendations, such as evacuation or sheltering in place, is much more straightforward than other threats. Decades of advancements in nuclear power facility preparedness, nuclear weapons testing, and Cold War preparedness initiatives have established a robust body of preparedness guidance and information.

Public Health and Protective Actions

The primary roles of public health agencies in the aftermath of a nuclear or radiological incident are to assess public health risks, develop recommendations to optimize protection of public health, and communicate recommendations to protect the public.

There are Environmental Protection Agency (EPA) standards for workers and for the general public that originated with planning for response to nuclear power facility accidents. They are called Protective Action Guides (PAGs). The most recent EPA PAGs for nuclear incidents were published in 1992. However, following the terrorist attacks of 2001, a series of national exercises were conducted to identify gaps and opportunities to improve the federal response to a variety of scenarios. It was discovered through these exercises, the changing nature of nuclear and radiological threats, and information emerging from the Chernobyl disaster that the existing guides were insufficient.

The Top Officials 2 Exercise was conducted in May 2003. It included a fictitious scenario of a foreign terrorist organization attacking the Seattle, Washington region with a RDD and the Chicago metropolitan area with Pneumonic Plague (*Yersinia pestis*). Leading up to the full-scale exercise, there were a number of smaller exercises including intelligence sharing, a cyber-attack, and credible terrorist threats against several locations (Department of Homeland Security, 2003). These exercise activities led to the identification of important gaps including the lack of clear guidance on long-term site restoration and clean-up. Other identified gaps included the need for more specific guidance for INDs and RDDs and additional guidelines for drinking water. These findings underscore the importance of full-scale exercises in identifying important gaps that are missed during earlier efforts and aside from an exercise would only surface during an actual event.

The PAGs include recommendations for the general public and separate recommendations for first responders. The triggers used for deciding when to shelter or evacuate populations at risk are all based on the PAGs. For example, during a nuclear or radiological incident, when first responders are balancing their moral obligations to protect the public against their personal risks, it is not a time to begin the development of a framework to facilitate a rational decision-making process. That is already done with the PAGs. Table 9–4 summarizes the suggested maximum dose for response workers under various conditions (Department of Homeland Security, 2006).

Table 9–4 Response Worker Protective Action Guidelines

Dose	Activity	Condition
5 rem (50 mSv)	All occupational exposures	All reasonably achievable actions taken to minimize dose
10 rem (100 mSv)	Protecting valuable property	Lower dose not practicable
25 rem (250 mSv)	Lifesaving or protection of large populations	Lower dose not practicable
>25 rem (>250 mSv)	Lifesaving or protection of large populations	Voluntary basis/fully aware of risks

Adapted from Protective Action Guides for Radiological Dispersal Device and Improvised Nuclear Device Incidents; Notice, Response Worker Guidelines. Available at: http://homer.ornl.gov/oepa/rules/71/71fr174.pdf.

There are also recommended initial actions for the general public for the time prior to the availability of radiological measurements and associated protective action guidance. If there is a suspicious explosion or if local authorities announce the possibility of a radiological release, it is best to cover your nose and mouth with a piece of cloth and take immediate cover inside the nearest building. Those taking shelter inside buildings should ensure that the doors and windows are closed and that the heating, ventilation, and air conditioning system is shut down. If an announcement is made to evacuate an area where you are inside a building, cover your nose and mouth and evacuate to another building at a safe distance. If you think you have been exposed to radioactive particles, it is important to get out of contaminated clothing as soon as possible. The biggest problem facing those in the immediate area of a suspicious blast or release is panic. In this time of calamity and confusion, the local authorities will have the responsibility to maintain a level head and remind the public of basic principles. For example, to your exposure to radiation, it is simply a matter of time, distance, and shielding.

The Basics of Nuclear and Radiological Protection

Time: Limiting the time of an exposure as much as possible.
Distance: Getting away from a blast site or release quickly will limit exposure.
Shielding: More material and higher density material between you and a radioactive source will reduce exposure. If inside a building, interior rooms or basements offer the best protection.

Medical Response

According to the Centers for Disease Control and Prevention, there are six areas that need to be considered by healthcare organizations as they prepare for nuclear and radiological incidents. These areas include notification and communication, triage, patient management, healthcare worker protection, surveillance, and community planning (Centers for Disease Control and Prevention, 2003).

Notification and communication includes an understanding of existing plans and procedures, how each organization fits into the Incident Command Structure, and where to find the expertise needed to answer detailed clinical and technical questions. It is essential that consistent internal and external communications are established, especially with the media. Public confidence in the health and medical response has far reaching ramifications. The fastest way to undermine and lose that assurance is through inconsistent messages and actions.

Triage needs to be planned in advance and practiced through exercises. The key to successful triage is the understanding of the proper assumptions and correct processes. Many injured and exposed individuals will triage and transport themselves. The patient surge and triage process is managed best by establishing a separate assessment or triage area away from the emergency department of a facility. The surge will occur on top of the typical demand for services and may include burns, trauma, external contamination in need of decontamination and internal contamination in need of the rapid administration of radiological pharmaceuticals.

Successful patient management following a radiological incident will depend on pre-event planning and exercises (Figure 9–10). It includes the organization of the physical layout with clear signage to direct and manage a large influx of people seeking care. A multistep process is needed that includes an initial triage, decontamination process, treatment of injuries, identification and treatment of ARS, management of clinical specimens, and appropriate referral of those with special needs, such as pregnant women and those needing mental health services.

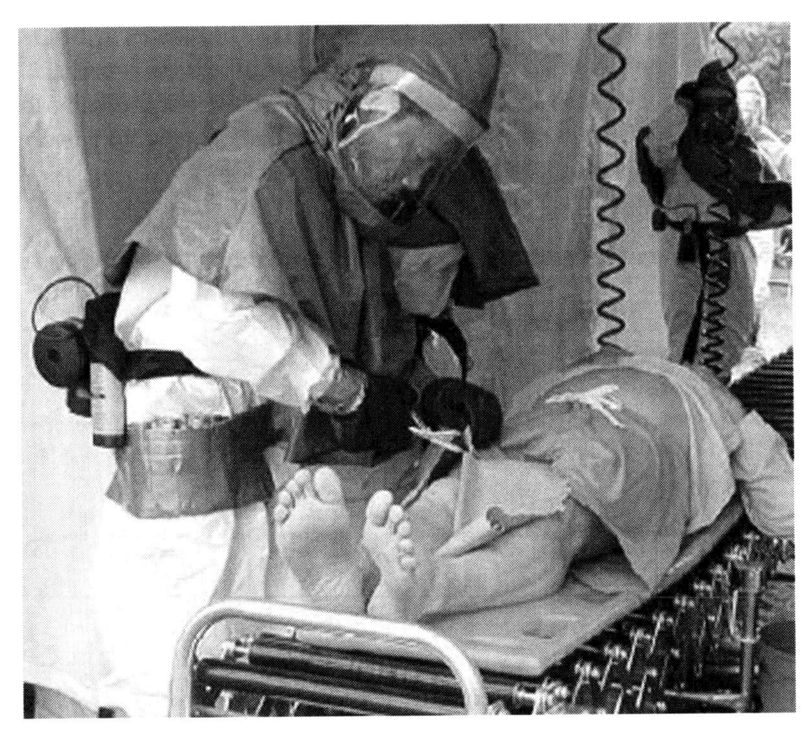

FIGURE 9–10 A healthcare provider cutting off a patient's contaminated clothing during a radiological exercise.

Source: Occupational Safety and Health Administration. Available at: http://www.osha.gov/dts/osta/bestpractices/html/images/hospital_firstreceivers_img1.jpeg.

Healthcare provider protection begins prior to an event. Personal protective equipment is essential for those in contact with potentially contaminated patients and an ample supply should be maintained at all times. Training will also play a key role in healthcare worker protection. Not all employees will receive the same training. Although they all should have an awareness level of nuclear and radiological training to gain an understanding of what the possible scenarios include, some will need more detailed training on decontamination, use of detection equipment, and detailed implementation of the response plan. Practice through exercises is essential to reinforce training and identify gaps in planning.

Surveillance is simply the collection and analysis of information on those seeking care after an incident. This is essential in the first hours and days following an incident to inform the decision-making on initial action and treatment recommendations. In the long term, it is important for follow-up activities, additional treatment suggestions, and long-term study of the health outcomes so lessons learned can be applied toward future events. The minimal information needed includes patient contact information, demographics, location at the time of the incident, and initial actions taken.

Community Planning is not about the final planning document. It is about relationships in the context of disaster scenarios. Healthcare organizations must cultivate the relationships needed to facilitate timely delivery of resources needed during a radiological disaster. These resources include additional supplies, equipment, staffing, and pharmaceuticals. The plan is not worth the paper it is written on, if it is not a reflection of the relationships built during the planning process.

Unique Pharmaceutical Issues

In the United States, when the local medical infrastructure is overwhelmed by a nuclear or radiological incident, they should notify their local public health agency who can request support through the state. If the state cannot meet the needs, they will request federal support. This may include the Strategic National Stockpile (SNS). Many preparedness planners and responders quickly think of the stockpile as a source of pharmaceuticals for mass prophylaxis following a bioterrorism attack or as the Chempack program that places nerve agent antidote in communities across the nation. Some are not as aware of the SNS resources available to support a radiological incident response.

The SNS has a variety of critical radiological pharmaceutical resources including (U.S. Department of Health and Human Services, 2008)

- Chelating agents
- Calcium diethylenetriamene pentaacetate
- Zinc diethylenetriamene pentaacetate
- These agents can be used to treat internal contamination of radiological agents that are transuranic heavy metals including plutonium, americium, and curium.
- Prussian blue to treat internal contamination of cesium and thallium
- Potassium iodide to treat internal contamination of radioactive iodine (I-131)
- Growth factors/cytokines for white blood cells to aid in the recovery of suppressed bone marrow

In most cases, these drugs need to be administered as quickly as possible and some require an "Emergency Use Authorization" from the Food and Drug Administration. That is another reason why preplanning these scenarios in detail is essential.

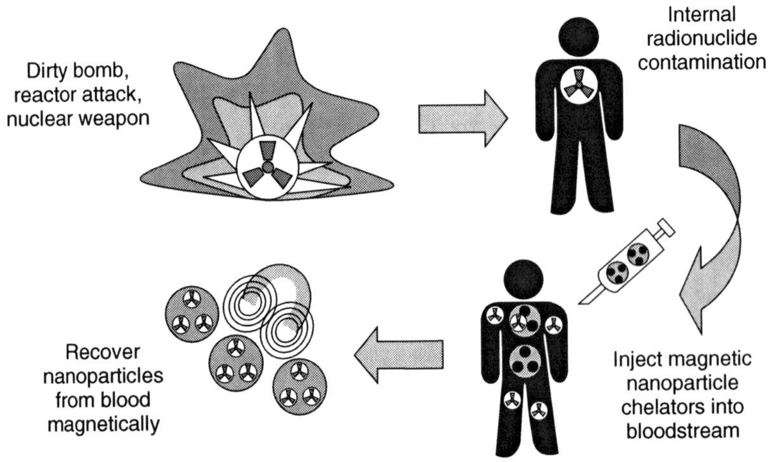

FIGURE 9–11 New biodegradable nanospheres treatment technologies.
Source: U.S. Armed Forces Radiobiology Research Institute. Available at: http://www.anl.gov/Media_Center/News/2006/photo/061020_dirty_bomb_treatment-large.jpg.

There are also several promising pharmaceutical advances on the horizon that will augment the existing treatment options. One particularly interesting advance uses nanotechnology. Biodegradable nanospheres between 100 and 5000 nm in diameter are delivered intravenously. One nanometer is one-billionth of a meter. They are small enough to pass through small vessels but large enough to avoid being filtered out of the bloodstream by the kidneys. They have surface proteins or other media that bind with the targeted radiological or biological toxins in the bloodstream but will not bind with white blood cells (U.S. Department of Energy, 2006). After they circulate and collect the offending threat from the bloodstream, a small shunt is inserted into an artery and the circulating blood is put past a powerful magnet that removes the nanospheres (Figure 9–11).

New Method of Assassination?

Alexander Litvinenko died a mysterious death in November 2006. He was a former KGB agent who moved to London and became an author who leveled allegations of corruption and murder toward his former colleagues in Russia. He wrote books and articles charging Russian Secret Service Agents with acts of terrorism to promote the rise of Vladimir Putin to the Presidency. He was only 43 years old as his health diminished over the course of several weeks in the hospital. Although there were difficulties in diagnosing him, it was eventually determined that he had been poisoned with polonium 210, a powerful alpha radiation emitter. Just before his death, he claimed that Putin was responsible for his murder (Elsen et al., 2007). This assassination is unique and could be the beginning of a terrible new method of murder by radiological "poisoning."

Recovery

The long-term issues associated with a nuclear or radiological incident may be far worse than the short-term impact. The persistent nature of many radiological sources could result in the need to abandon a large affected area for years. Consider the devastation of Hurricane Katrina and the slow recovery of New Orleans. Many evacuated residents have chosen to never return and the city will never be the same. In that case, the people have a choice. In the case of a dirty bomb or other persistent radiological hazard, there may be no choice. A large area, even of the downtown district in a major city, could become uninhabitable for years following a dirty bomb or other radiological release.

This was certainly on the minds of the DHS interagency work group developing clean-up guidelines in 2006. Again, this was a change associated with the PAGs. The problem is that there are so many potential scenarios with such a broad range of outcomes that it makes it difficult to establish numerical standards. The guidelines instead refer to a variety of other sources and leave the decision to local, state, and federal responders to decide on a case by case basis. While this sounds like a reasonable approach, the International Commission on Radiation Protection is among those considered as providing acceptable guidance. They have supported a long-term, acceptable clean-up standard of 10,000 millirems per year. That is 30 times more than an average person receives from all natural and manmade sources each year (American National Standards Institute, 2006). The mainstream media jumped on this, comparing it to 1600 chest X-rays, the current Nuclear Regulatory Commission public exposure limit of 100 millirems, and the estimated cancer rate of one in every four people who are exposed to that level (Hebert, 2006). If this was the media response to a recommendation that defaulted to other sources of expertise generally regarded as competent to make the recommendations, what will the media and public response be when actual names and faces are associated with an abandoned area? With ongoing questions concerning "how clean is clean," there is a high likelihood that the long-term impact of a dirty bomb or other dispersion of a persistent radiological threat is going to make the affected area uninhabitable for an extended period of time. This may or may not be due to actual health threats.

While the environmental impact is being debated, the human health impact will be a source of ongoing controversy as well. Those exposed and many more who never were but are still concerned (the "worried well") will suffer the psychological stress associated with the possibility of a looming terminal diagnosis. As with Chernobyl, there will be individuals for years to come placing the responsibility for a variety of health conditions on the incident. While some will have legitimate claims, it is unfortunate that many who do may be lost in the noise of those whose claims are not legitimate. Should this scenario occur, it will be an ongoing challenge for years, and possibly for generations. The only way to minimize the chronic social impact is to manage it well from the moment it occurs, including the registration and close monitoring of those truly at risk.

Summary

The threat of a nuclear or radiological disaster is real and growing. As technology continues to move forward, more dangerous information and materials are likely to become available to rogue nations and terrorists in the future. Although most of our nuclear and radiological preparedness is based on transportation accident and nuclear facility accident scenarios, in recent years many initiatives have begun addressing the unique aspects of an intentional nuclear or radiological attack. We are fortunate that our history with industrial and military applications has established a preparedness foundation. However, much work remains in developing the tools needed for effective management of these incidents in the future.

While we have a firm grasp on nuclear and radiological principles, our gaps are in the health and medical issues. First responders can easily respond and characterize the threats. What we do about them is still wrought with questions. New treatment approaches and pharmaceuticals are needed to effectively care for exposed individuals. More training and better mass care approaches are needed in healthcare. Public health needs more effective processes to identify and track those exposed and also needs key environmental health questions answered so they can confidently chart a course for managing the local public health threats that may persist.

Websites

American College of Radiology: www.acr.org/.

Armed Forces Radiobiology Research Institute: www.afrri.usuhs.mil/.

Armed Forces Radiobiology Research Institute, AFRRI Pocket Guide: Emergency Radiation Medicine Response, September 2007: www.afrri.usuhs.mil/www/outreach/pdf/pcktcard.pdf. Other AFRRI Publications: www.afrri.usuhs.mil/outreach/infoprod.htm.

Centers for Disease Control and Prevention, Radiation Emergencies: www.bt.cdc.gov/radiation/.

Department of Energy: www.energy.gov/.

Department of Health and Human Services, Radiation Event Medical Management: www.remm.nlm.gov/.

Department of Homeland Security, Working Group on Radiological Dispersal Device Preparedness, Medical Preparedness and Response Sub-Group: www1.va.gov/emshg/docs/Radiologic_Medical_Countermeasures_051403.pdf.

Environmental Protection Agency, Protective Action Guidelines: www.epa.gov/radiation/rert/pags.html.

Environmental Protection Agency, Radiation Protection: www.epa.gov/radiation/index.html.

Food and Drug Administration, Emergency use authorization of Medical Products: www.fda.gov/oc/guidance/emergencyuse.html.

Food and Drug Administration, Radiological Health Program: www.fda.gov/cdrh/radhealth/.

Health Physics Society: www.hps.org/.

International Atomic Energy Agency: www.iaea.org/.

National Institutes of Health, Fact Sheet, What we know about radiation: www.nih.gov/health/chip/od/radiation/.

Nuclear Regulatory Commission, Emergency Preparedness: www.nrc.gov/about-nrc/emerg-preparedness.html.

Nuclear Regulatory Commission, Radiation Basics: www.nrc.gov/about-nrc/radiation/health-effects/radiation-basics.html.

Oak Ridge Institute, Radiation Emergency Assistance Center/Training Site: www.orise.orau.gov/reacts/.

U.S. Federal Radiological Monitoring and Assessment Center: www.nv.doe.gov/nationalsecurity/homelandsecurity/frmac.htm.

U.S. National Response Team: www.nrt.org/.

References

American National Standards Institute. (2006). Critics call DHS dirty bomb cleanup guidelines too weak. ANSI News and Publications. January 4, 2006. www.ansi.org/news_publications/news_story.aspx?menuid=7&articleid=1116.

Arms Control Association. (2007). Nuclear weapons: who has what at a glance. Strategic Arms Control Policy Fact Sheet. October 2007. www.armscontrol.org/factsheets/Nuclearweaponswhohaswhat.

Centers for Disease Control and Prevention. (2003). Interim guidelines for hospital response to mass casualties from a radiological incident. December 2003. www.bt.cdc.gov/radiation/pdf/MassCasualtiesGuidelines.pdf.

Centers for Disease Control and Prevention. (2005a). Acute radiation syndrome. A fact sheet for physicians. March 18, 2005. www.bt.cdc.gov/radiation/pdf/arsphysicianfactsheet.pdf.

Centers for Disease Control and Prevention. (2005b). Cutaneous radiation injury. A fact sheet for physicians. June 29, 2005. www.bt.cdc.gov/radiation/pdf/criphysicianfactsheet.pdf.

Centers for Disease Control and Prevention. (2005c). Prenatal radiation exposure. A fact sheet for physicians. March 23, 2005. www.bt.cdc.gov/radiation/pdf/prenatalphysician.pdf.

Department of Homeland Security. (2003). Top Officials (TOPOFF) exercise series: TOPOFF 2, after action summary report for public release. December 19, 2003. www.dhs.gov/xlibrary/assets/T2_Report_Final_Public.doc.

Department of Homeland Security. (2006). Protective action guides for radiological dispersal device (RDD) and improvised nuclear device (IND) incidents. *Fed Regist* 71(1):174–196. http://homer.ornl.gov/oepa/rules/71/71fr174.pdf.

Elsen, J., Alexander, V., & Litvinenko. (2007). Times Topics, People. May 31, 2007. http://topics.nytimes.com/top/reference/timestopics/people/l/alexander_v_litvinenko/index.html.

Figley, M. M. (1980). Introduction of SI units. *Am J Roentgenol* 134(1):208. www.ajronline.org/cgi/reprint/134/1/208.pdf.

Hebert, H. J. (2006). Government has dirty bomb cleanup guide. *San Francisco Chronicle*. January 4, 2006. www.sfgate.com/cgi-bin/article.cgi?f=/n/a/2006/01/03/national/w155750S54.DTL&type=health.

International Atomic Energy Agency. (2005). *Chernobyl's 700,000 "Liquidators" struggle with psychological and social consequences*. Staff Report. August 2005. www.iaea.org/NewsCenter/Features/Chernobyl-15/liquidators.shtml.

Jaworowski, Z. (2000–2001). The truth about Chernobyl is told. *21st Century Science and Technology Magazine*. Winter 2000–2001. www.21stcenturysciencetech.com/articles/chernobyl.html.

Library of Congress. (1996). Revelations from the Russian archives. January 1996. www.loc.gov/exhibits/archives/cher.html.

Norris, R. S., & Kristensen, H. M. (2006). NRDC nuclear notebook: global nuclear stockpiles 1945–2006. *Bull At Sci* 62(4):64–67. http://thebulletin.metapress.com/content/c4120650912x74k7/fulltext.pdf.

Norris, R. S., & Kristensen, H. M. (2007a). NRDC nuclear notebook: Russian nuclear forces. *Bull At Sci* 63(2):61–64. http://thebulletin.metapress.com/content/d41x498467712117/fulltext.pdf.

Norris, R. S., & Kristensen, H. M. (2007b). NRDC nuclear notebook: U.S. nuclear forces. *Bull At Sci* 63(1):79–82. http://thebulletin.metapress.com/content/d41x498467712117/fulltext.pdf.

Patterson, W. C. (1986). Chernobyl: worst but not first. *Bull At Sci* 42:43–45.

Purdue University. (2008). Cell phone sensors detect radiation to thwart nuclear terrorism. *Science Daily*. January 24, 2008. www.sciencedaily.com/releases/2008/01/080122154415.htm.

The Chernobyl Forum. (2005). Chernobyl's legacy: health, environmental and socio-economic impacts and recommendations to the governments of Belarus, the Russian Federation and Ukraine, The Chernobyl Forum: 2003–2005. September 2005. http://chernobyl.undp.org/english/docs/chernobyl.pdf.

United Nations. (1988a). *Sources, effects and risks of ionizing radiation: United Nations Scientific Committee on the effects of atomic radiation*. 1988 report to the General Assembly, with annexes, Appendix: acute radiation effects of victims of the Chernobyl nuclear power plant accident. United Nations, United Nations Publication, Sales Publication E.88.IX.7. United Nations, New York. www.unscear.org/docs/reports/1988/1988r_unscear.pdf.

United Nations. *(1988b). Sources, effects and risks of ionizing radiation: United Nations Scientific Committee on the effects of atomic radiation*. 1988 report to the General Assembly, with annexes, Annex D: exposures from the Chernobyl accident. United Nations, United Nations Publication, Sales Publication E.88.IX.7. United Nations, New York. www.unscear.org/docs/reports/1988/1988i_unscear.pdf.

United Nations Office for Disarmament Affairs. (1968). Multilateral arms regulation and disarmament agreements, Treaty on the Non-Proliferation of Nuclear Weapons. http://disarmament.un.org/TreatyStatus.nsf/.

United States Nuclear Regulatory Commission. (2006). Backgrounder on Chernobyl nuclear power plant accident. April 2006. www.nrc.gov/reading-rm/doc-collections/fact-sheets/chernobyl-bg.pdf.

U.S. Department of Energy, Argonne National Laboratory. (2006). Biodegradable nanospheres offer novel approach for treatment of toxin exposure, drug delivery. Argonne National Laboratory News Release. October 20, 2006. www.anl.gov/Media_Center/News/2006/CMT061020.pdf.

U.S. Department of Health and Human Services. (2008). Radiation event medical management. Strategic national stockpile. www.remm.nlm.gov/sns.htm.

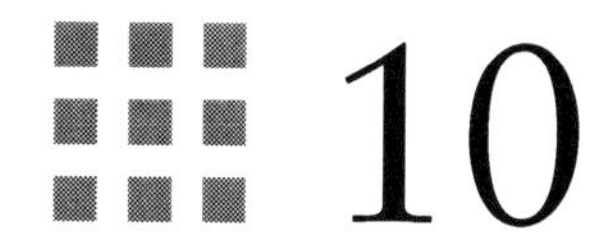

10 Pandemic Influenza

Objectives of This Chapter

- Differentiate between seasonal and pandemic influenza.
- Identify the proteins that changed during genetic mutation of a virus.
- Describe the pharmaceutical options available to prevent influenza or reduce severity of disease.
- List the World Health Organization phases of pandemic response.
- Explain the difference between the World Health Organization Phases and the United States Response Stages.
- Describe how an outbreak peak curve can be compressed and what benefits are derived from it.
- List what individuals can do to reduce the risk of influenza.
- Define nonpharmaceutical interventions and give examples.
- Explain how the Pandemic Severity Index is determined and how it is used.
- Describe the limitations of a government response to a pandemic.

They were a young, adventurous, and hardworking couple with big plans ahead. Simon Wickstrom was 21 years old when he married Selma who was just 16. Simon had emigrated from Sweden to Minnesota and in 1898 he built a farmhouse for his new bride in Minnesota. It was not long before Selma gave birth to their first child, Ella. She was just 17 when their first child arrived. Many more children followed. Seven more arrived in the coming years, including Ralph, Lester, Carrie, Ted, Florence, Elmer, and Alvin (Figure 10–1). It was a hard life filled with long days of difficult work. As their family grew, Selma had her older children, Ella and Ralph, to help with the chores.

November 1918 should have been a time for joyful celebration for families like the Wickstroms. The holidays were approaching and with the signing of the armistice ending World War I, peace was on the horizon. Though the "Great War" was ending, a new battle was emerging. It would eventually kill far more people than any war in history. The influenza pandemic of 1918–1919 became the worst outbreak in modern history infecting about one-third of the global population (Burnet and Clark, 1942; Frost, 1920). The case-fatality rate for most influenza pandemics is generally less than 0.1%. However, the 1918 influenza strain had case-fatality rates estimated to be greater than 2.5% (Marks and Beatty, 1976; Rosenau and Last, 1980). As a result, the global death estimates from this pandemic are about 50 million (Johnson and Mueller, 2002).

The influenza pandemic reached across every continent and caused devastating illnesses even in the most remote regions. The Wickstrom family of rural Minnesota was no exception. The infection swept through the family sickening the parents and all eight children. Those who were less ill cared for the others. The eldest daughter, 19-year-old

FIGURE 10–1 The Simon Wickstrom Family in Synnes Township, Minnesota. Photo courtesy of the U.S. Department of Health and Human Services, Centers for Disease Control and Prevention. Available at: http://www.pandemicflu.gov/storybook/images/courage_deokal.jpg.

Ella, took care of her father and the other children, while her mother Selma and one of her brothers maintained the farm. In late November, 13-year-old Lester died, followed 5 days later by his 40-year-old father, Simon. The children were given little time to say goodbye to their brother and father. Their bodies were placed in wooden coffins and hastily buried by the local undertaker. Selma became a single mother at the age of 36. She and the surviving children, ranging in age from 1–19, were left to carry on in that chilling Minnesota winter (Centers for Disease Control and Prevention, 2006). These kinds of losses played out millions of times around the world in 1918 and 1919 and left countless legacies of suffering and courage, the depth and breadth of which will never be fully known.

Pandemic Definitions

Antibiotic (also antimicrobial): A drug produced by bacteria or fungi that destroys or prevents the growth of other bacteria and fungi.

Antiviral: A drug that is used to prevent or cure a disease caused by a virus by interfering with the ability of the virus to multiply or spread from cell to cell.

Asymptomatic: Presenting no symptoms of disease.

Avian flu (also AI or bird flu): A highly contagious viral disease with up to 100% mortality in domestic fowl caused by influenza A virus subtypes H5 and H7. Low-pathogenic AI causes few problems and is carried by many birds with no resulting problems. Highly pathogenic AI kills birds and if transmitted to humans, can also be fatal. There is little or no human immunity but humans are rarely affected.

Community-based measures to increase social distance include measures applied to whole neighborhoods, towns, or cities (e.g., snow days, establishment of fever clinics, and community-wide quarantine).

Containment measures that apply to use of specific sites or buildings include cancellation of public events (e.g., concerts, sports events, and movies), closure of office buildings, apartment complexes, or schools, and closure of public transit systems. These measures may also involve restricting entrance to buildings or other sites (e.g., requiring fever screening or use of face masks before entry).

Drift: The process in which influenza virus undergoes normal mutations. The amount of change can be subtle or dramatic, but eventually as drift occurs, a new variant strain becomes dominant. This process allows influenza viruses to change and reinfect people repeatedly through their lifetime and is the reason influenza virus strains in vaccine must be updated each year. See also "shift."

Epidemic: A disease occurring suddenly in humans in a community, region, or country in numbers clearly in excess of those that may be typical.

H5N1: A variant of avian influenza, which is a type of influenza virulent in birds. It was first identified in Italy in the early 1900s and is now known to exist worldwide. There are both low and highly pathogenic variants in different regions of the world.

Hemagglutinin: An important surface structure protein of the influenza virus, that is, an essential gene for the spread of the virus throughout the respiratory tract. This enables the virus to attach itself to a cell in the respiratory system and penetrate it. It is referred to as the "H" in influenza viruses. See also "neuraminidase."

HPAI: Highly pathogenic form of avian influenza. Avian flu viruses are classified based upon the severity of the illness and HPAI is extremely infectious among humans. The rapid spread of HPAI, with outbreaks occurring at the same time, is of growing concern for human health as well as for animal health. See also "LPAI."

Influenza: A serious disease caused by viruses that infect the respiratory tract.

Isolation: A state of separation and restriction of movement between persons or groups to prevent the spread of disease. Isolation measures can be undertaken in hospitals or homes, as well as in alternative facilities.

LPAI: Low pathogenic form of avian influenza. Most avian flu strains are classified as LPAI and typically cause little or no clinical signs in infected birds. However, some LPAI virus strains are capable of mutating under field conditions into HPAI viruses. See HPAI.

Mutation: Any alteration in a gene from its natural state. This change may be disease causing or a benign, normal variant.

Neuraminidase: An important surface structure protein of the influenza virus that is an essential enzyme for the spread of the virus throughout the respiratory tract. It enables the virus to escape the host cell and infect new cells. It is referred to as the "N" in influenza viruses. See also "hemagglutinin."

Pandemic: The worldwide outbreak of a disease in humans numbers is clearly in excess of normal.

Pathogenic: Causing disease or capable of doing so.

Prepandemic vaccine: A vaccine created to protect against currently circulating H5N1 avian influenza virus strains with the expectation that it would provide at least some protection against new virus strains that might evolve.

Prophylactic: A pharmaceutical or a procedure that prevents or protects against a disease or condition (e.g., vaccines, antibiotics).

Quarantine: A time period of separation or restriction of movement decreed to control the spread of disease. Before the era of antibiotics, quarantine was one of the few available means of halting the spread of infectious disease. It is still employed today as needed. Individuals may be quarantined at home or in designated facilities.

Seasonal flu: A respiratory illness that can be transmitted from person to person. Most people have some immunity, and a vaccine is available. This is also known as the common flu or winter flu.

Shift: The process in which the existing H (hemagglutinin) and N (neuraminidase) are replaced by significantly different H and Ns. These new H or H/N combinations are perceived by human immune systems as new, so most people do not have preexisting antibody protection to these novel viruses. This is one of the reasons that pandemic viruses can have such a severe impact on the health of populations. See also "drift."

Snow days: Days on which offices, schools, and transportation systems are closed or cancelled, as if there were a major snowstorm. This approach may be recommended to reduce disease transmission.

Virus: Any of various simple submicroscopic parasites of plants, animals, and bacteria that often cause disease and that consist essentially of a core of RNA or DNA surrounded by a protein coat. Unable to replicate without a host cell, viruses are typically not considered living organisms.

Widespread or community-wide quarantine refers to the closing of community borders or the erection of a real or virtual barrier around a geographic area (a cordon sanitaire) with prohibition of travel into or out of the area.

Basic Facts about Pandemic Influenza

Influenza, referred to as the flu, is a highly infectious respiratory disease caused by the influenza virus. It is a segmented, single-stranded, RNA virus that is part of the Orthomyxoviridae family (Figure 10–2). The name influenza comes from the Italian form of the Latin word “influentia.” This was because influenza illnesses were believed to result from occult “influences.” The Orthomyxoviridae family name is derived from the Greek words “orthos” or straight and “myxo” or mucus (International Committee on Taxonomy of Viruses, 2006). The name of a specific influenza strain is usually based on where it is initially discovered. This includes references to the geographic location (Hong Kong flu) or animal reservoir (Bird flu).

Influenza pandemics occur when a dramatic reassortment of the genetic structure of the virus makes it unrecognizable to the human immune system. Throughout the twentieth century there were three pandemics, including 1918, 1957, and 1968 (Cox and Subbarao, 2000). The severity of seasonal or pandemic influenza is caused by the genetic structure of the virus. There are three types of influenza viruses including A, B, and C. Influenza A is the most dangerous for humans and influenza C is the least dangerous. The specific structure of the influenza virus is expressed by the proteins it carries. Hemagglutinin (HA) proteins of the virus are the components that enable it to penetrate a foreign body. There are 16 different proteins possible. Neuraminidase (NA) is another key protein of the virus. It has nine varieties. These two proteins are the defining factors for each strain of influenza and are the basis for the “H_N_” designation. For example, there have been recent concerns over H5N1 as a potential candidate for a future influenza pandemic. However, there is one more important distinction between the various strains of viruses. Although a virus may have the same genetic assortment, for example Influenza A, H5N1, there are both low-pathogenic and high-pathogenic strains. For several years, a low-pathogenic avian influenza (LPAI) H5N1 has been in North America, while there has been a highly pathogenic avian influenza (HPAI) H5N1 in Asia. Even though they are very similar, the “low path” North American strain is not a human health concern, while the “high path” Asian strain is raising significant international concerns that it could become the source of the next large human influenza pandemic.

The changes in the virus are caused by two types of mutations. Antigenic drift is a constant process of genetic change that naturally occurs as a virus replicates itself. Small imperfections happen as copies are made and they cause a genetic drift, or slow process of change to occur. Occasionally, an antigenic shift occurs. This is a dramatic reassortment of the genetic structure of the virus when it is combined with other influenza strains and exchanges genetic material (Webster and Govorkova, 2006). This usually occurs in animals such as swine that carry both human and animal strains of influenza virus and provide an ideal reservoir for various strains to mix and shift their genetic structure.

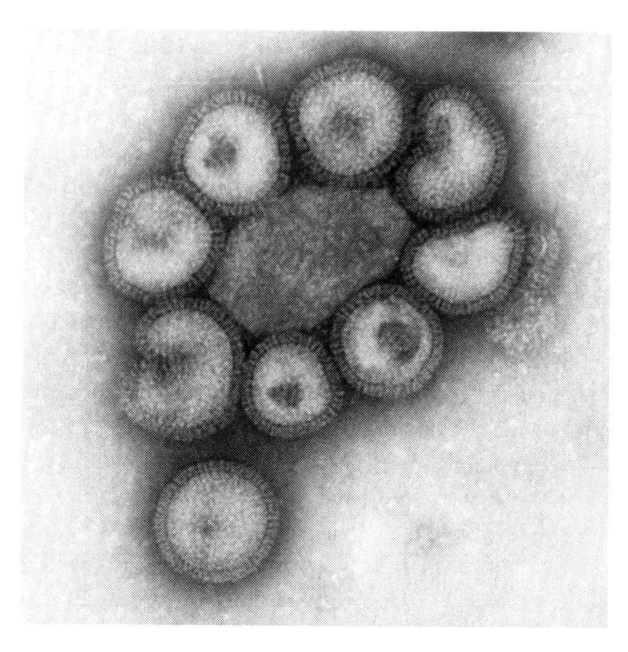

FIGURE 10–2 This negative-stained transmission electron micrograph (TEM) depicts the ultrastructural details of a number of influenza virus particles, or "virions." Photo by Cynthia Goldsmith, Courtesy of CDC/Public Health Image Library PHIL ID# 10072.

Table 10–1 The Differences Between Seasonal and Pandemic Influenza

Seasonal Influenza	Pandemic Influenza
Annual occurrence	Rare occurrence (several times per century)
Some natural human immunity	Little or no human immunity
Weak immune systems at increased risk	Those at increased risk have a strong immune system
Vaccine available before flu season	Vaccine not available for several months
Adequate antivirals for those at increased risk	Inadequate supply of antivirals for those at risk
Healthcare system is adequate to provide care	Healthcare system overwhelmed
Mild, nonlife-threatening illness in most people	Severe, life-threatening illness in many more
Deaths in the U.S. in the tens of thousands	Deaths in the U.S. in the hundreds of thousands
Does not normally close schools	School closures likely
Does not cause travel restrictions among the healthy	Travel restrictions likely
Mild impact on the economy and business continuity	Severe impact on the economy and business continuity

Adapted from the Department of Health and Human Services, pandemicflu.gov site: http://www.pandemicflu.gov/general/season_or_pandemic.html.

Health Threat

In a single year, between 5% and 20% of the entire U.S. population is infected with influenza. The majority of cases are mild but many are severe. About 200,000 of those infected are hospitalized from influenza and related complications every year and about 36,000 died (Centers for Disease Control and Prevention, Influenza (flu): key facts about seasonal influenza, www.cdc.gov/flu/pdf/keyfacts.pdf). Influenza symptoms typically include fever, muscle pain, runny nose, and cough. Those who die from seasonal influenza usually have weaker immune systems and other complications like pneumonia. The primary differences during an influenza pandemic are the disease severity and the populations affected (See Table 10–1). Not only will the case-fatality ratio be excessive during a pandemic, but often those most severely affected are completely different demographically. Rather than populations with weaker immune systems having more severe illness, those with healthier immune systems may be more susceptible. When the human immune system is fighting an infection, cytokines are responsible for activating the immune functions and also stimulating production of more cytokines. It is a looped system that keeps the immune system balanced to fight illness effectively while minimizing damage to healthy cells. Some highly pathogenic strains of influenza produce a "cytokine storm" where cytokines and immune cells become hyperstimulated and damage the healthy cells of an infected respiratory tract.

During the 1918 pandemic, many between the ages of 18 and 40 with healthier immune systems had more severe illness due to cytokine stimulation. As the immune system began damaging the healthy cells of the respiratory tract, it led to acute respiratory distress syndrome (ARDS) (Kobasa et al., 2004; Osterholm, 2005). This condition killed many within several days. Others succumbed over a longer period of time due to bacterial infections of the lungs. These infections were ushered in by viral damage to the epithelial cells of the respiratory tract. Although the epithelial cells usually sweep the respiratory tract clean, as the influenza viruses damage them, bacteria from the mouth can make their way down the respiratory tract and cause secondary illnesses.

Because influenza is caused by a virus, there are only two pharmaceutical options. A vaccine is needed to prevent or to reduce the severity of influenza (Figure 10–3). Antiviral drugs may also reduce the severity and length of the illness. Antiviral drugs are the current focus of pandemic preparedness efforts because vaccine production requires access to the specific pathogenic organism. Influenza viruses quickly mutate making them evasive targets for vaccine development. Despite these historical limitations, a new prepandemic H5N1 vaccine has been approved and is in production. Many stockpiles in nations around the world are starting to include this new vaccine with the hope that it will offer sufficient protection against an H5N1 pandemic. It is not known if it will offer enough protection or if a future H5N1 mutation could make that vaccine ineffective. If the current influenza vaccine production capacity is used to quickly make a new vaccine from a newly discovered influenza strain, the global influenza vaccine production capacity is about 500 million doses per year. The first shipments of vaccine to leave those facilities would take several months to develop. The vaccine is cultured in eggs using a time-consuming process (Morse et al., 2006). This production schedule cannot be changed using the current technologies and manufacturing approaches. New production methods using cell cultures are starting to emerge and will likely speed this production process in the future.

Antiviral drugs are the most effective pharmaceutical options currently available for managing an avian influenza pandemic (Schunemann et al., 2007). Although there are multiple drugs approved by the U.S. Food and Drug Administration for treatment and prevention of influenza, in recent years there are strains of influenza emerging that show resistance to many of the older antiviral drugs (Centers for Disease Control and Prevention, 2006). As a result, there are currently two newer antiviral drugs serving as the primary choices for antiviral stockpiling efforts. Zanamivir (Relenza) and oseltamivir phosphate (Tamiflu) were both approved in 1999 for the prevention and treatment of acute uncomplicated illness because of influenza A and B (Food and Drug Administration, Center for Drug Evaluation and Research, 2008).

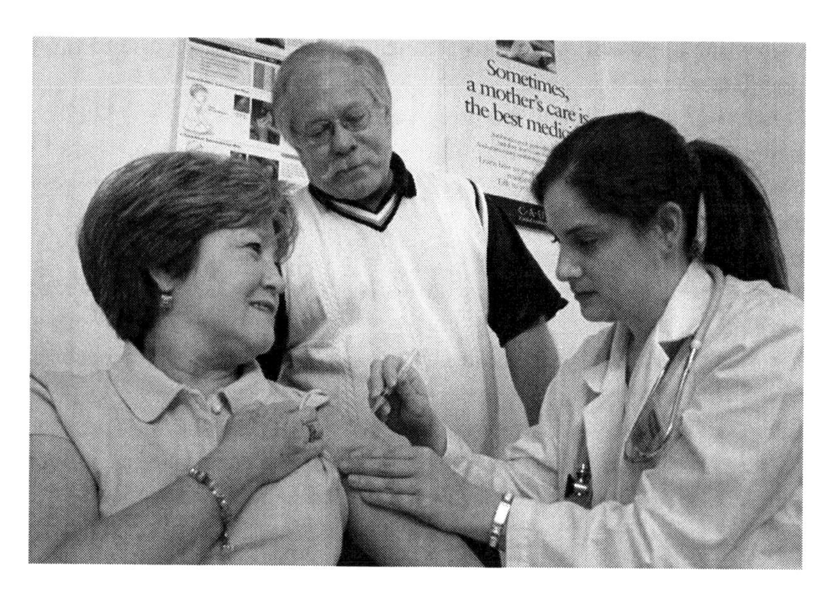

FIGURE 10–3 Influenza vaccination is the first line of defense. Photo courtesy of CDC/Public Health Image Library, James Gathany, PHIL ID# 9306.

There is also an important role for antimicrobial drugs in pandemic influenza preparedness and response. Secondary bacterial infections are a common cause of death among those who die from influenza-related complications. In fact, about one-quarter of those who die from seasonal influenza are found with coinfections from bacterial infections as major factors in mortality (Bhat et al., 2005; Lim et al., 2001). It is important to determine the organism's susceptibility to the antimicrobial drug selected. It is also crucial to effectively triage those needing antimicrobial therapy so individuals with weak immune function receive priority treatment (Gupta et al., 2008).

Prevention

There are several measures that may be instituted to reduce the risk of influenza infection. These include both pharmaceutical and nonpharmaceutical measures. The primary pharmaceutical measure for influenza is vaccination. As stated earlier, the problem with pandemic influenza is that a rapid, drastic shift has occurred in the virus that does not allow sufficient time to incorporate the new strain into a vaccine before the pandemic is already spreading. There will likely be a delay of several months using the current production processes. The other pharmaceutical options are antiviral drugs. These drugs may be the only pharmaceutical line of defense in the initial weeks and months of an influenza pandemic before a vaccine can be produced.

Top Ten Individual Influenza Prevention Tips

1. Wash your hands often. It decreases your risk for all sorts of illnesses including influenza.
2. Clean frequently touched surfaces often. This includes door handles, hand rails, etc.

3. Keep your hands off your face. Touching your eyes, nose, and mouth may inoculate you with a virus.
4. Cover your mouth when you cough or sneeze.
5. Stay away from those who are ill.
6. Stay away from crowded congregate settings when there is illness in the community.
7. Stay home when you are ill.
8. Avoid unnecessary travel when illness is prevalent. Take extra precautions if traveling.
9. Get a pneumococcal vaccination. It may prevent a secondary bacterial infection if you get the flu.
10. Get a flu shot every year. Even if the current vaccine does not protect specifically against avian influenza, it offers some protection against influenza and may reduce the overall number of flu cases during a pandemic.

Adapted from the New York City Department of Health and Mental Hygiene, Pandemic Influenza Preparedness and Response Plan, pp. 202–203. Available at: http://www.nyc.gov/html/doh/downloads/pdf/cd/cd-panflu-plan-06.pdf.

Antiviral medications may be used early in a pandemic for close contacts of initial cases and to contain disease clusters. Close contacts may include family members, friends, coworkers, healthcare providers, and fellow travelers on aircraft, buses, etc. The use of antiviral drugs may contain some disease clusters. They may also prevent cases in high-risk facilities such as long-term care or in less populated, well-defined regions like military installations or small towns. The use of antiviral drugs for prevention must be weighed against their use for treatment. Early in the arrival of a pandemic to a new region, it may be best to use antiviral medications to prevent infections. However, once influenza cases begin to spread across a community, priority antiviral allocation should be given to those with severe illness to improve their outcomes.

Nonpharmaceutical prevention measures may offer the best option for some regions and the only option for many more. With billions of people at risk during a pandemic, there will not be adequate pharmaceutical measures to go around in the foreseeable future (World Health Organization, 2005). There are three primary nonpharmaceutical interventions that offer the best prevention. These include isolation, quarantine, and social distancing.

During a pandemic, hospitals will be full and many who are able to care for themselves or who have an adequate support system to provide care may isolate themselves from others to reduce the risk of additional transmission. There are also those who are not yet ill but may have been exposed to the virus. They may choose self-quarantine. Others with no illness or known exposure may choose social distancing. The concept is the same for all three of these responses. In the context of prevention, the primary distinction between isolation, quarantine, and social distancing is the health status of the person taking action to keep away from others. In addition, basic community infection control measures should be taken. This includes promoting good hand hygiene and strict respiratory precautions for those that may be ill.

Lessons from The Three Pandemics of the Last Century

1. No two pandemics are alike. There are differences in mortality, severity, and patterns of spread.
2. The need for rapid hospital surge capacity is a common factor in all pandemics.
3. Pandemic viral strains are more lethal and often cause severe disease in unlikely age groups, such as young adults.
4. Pandemics occur in waves. Age groups and geographical areas not affected initially are more likely to be vulnerable in subsequent waves. Subsequent waves are often more severe.
5. Virological surveillance is often the key to alerting public health agencies to a coming pandemic by isolating and characterizing the virus, and making it available to vaccine manufacturers.
6. Over the centuries, most pandemics have originated in parts of Asia where dense populations of humans live in close proximity to ducks and pigs.
7. Historically, public health interventions may delay international spread of pandemics, but cannot stop them. Quarantine and travel restrictions have shown little effect.
8. Delaying spread can flatten the epidemiological peak, thus distributing cases over a longer period of time. Having fewer people ill at a given time increases the likelihood that medical and other essential services can be maintained and improves surge capacity.
9. The impact of vaccines on a pandemic remains to be demonstrated. In 1957 and 1968, vaccine manufacturers responded with too little, too late to have an impact.
10. Countries with domestic manufacturing capacity will be the first to receive vaccines.
11. The tendency of pandemics to be more severe in later waves may extend the time before large supplies of vaccine are needed to prevent severe disease in high-risk populations.
12. Regions with good annual influenza vaccination programs will have experience in the logistics of vaccine administration to at least some groups requiring priority protection during a pandemic.

Adapted from Avian influenza: Assesing the pandemic threat. World Health Organization. pp. 31–33. Available at: http://www.who.int/csr/disease/influenza/H5N1-9reduit.pdf.

In the Fall of 2005, the White House released a national strategy for pandemic preparedness and response. It includes three steps. These steps include delaying or stopping a pandemic before it reaches the United States, reducing the spread of illness and mitigating suffering and death, and sustaining society, critical infrastructure, and the economy. The strategy also defines three pillars on which these activities will be sustained. They include preparedness and communication, surveillance and detection, and response and containment (Homeland Security Council, 2005). With this strategic vision, pandemic preparedness funding and associated requirements began arriving in 2006 for state and local governments to initiate their plans and begin training and exercising for a pandemic. Since then, great strides have been made in pandemic preparedness initiatives. The media focus in 2006 on the growing pandemic threat resulted in widespread participation of many public and private organizations in preparedness initiatives that they would normally not engage in or commit resources toward.

Immediate Actions

The World Health Organization (WHO) has established six alert phases for pandemic influenza. This is designed to give the world a common reference for communicating the current pandemic risk level. With each increase in pandemic phases, a variety of response measures are instituted. The WHO phases address the need for triggers. As events unfold leading up to a pandemic, a common frame of reference to understand when to take action is essential. There is no way to predict the timeframe for changes between these phases. It may take days or years to move up or down from one phase to another. That is why coordinated pandemic preparedness planning is so essential. The interpandemic level includes Phase 1 with the lowest human risk and Phase 2 where a new virus is discovered in animals that is not yet transmitted to humans but posing an increased risk. The Pandemic Alert level indicates that humans are beginning to become infected with a new virus. The degree to which infections are spreading among people is broken into the next three phases. Phase 3 indicates transmission from animals to human with little or no transmission from person to person. For several years, this has been the alert level for H5N1. Phase 4 indicates that person-to-person transmission is increasing but not yet reaching a point of being significant. Phase 5 is the point where the person-to-person transmission becomes common but is still localized. The final Pandemic level is Phase 6. It signifies an efficient and sustained person-to-person transmission of the illness.

The U.S. Government response actions are comprised of six stages. However, these stages do not run concurrently with the WHO phases (See Table 10–2). The phases for U.S. Government actions are based on the specific threat to the U.S. population. As there is no immediate public health threat until person-to-person transmission has been established, that is where the U.S. Government response begins.

The goals of the first stage of the U.S. Government pandemic response are to quickly confirm any overseas outbreaks with pandemic potential and support the response to control them. This includes support such as rapid response teams to help investigate them, enhancing local laboratory capacity, and supporting clinical activities. There may also be traveler screening or restrictions instituted for the affected region in an effort to contain any outbreak before it spreads. At the same time support is being provided to the affected region, measures are being taken across the United States to prepare for the possible arrival of a pandemic. This includes prepositioning antiviral stockpiles and the possible use of any available prepandemic vaccine for those at high risk.

Once a human outbreak is confirmed, the WHO phase is elevated to Phase 4 if it is a small, isolated cluster of human cases or to Phase 5 if it is a larger cluster but still limited to a defined region. Either of those phases will result in a Stage 2 U.S. response.

Table 10–2 Correlation Between the World Health Organization Pandemic Phases and the United States Federal Response Stages

WHO Phases		Federal Government Response Stages	
Interpandemic period			
1	No new influenza virus subtypes have been detected in humans. An influenza virus subtype that has caused human infection may be present in animals. If present in animals, the risk of human disease is considered to be low.	0	New domestic animal outbreak in at-risk country
2	No new influenza virus subtypes have been detected in humans. However, a circulating animal influenza virus subtype poses a substantial risk of human disease.		
Pandemic alert period			
3	Human infection(s) with a new subtype, but no human-to-human spread, or at most rare instances of spread to a close contact.	0	New domestic animal outbreak in at-risk country
		1	Suspected human outbreak overseas
4	Small cluster(s) with limited human-to-human transmission but spread is highly localized, suggesting that the virus is not well adapted to humans.	2	Confirmed human outbreak overseas
5	Larger cluster(s) but human-to-human spread still localized, suggesting that the virus is becoming increasingly better adapted to humans, but may not yet be fully transmissible (substantial pandemic risk).		
Pandemic period			
6	Pandemic phase increased and sustained transmission in general population.	3	Widespread human outbreaks in multiple locations overseas
		4	First human case in North America
		5	Spread throughout United States
		6	Recovery and preparation for subsequent waves

Source: U.S. Department of Health and Human Services, Community Strategy for Pandemic Influenza Mitigation.[22]

The actions to contain the outbreak outside the nation's borders will be increased as the National Response Framework ramps up in support of international control measures. At the same time, screening measures will be increased according to the threat. Antiviral and vaccination prioritization will be revised as necessary according to the epidemiology observed.

The final WHO declaration, Phase 6, is an indication that the pandemic has arrived. There are four U.S. response stages during this final WHO phase that are all based on the status of the pandemic on the North American continent. Stage 3 is when there are widespread outbreaks in other nations but not in North America. During this stage, everything that can be done to prevent or delay the arrival of the pandemic will be done. At the same time, meticulous surveillance activities will be in progress to identify the pandemic arrival. All available antiviral medications, vaccines, antibiotics for secondary infections, etc. will be prepared while healthcare facilities arrange for additional patient surge. Stage 4 signals the arrival of the first case in North America. All available resources will focus on rapid identification and containment of the initial case or cases. State and local pandemic plans will be activated and domestic travel will be limited to essential trips only. Stage 5 signals the spread of the pandemic across the United States. It is likely to come in waves hitting some areas severely while missing others completely. There is no way to predict how it will unfold. The only thing that is certain is the lack of government capacity to respond to a widespread pandemic. There are not sufficient resources available for the federal government to be expected to manage a pandemic. In fact, a pandemic will render all levels of government incapable of meeting the needs within their jurisdiction. Individual and family preparedness, a strong public health infrastructure, and clear communication of nonpharmaceutical interventions and pharmaceutical options will be the keys to managing a pandemic in the United States.

> *"Any community that fails to prepare with the expectation that the federal government will, at the last moment, be able to come to the rescue will be tragically wrong. Not because the federal government lacks will, not because we lack wallet, but because there is no way in which 5000 different communities can be responded to simultaneously, which is a unique characteristic of a human pandemic."*
>
> Mike Leavitt, Health and Human Services Secretary (Manning, 2006)

Once a pandemic begins to sweep across a community, it will likely overwhelm local acute care facilities and cause a ripple effect of patient surge that will strain services across

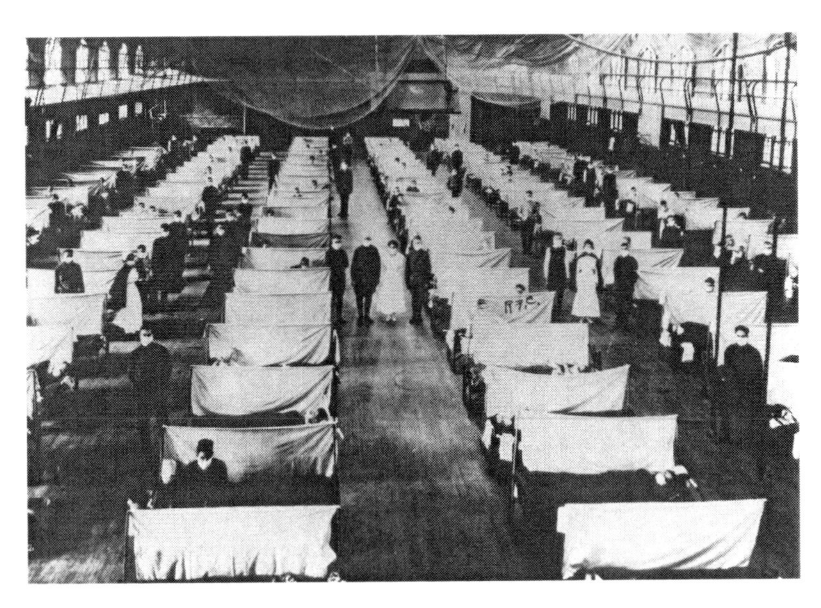

FIGURE 10–4 Mass care during the 1918 flu pandemic.

Source: Department of Health and Human Services. Available at: http://1918.pandemicflu.gov/pics/photos/Iowa_Flu2.jpg.

a region, a state, and eventually the nation. To save as many lives as possible, the healthcare system must remain functional and delivering the best care possible under whatever conditions arise (Toner et al., 2006). The strategy to accomplish this is to institute community measures that will delay and compress the outbreak peak (See Figure 10–5). As interventions are instituted, some cases of illness will be delayed and others prevented. As a result, the cases in a given region may be spread over a longer period of time. Even though some who delay their illness eventually become ill, as the cases are spread over a more manageable timeframe, it keeps the peak number of cases more manageable for providers and makes it easier to sustain essential services.

Although there are a variety of community measures that can be taken during a pandemic to reduce the risk of exposure and subsequent cases of illness, a balance is needed when instituting interventions (See Figure 10–7). No intervention is without some economic or social cost. If individuals are isolated or quarantined, there must be a support system in place to provide everything they need. This not only includes essentials such as groceries and prescriptions, but also includes services like child care for ill parents of small children and other services. There must be economic protection to allow those who are ill to work from home or at least to have some assurance that taking sick days will not result in job loss. A variety of consequences must also be considered when closing schools and day care centers or when discouraging public gatherings. There are substantial economic outcomes to these measures. If children are not in school or day care, some parents will be unable to work. If mass transit and congregate gatherings are discouraged, a variety of industries will suffer. These interventions may be important to public health but must be used judiciously. If they are used too soon or taken to an inappropriate extreme, the political and social consequences may be too burdensome and the public may not follow them when they are actually needed. On the other hand, if interventions are established too late, they may be ineffective.

To establish appropriate triggers for these mitigation strategies, the case-fatality ratio is used. As a pandemic unfolds, it will be quickly evident how many cases are observed and

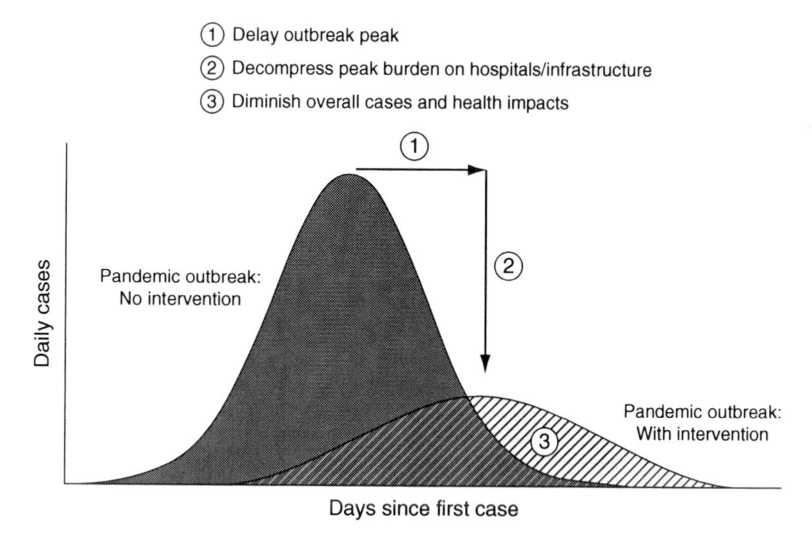

FIGURE 10–5 The goal of community mitigation measures during a pandemic is to compress the peak number of cases and keep the number of ill at any given time as manageable as possible.

Source: U.S. Department of Health and Human Services, Community Strategy for Pandemic Influenza Mitigation (U.S. Department of Health and Human Services, 2007).

how many of those cases have a fatal outcome. If it is less than 0.1%, few interventions are recommended. If the case-fatality ratio exceeds 1%, all of the interventions that can be instituted are recommended. This is based on a Pandemic Severity Index determined by the case-fatality ratio. Categories 1 through 5 correlate with specific interventions that are recommended based on the severity of the pandemic (Figure 10–6).

Nonpharmaceutical Interventions

1. Isolation of the sick at home or in healthcare settings. Treatment with antiviral medication as appropriate.
2. Voluntary quarantine of households with confirmed or probable influenza. Prophylactic treatment as appropriate.
3. Closure of daycares, K–12 schools, and colleges. This should be followed by reducing child contacts through social distancing.
4. Social distancing of adults including cancellation of large gatherings and altered work schedules.

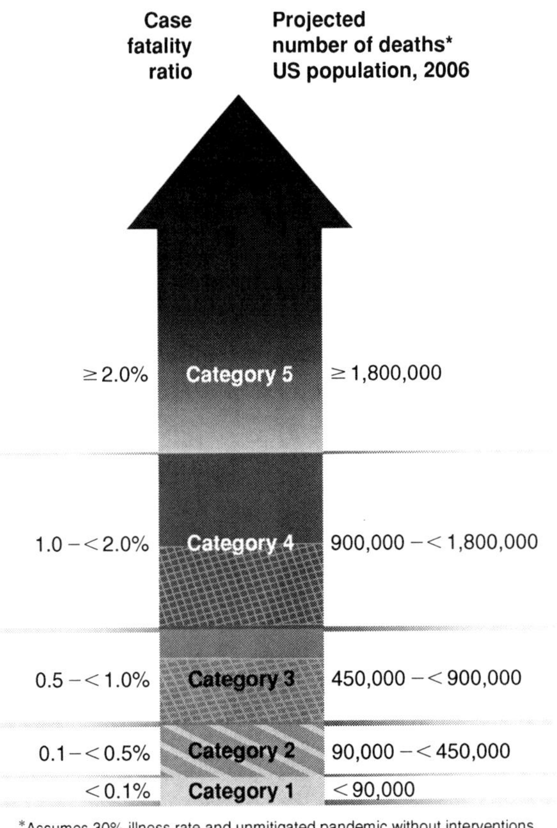

FIGURE 10–6 Pandemic Severity Index.

Source: U.S. Department of Health and Human Services, Community Strategy for Pandemic Influenza Mitigation.[22]

	Pandemic severity index		
Interventions* by Setting	1	2 and 3	4 and 5
Home **Voluntary isolation** of ill at home (adults and children); combine with use of antiviral treatment as available and indicated	**Recommend†§**	**Recommend†§**	**Recommend†§**
Voluntary quarantine of household members in homes with ill persons¶ (adults and children); consider combining with antiviral prophylaxis if effective, feasible, and quantities sufficient	**Generally not recommended**	**Consider****	**Recommend****
School **Child social distancing**			
– dismissal of students from schools and school based activities, and closure of child care programs	**Generally not recommended**	**Consider: ≤4 weeks††**	**Recommend: ≤12 weeks§§**
– reduce out-of-school social contacts and community mixing	**Generally not recommended**	**Consider: ≤4 weeks††**	**Recommend: ≤12 weeks§§**
Workplace/Community **Adult social distancing**			
– decrease number of social contacts (e.g., encourage teleconferences, alternatives to face-to-face meetings)	**Generally not recommended**	**Consider**	**Recommend**
– increase distance between persons (e.g., reduce density in public transit, workplace)	**Generally not recommended**	**Consider**	**Recommend**
– modify postpone, or cancel selected public gatherings to promote social distance (e.g., postpone indoor stadium events, theatre performances)	**Generally not recommended**	**Consider**	**Recommend**
– modify work place schedules and practices (e.g., telework, staggered shifts)	**Generally not recommended**	**Consider**	**Recommend**

FIGURE 10–7 Interventions recommended based upon the Pandemic Severity Index.

Source: U.S. Department of Health and Human Services, Community Strategy for Pandemic Influenza Mitigation.[22]

The challenges of managing a pandemic are immense, and the limitations of the government response at all levels are known. The nongovernmental resources from businesses, faith-based organizations, and others will certainly play a major role in a community pandemic response. Even communities with relatively broad preparedness participation from these organizations have rarely identified and characterized the resources and support that these organizations may offer during a pandemic crisis. It is also not known what policies may be instituted by these organizations that will support or counter nonpharmaceutical interventions. It is impossible to effectively strategize with the substantial gaps that exist in our current planning assumptions. This is why the involvement of these organizations in preparedness activities will play an important role in an effective pandemic response.

Recovery Challenges

It seems that pandemic preparedness and response is so ominous and complex that when a WHO Phase 6 is reached, the planning stops. It is important to consider the post influenza pandemic scenarios and extend the process well beyond where many plans currently stop. For example, if there are significant fatalities within a certain demographic, special challenges will emerge. If there are substantial numbers of elderly losses, the secondary impact may be the loss of primary caregivers for children in some urban communities. If the losses are more among the working population, there are essential businesses and services that may be short-handed and unable to perform indispensable functions. If healthcare workers suffer excessive losses, the availability or standard of care may be compromised. Even when the pandemic has finished, critical infrastructure may take a long time to return to business as usual. Although there are too many unknowns to predict what special challenges may result from a pandemic, it is important to push forward in the planning, training, and exercise processes to include a recovery component that discusses the possibilities. This should include consideration for a variety of scenarios as well as the mental health impact and other long-term issues that will likely result.

Summary

Pandemic influenza has occurred at fairly regular intervals throughout history and will be repeated. There are significant differences between each of those historical outbreaks and there will certainly be unique, unanticipated aspects to any future pandemic. Many gaps remain in our preparedness activities and things have changed in our society that may work against us in a pandemic. The 2006 media attention on the growing threat of avian influenza prompted interest and involvement by many organizations that are not usually engaged in preparedness. Unfortunately, as time passes, there is clearly a diminishing sense of urgency and much of the previous momentum is being lost. As a society, we are also seeing an erosion of social capital. Neighbors do not communicate with each other like they have in the past and most people are not as involved in their communities as their parents or grandparents. This neighborhood erosion will weaken the ability of communities to identify those at increased risk and respond effectively. We have also lost some of the important, age old skills that our ancestors had. In 1918, infectious disease was still a leading cause of mortality and people knew how to care for others. Our modern society has provided better access to healthcare for most people, but as a result many parents do not know how to care for a feverish child. They simply call the pediatrician and follow the instructions provided. If a pandemic is severe and that quick access to answers and care is not available, there are many who simply lack the knowledge and skill to provide basic care for others.

We also have several things working in our favor to reduce the impact of a pandemic. Tremendous strides have been made in pharmaceuticals, critical care, and public health interventions. The World Health Organization and public health organizations around the world are collaborating closely on monitoring the progress of H5N1 in both the human and animal populations. Unprecedented international coordination will occur if and when the pandemic phase is elevated. In the United States, the recent investment in public health preparedness following the terrorist attacks of 2001 and more recent investments specifically in pandemic preparedness have filled several gaps in a public health infrastructure that had been eroding for decades. There is some good news in pandemic preparedness.

Websites

Centers for Disease Control and Prevention, Avian Influenza (Bird Flu): www.cdc.gov/flu/avian/.

Centers for Disease Control and Prevention, Prevention and Control of Influenza: Recommendations of the Advisory Committee on Immunization Practices (ACIP), 2008: www.cdc.gov/mmwr/preview/mmwrhtml/rr5707a1.htm.

Centers for Disease Control and Prevention, Resources for Pandemic Flu: www.cdc.gov/flu/Pandemic/.

Center for Infectious Disease Research and Policy, Pandemic Influenza: www.cidrap.umn.edu/cidrap/content/influenza/panflu/.

European Influenza Surveillance: www.eiss.org/.

Food and Agriculture Organization of the United Nations, Avian Influenza: www.fao.org/avianflu/en/index.html.

New England Journal of Medicine, Avian Influenza A (H5N1) Infection in Humans (2005): content.nejm.org/cgi/content/full/353/13/1374.

United Kingdom, Department for Environment, Food, and Rural Affairs, Avian Influenza: www.defra.gov.uk/animalh/diseases/notifiable/ai/index.htm.

U.S. Department of Agriculture, Avian Influenza (Bird Flu): www.usda.gov/wps/portal/usdahome? navtype=SU&navid=AVIAN_INFLUENZA.

U.S. Department of Health and Human Services, The Great Pandemic: The United States in 1918–1919: http://1918.pandemicflu.gov/index.htm.

U.S. Department of Health and Human Services, One-stop access to U.S. Government avian and pandemic flu information: www.pandemicflu.gov/.

U.S. Department of Homeland Security, Pandemic Influenza Preparedness, Response, and Recovery: Guide for Critical Infrastructure and Key Resources: www.pandemicflu.gov/plan/pdf/cikrpandemicinfluenzaguide.pdf.

U.S. Department of the Interior, Avian Influenza: www.doi.gov/issues/avianflu.html.

U.S. Food and Drug Administration, Flu Information: www.fda.gov/oc/opacom/hottopics/flu.html.

U.S. Geological Survey, Avian Influenza: www.nwhc.usgs.gov/disease_information/avian_influenza/index.jsp.

White House, National Strategy for Pandemic Influenza Implementation Plan, One Year Summary: www.whitehouse.gov/homeland/pandemic-influenza-oneyear.html.

Wildlife Conservation Society, Global Avian Influenza Network for Surveillance: www.gains.org/.

World Health Organization, Epidemic and Pandemic Alert and Response, Avian Influenza: www.who.int/csr/disease/avian_influenza/en/.

World Health Organization, Epidemic and Pandemic Alert and Response, Current WHO Phase of Pandemic Alert: www.who.int/csr/disease/avian_influenza/phase/en/index.html.

World Health Organization, Epidemic and Pandemic Alert and Response, Global Influenza Surveillance: www.who.int/csr/disease/influenza/influenzanetwork/en/.

World Health Organization, Avian Influenza: www.who.int/topics/avian_influenza/en/.

World Health Organization, Influenza: www.who.int/topics/influenza/en/.

References

Bhat, N., Wright, J. G., Broder, K. R., et al. (2005). Influenza-associated deaths among children in the United States, 2003–2004. *N Engl J Med* 353:2559–2567.

Burnet, F., & Clark, E. (1942). *Influenza: A survey of the last 50 years in the light of modern work on the virus of epidemic influenza.* Melbourne, Australia: MacMillan.

Centers for Disease Control and Prevention (2006). Health alert: CDC recommends against the use of amantadine and rimantadine for the treatment or prophylaxis of influenza in the United States during the 2005–06 influenza season. www.cdc.gov/flu/han011406.htm.

Centers for Disease Control and Prevention (2006). *Pandemic influenza storybook: personal recollections from survivors, families, and friends*. Angela J. Deokar. CDC, Department of Health and Human Services. www.pandemicflu.gov/storybook/stories/courage/otto/index.html.

Cox, N. J., & Subbarao, K. (2000). Global epidemiology of influenza: past and present. *Ann Rev Med* 51: 407–421.

Food and Drug Administration, Center for Drug Evaluation and Research (2008). Influenza (flu) antiviral drugs and related information. www.fda.gov/cder/drug/antivirals/influenza/default.htm.

Frost, W. H. (1920). Statistics of influenza morbidity. *Public Health Rep* 35:584–597.

Gupta, R. K., George, R., & Nguyen-Van-Tam, J.S. (2008). Bacterial pneumonia and pandemic influenza planning. *Emerg Infect Dis* 14:1187–1192. www.cdc.gov/EID/content/14/8/1187.htm.

Homeland Security Council (2005). *National strategy for pandemic influenza*. Washington, DC: The Whitehouse. www.whitehouse.gov/homeland/pandemic-influenza.html.

International Committee on Taxonomy of Viruses (2006). Index of viruses—*Orthomyxoviridae*. In: Büchen-Osmond. C., ed. *ICTVdB—the universal virus database, version 4*. New York: Columbia University. www.ncbi.nlm.nih.gov/ICTVdb/Ictv/fs_index.htm.

Johnson, N. P., & Mueller, J. (2002). Updating the accounts: global mortality of the 1918–1920 "Spanish" influenza pandemic. *Bull Hist Med* 76:105–115.

Kobasa, D., Takada, A., Shinya, K., et al. (2004). Enhanced virulence of influenza A viruses with haemagglutinin of the 1918 pandemic virus. *Nature* 431:703–707.

Lim, W.S., Macfarlane, J. T., Boswell, T. C., et al. (2001). Study of community acquired pneumonia aetiology (SCAPA) in adults admitted to hospital: implications for management guidelines. *Thorax* 56:296–301.

Manning, A. (2006). U.S. expects bird flu this year. *USA Today*. March 20, 2006. www.usatoday.com/news/health/2006-03-20-bird-flu_x.htm.

Marks, G., & Beatty, W. K. (1976). *Epidemics*. New York: Scribners.

Morse, S. S, Garwin, R. L., & Olsiewski, P.J. (2006). Next flu pandemic: what to do until the vaccine arrives? *Science* 314:929.

Osterholm, M. T. (2005). Preparing for the next pandemic. *N Engl J Med* 352(18):1839–1842. http://content.nejm.org/cgi/content/full/352/18/1839.

Rosenau, M. J., Last, J. M. (1980). *Maxcy-Rosenau preventative medicine and public health*. New York: Appleton-Century-Crofts.

Schünemann, H. J., Hill, S. R., Kakad, M., et al. (2007). WHO rapid advice guidelines for pharmacological management of sporadic human infection with avian influenza (H5N1) virus. *Lancet Infect Dis* 7:21–31.

Toner, E., Waldhorn, R., Maldin, B., et al. (2006). Hospital preparedness for pandemic influenza. *Biosecur Bioterror* 4(2):207–217.

U.S. Department of Health and Human Services (2007). Community strategy for pandemic influenza mitigation. www.pandemicflu.gov/plan/community/community_mitigation.pdf.

Webster, R. G., & Govorkova, E. A. (2006). H5N1 influenza—continuing evolution and spread. *N Eng J Med* 21:2174–2177.

World Health Organization (2005). Avian influenza: assessing the pandemic threat. www.who.int/csr/disease/influenza/WHO_CDS_2005_29/en/index.html.

11 Thunderstorms and Tornadoes

Objectives of This Chapter

- Describe how a thunderstorm is formed.
- Explain the significance of thunderstorms relative to other U.S. natural disasters.
- List the hazards produced by thunderstorms.
- Recognize the levels of the Fujita Scale.
- Describe the types of injuries seen with tornadoes.
- Explain how lightning is formed.
- Describe the morbidity and mortality of lightning.
- List mitigation measures for tornadoes.
- Describe the immediate actions that should be taken during a tornado warning.
- Recognize the hazards posed during recovery.

Introduction—The 1953 Beecher Tornado

Throughout the history of severe storms, there have been miraculous stories of survival. One such story came from a family in Beecher, near Flint, Michigan, who survived a direct hit from a massive tornado in 1953. It was one of eight that hit the region in one night killing 116 and injuring 844, and it remains the deadliest storm in the history of Michigan (Figure 11–1). The following account has been adapted from the firsthand account of Michael Naruta (National Oceanic and Atmospheric Administration, National Weather Service, 2003).

An intense storm built throughout the evening as Zena and her three children (Tom, Mike, and Gordon) huddled together in their farmhouse. They had been through severe thunderstorms before but this one was different. The lightning was particularly intense as the storm built. The lights went out and Zena picked up a flashlight and gathered the kids in the kitchen. Suddenly the wind picked up and became louder. The house creaked and the windows rattled as the noise grew to a loud roar. It sounded like a freight train as the windows cracked and then exploded. A swarm of paper, dishes, and other belongings filled the living room and kitchen as the plaster walls began to crumble. Tom, age 15, impulsively ran upstairs to his bedroom and pulled the covers over his head. As the roof was peeled back and the walls collapsed, Tom and his bed were thrown into the front yard. Zena and the other two children had no place to go as the walls began to crumble. They were knocked to the floor and buried under the debris. Fortunately, they fell next to the kitchen table. The sturdy old table created a void spot among the rubble where Zena and the children were spared from the full weight of the collapsing farmhouse.

FIGURE 11–1 One of the eight tornadoes that hit the Flint, Michigan, region on June 8, 1953.
Source: National Oceanic and Atmospheric Administration. Available at: http://www.crh.noaa.gov/dtx/1953beecher/images/erie.jpg.

Everything was suddenly quiet. Tom picked himself up and approached the debris pile that was his home minutes earlier. Beneath the rubble, Zena heard the approaching sound of Tom calling out to them. She could feel the flashlight next to her and turned it on. Through the dust, she could see small openings in the debris. She called out to Tom as she shone the light through the breaks in the debris. As Tom began moving the rubble aside, fresh air flooded the space and they could see stars above them. Zena handed out her youngest child, Gordon. The 6-month-old was uninjured. Then his older brother, Mike was worked free and then Zena. There was a rush of relief and joy as she realized that they had all survived. Tom had injuries to his hands and feet. He was barefoot as he worked to dig his family from the remnants of their house. Zena was bleeding from injuries to her head and neck. She had shielded the children with her body and was pelted by flying glass and fragments. As the lightning flashed in the distance, it gave them just enough light to see that they had lost everything.

Thunderstorm and Tornado-Related Definitions

Downburst: A strong downdraft resulting in an outward burst of damaging winds on or near the ground. Downburst winds can produce damage similar to a tornado.

Downdraft: A small-scale column of air that rapidly sinks toward the ground, usually accompanied by precipitation. A downburst is the result of a strong downdraft.

Fujita Scale (F Scale): A scale describing wind damage intensity. All tornadoes, and some severe windstorms, are assigned a number from the scale according to wind speed and damage caused.

Funnel cloud: A condensation funnel extending from the base of a thunderstorm, associated with a rotating column of air that is not in contact with the ground (not yet a tornado).

Hook (hook echo): A radar reflectivity pattern characterized by a hook-shaped extension of a thunderstorm echo, usually in the right-rear part of the storm relative to direction of motion. A hook often is associated with favorable conditions for tornado development.

Microburst: A small, concentrated downburst affecting an area less than 2.5 miles (4 km) across. Most only last about 5 minutes but some may last longer.
Multiple-vortex: A tornado with two or more funnels or debris clouds present at the same time, often rotating around a common center or each other. Multiple-vortex tornadoes can be especially dangerous.
Severe thunderstorm: A thunderstorm that has winds of at least 50 knots (57.5 mph/92.5 km/hour) and produces hail 0.75 inches (1.9 cm) or tornadoes.
Severe thunderstorm warning: Issued by the National Weather Service to warn the public, emergency management, and response agencies when a severe thunderstorm is expected or is occurring. The warning will include where the storm is occurring, its direction of movement, and the primary threats from the storm.
Severe thunderstorm watch: Issued by the National Weather Service when conditions are favorable for the development of severe thunderstorms in a defined area.
Shear (wind shear): Variation in wind speed and/or direction over a short distance. Shear usually refers to vertical wind shear, that is, the change in wind with height.
Shelf cloud: A low, horizontal, wedge-shaped cloud associated with a thunderstorm front.
Squall line: A line or band of active thunderstorms.
Straight-line winds: The storm winds not associated with rotation, used to differentiate from tornadic winds.
Supercell: A severe thunderstorm with updrafts and downdrafts that are nearly balanced allowing the storm to sustain for hours; they often produce large hail and tornados.
Tornado: A rapidly rotating column of air that emerges from thunderstorm, making contact with the ground. It has a destructive circulation that causes damage and injury when it reaches the ground. However, the visible portion might not extend all the way to the ground.
Tornado warning: Issued by the National Weather Service to warn the public, emergency management, and response agencies when a tornado is likely to occur or is occurring. The warning will include where the storm was occurring and its direction of movement.
Tornado watch: Issued by the National Weather Service when conditions are favorable for the development of severe thunderstorms and possible tornados in the area.
Wall cloud: An area of clouds that extends under a severe thunderstorm; if a wall cloud rotates, it sometimes produces a tornado.
Waterspout: A tornado that reaches the surface of the water.

Thunderstorms and Tornadoes

A thunderstorm is formed when moisture and rising warm air encounters weather factors providing lift. The lift is usually provided by a warm or cold front but can also be provided by mountain or sea breezes. They can occur as a single thunderstorm cloud, a cluster, or a line. It is considered a "severe" thunderstorm when the winds gust to more than 50 knots (57.5 mph/92.5 km/hour) and produces hail 0.75 inches (1.9 cm) or tornadoes. Thunderstorms average about 15 miles (24 km) in diameter and last about 30 minutes. They carry a variety of hazards including lightning, hail, heavy rain, strong winds, and tornadoes (National Oceanic and Atmospheric Administration, National Weather Service, 2008). The United States has more severe weather events than any other nation on the Earth. About 10,000 thunderstorms, 5000 floods, 1000 tornadoes, and an average of two deadly hurricanes hit the United States annually resulting in about 500 fatalities and $14 billion in damage. In fact, about 90% of all federally declared disasters are storm related (National Oceanic and Atmospheric Administration, National Weather Service, 2008).

FIGURE 11–2 A severe thunderstorm in South Dakota, July 20, 2002.

Source: National Oceanic and Atmospheric Administration, National Weather Service. Available at: http://www.crh.noaa.gov/Image/unr/07-20-02/bk_ltg2.jpg.

Lightning poses a serious safety threat. Positive and negative charges build inside the storm clouds as the moisture separates into water and ice inside the clouds at high altitudes. Positive charges build in the ice particles while negative charges build in the water droplets. When enough energy builds between the two, lightning is discharged as a bolt of light. It can occur within a storm cloud or from cloud to cloud but poses a substantial health risk when it strikes the ground. The lightning produced by a thunderstorm has an incredible reach. Even when lightning hazards are not apparent, they can strike up to 10 miles from the storm's rainfall (Holle et al., 1995). The bolt can reach a temperature of nearly 50,000°F and dissipate in a split second. It is estimated that across the United States about 30 million lightning strikes hit the ground every year Krider and Uman, 1995). The instant heating and cooling of the air around lightning causes thunder.

These storms can produce a variety of hazards, including heavy rain, lightning, hail, strong winds, and tornadoes. The rain can contribute to riverine or flash floods and is the most lethal aspect of thunderstorms. Flooding is addressed in detail in Chapter 6. The winds produced are either straight line or rotational. The rotational winds in the form of tornadoes are the most notorious, although straight line winds can drop from a thunderstorm cloud at over 100 mph. These powerful downbursts can cause damage similar to a tornado and are particularly dangerous to aircraft.

Tornadoes are a destructive product of thunderstorms. Although tornadoes are possible almost anywhere, the vast majority occur in the United States. They can have winds from 100 mph to over 300 mph. Some can reach over a mile wide and stay on the ground for many miles. Once they are on the ground, they will typically move forward at about 30 mph. The University of Chicago professor, T. Theodore Fujita, developed the Fujita Tornado Damage Scale in 1971 to characterize the severity of tornadoes (Table 11–1). It does not use actual wind measurements but estimates are made based on the observed damage. An updated or enhanced F Scale has recently been developed using

Table 11–1 Fujita Tornado Damage Scale

Scale	Wind Estimate	Typical Damage
F0 (weak)	40–72 mph	Light damage: some damage to chimneys; branches broken off trees; shallow-rooted trees pushed over; sign boards damaged
F1 (weak)	73–112 mph	Moderate damage: roof damage; mobile homes pushed off foundations or overturned; moving autos blown off roads
F2 (strong)	113–157 mph	Considerable damage: roofs torn off; mobile homes demolished; trains overturned; large trees uprooted; object missiles generated; cars lifted
F3 (strong)	158–206 mph	Severe damage: roofs and some walls torn off well-constructed houses; trains overturned; most trees uprooted; heavy cars lifted off the ground and thrown
F4 (violent)	207–260 mph	Devastating damage: well-constructed houses leveled; structures with weak foundations some distance; cars thrown and large missiles generated
F5 (violent)	261–318 mph	Incredible damage: strong frame houses leveled; automobile-sized missiles fly in excess of 100 m (109 yd); trees debarked; incredible devastation

Adapted from National Oceanic and Atmospheric Administration, National Weather Service, Fujita Tornado Scale. Available at: http://www.spc.noaa.gov/faq/tornado/f-scale.html.

8 levels of damage to 28 indicators including a variety of structures (National Oceanic and Atmospheric Administration, National Weather Service, 2007). However, the original F0–F5 categories are not drastically changed.

Effects on the Human Populations

The impact of these storms on morbidity and mortality is tremendous and diverse. While flooding is the major cause of death, high winds and tornadoes injure many more people every year from collapsed structures and traumatic missile injury from flying debris. In the United States, tornadoes injure about 1000 people each year and kill about 50 (Lillibridge, 1997). Many of these injuries are due to the vulnerability of mobile homes, and storm shelters have proven to be an effective measure to reduce these injuries (CDC, 1992; Eidson et al., 1990; Liu et al., 1996). Reducing tornado-related morbidity and mortality relies upon access to storm shelters (Liu et al., 1996). It also depends on public education. During a devastating 1994 storm in Alabama, only 31% of those who heard the tornado warning actually took shelter (Office of Meteorology, National Weather Service, National Oceanic and Atmospheric Administration, 1996).

Although there are millions of lightning strikes around the world each day, relatively few cause injury. Although thousands of people are struck by lightning annually, a review of lightning strikes from 1959 to 1990 showed an annual U.S. average of just over 250 lightning injuries and about 90 deaths per year (Lopez and Holle, 1995). About a third of those struck by lightning die from their injuries and nearly 75% of them have permanent disability (Cooper, 1980). Sixty-three percent of those who die are lost within the first hour of their injuries (Duclos and Sanderson, 1990). The vast majority of these injuries (92%) occur between May and September and nearly three quarters (73%) occur in the afternoon and evening. About half (52%) of those injured by lightning are engaged in outdoor recreational activities and a quarter (25%) are engaged in outdoor work activities (Lopez and Holle, 1995). Nearly all of the injuries and fatalities are preventable.

Recent research has begun to reveal more subtle effects of thunderstorms. There is a growing body of literature showing increases in asthma attacks after thunderstorms. Specifically, storms with high winds and rain are associated with a 3% rise in asthma-related emergency department visits (Grundstein et al., 2008). It is believed that this is attributed to the rupture and release of pollen grains during the storms.

Public health and emergency management organizations should work more closely to communicate the health risks associated with these storms and raise public awareness through educational initiatives. They should focus mostly on individuals in mobile homes who are most vulnerable.

FIGURE 11–3 Tornado damage from a 1999 Oklahoma City tornado. Note the board driven through the car door.

Source: National Oceanic and Atmospheric Administration, National Weather Service. Available at: http://www.srh.noaa.gov/oun/storms/19990503/damage/mb990124.jpg.

Prevention

Before a thunderstorm or tornado, the essential preparedness activities described in Chapter 1 apply. You should have a plan, have a preparedness kit, and follow the instructions of authorities. The plan should include both a communication plan on how you will communicate with family or friends immediately after a storm and an action plan on how you will immediately react to an imminent thunderstorm threat. A preparedness kit at home is also essential. A thunderstorm can take out critical utilities over a huge area for days or even weeks.

Stay alert for dangerous weather conditions by keeping an eye out for threatening skies, hail, increasing winds, etc. When threatening weather is approaching, you should monitor radio or television for reports. It is also important to maintain a National Oceanic and Atmospheric Administration weather radio. These radios have grown into all-hazards warning systems. Not only do they alert areas at risk about dangerous weather, they are also now used for environmental announcements such as chemical spills and public safety issues like Amber Alerts.

Look for anything that could fall during high winds and cause damage or injury. Keep trees trimmed, particularly those near power lines, and remove dead trees and branches. There are simple preparedness rules to remember. The 30/30 lightning safety rule is a good example. If you see lightning and cannot count to 30 before hearing thunder, you are within range of lightning and should not be in a vulnerable place. It is best to go indoors. This level of risk applies until 30 minutes after you hear the last rumble of thunder. Vulnerable locations include any tall, isolated trees, hilltops, open fields, the beach, and the water. Small structures in open area are also susceptible. Anything metal, including bicycles, golf clubs, etc., should be avoided since they can attract lightning. It is a myth that rubber-soled shoes and rubber tires on vehicles can protect you. Although being in a car is safer than being in the open, it is still vulnerable. It is not a myth that it is dangerous to take a shower or bath during a storm. The plumbing and bathroom fixtures can conduct electricity and a surge of energy from lightning can reach you.

Immediate Actions

When a thunderstorm approaches, actions are dictated by circumstances. One of the most acute risks is lightning. If you are in the water or on a boat, get on dry land and seek shelter quickly. If in an open outdoor area, find a low lying place like a ravine or valley but be alert for flash floods. If on an outdoor playing field (soccer, baseball, etc.), activities should be suspended and shelter should be sought. Avoid things that may attract lightning, such as tall, isolated trees, etc. Those who have experienced lightning injuries have noticed that just before the strike, they could feel their hair beginning to stand on end. If you ever feel that sensation, immediately squat down and make yourself a small target. Do not lie flat on the ground.

If a tornado warning is issued and you are in a structure, put as many walls and as much depth between you and the winds as possible. If there is a basement or storm cellar, it is the first choice. Otherwise an interior hallway or bathroom is preferred. Stay away from corners, windows, doors, and outside walls. If in a vehicle, trailer, or mobile home, you should seek other shelter. If none is available, lie flat in a ditch or depression and cover your head with your hands. Most fatalities are caused by flying debris.

Response and Recovery Challenges

The risks associated with recovery from thunderstorms and tornadoes depend on the post-disaster landscape. Collapsed buildings carry risks of traumatic injury from the debris as well as the potential for hazardous chemical exposure from household chemicals spilling and mixing. In addition, the debris field can draw pests that can pose additional hazards. Recovery is a long and arduous process. More injuries sometimes occur as tension and frustration are combined with a hazardous environment.

Summary

Thunderstorms and tornadoes can pose an array of immediate risks to health and safety. As with other natural disasters, basic preparedness can make an enormous difference in the immediate and long-term impact. Risk-based preparedness initiatives should focus on the regions and populations at greatest risk. Oklahoma should have a more aggressive thunderstorm and tornado preparedness initiative than Utah. In public outreach, mobile home parks should receive more attention in public awareness initiatives than those in sturdy structures with basements. The key principles of awareness are understood by the public better than any other natural disaster threat. The problem is apathy that can only be overcome through constant reminders using a variety of approaches.

Websites

American Red Cross, Thunderstorm Preparedness (In English and Spanish): www.redcross.org/services/prepare/0,1082,0_247_,00.html.

American Red Cross, Tornado Preparedness (In English and Spanish): www.redcross.org/services/prepare/0,1082,0_248_,00.html.

Centers for Disease Control and Prevention, Tornado Preparedness: www.bt.cdc.gov/disasters/-tornadoes/.

Federal Emergency Management Agency, Thunderstorm Preparedness: www.fema.gov/hazard/-thunderstorm/index.shtm.

Federal Emergency Management Agency, Tornado Preparedness: www.fema.gov/hazard/tornado/index.shtm.

National Oceanic and Atmospheric Administration, National Weather Service, NOAA Weather Radio: www.weather.gov/nwr/.

National Oceanic and Atmospheric Administration, National Weather Service, Severe Weather: www.noaawatch.gov/themes/severe.php.

National Oceanic and Atmospheric Administration, National Weather Service, StormReady Program Site: www.stormready.noaa.gov/.

References

CDC. (1992). Tornado disaster—Kansas, 1991. *MMWR* 41:181–183.

Cooper, M. A. (1980). Lightning injuries: prognostic signs for death. *Ann Emerg Med* 9:134–138.

Duclos, P. J., & Sanderson, L. M. (1990). An epidemiological description of lightning-related deaths in the United States. *Int J Epidemiol* 19:673–679.

Eidson, M., Lybarger, J. A., Parsons, J. E., MacCormack, J. N., & Freeman, J. I. (1990). Risk factors for tornado injuries. *Int J Epidemiol* 19:1051–1056.

Grundstein, A., Sarnat, S. E., Klein, M., et al. (2008). Thunderstorm associated asthma in Atlanta, Georgia. *Thorax* 63:659–660.

Holle, R. L., Lopez, R. E., Howard, K. W., Vavrek, J., & Allsopp, J. (1995). Safety in the presence of lightning. *Semin Neurol* 15:375–380.

Krider, E. P., & Uman, M. A. (1995). Cloud-to-ground lightning: mechanisms of damage and methods of protection. *Semin Neurol* 15:227–232.

Lillibridge, S. R. (1997). Tornadoes. In: Noji, E., ed. *The public health consequences of disasters*. New York: Oxford University Press; pp.228–244.

Liu, S., Quenemoen, L. E., Malilay, J., Noji, E., Sinks, T., & Mendlein, J. (1996). Assessment of a severe weather warning system and disaster preparedness, Calhoun County, Alabama, 1994. *Am J Public Health* 86:87–89.

Lopez, R. E., & Holle, R. L. (1995). Demographics of lightning casualties. *Semin Neurol* 15:286–295.

National Oceanic and Atmospheric Administration, National Weather Service. (2003). 1953 Beecher Tornado, 50th anniversary. www.crh.noaa.gov/dtx/1953beecher/storiesNaruta.php.

National Oceanic and Atmospheric Administration, National Weather Service. (2007). Enhanced F scale for tornado damage. www.spc.noaa.gov/faq/tornado/ef-scale.html.

National Oceanic and Atmospheric Administration, National Weather Service. (2008). Severe weather. www.noaawatch.gov/themes/severe.php.

National Oceanic and Atmospheric Administration, National Weather Service. (2008). StormReady program site. www.stormready.noaa.gov/.

Office of Meteorology, National Weather Service, National Oceanic and Atmospheric Administration. (1996). *Tornado safety*. Washington, DC: National Oceanic and Atmospheric Administration, National Weather Service; Publication no. 1996-413-872.

12 Volcanoes

Objectives of This Chapter

- List the types of volcanoes.
- Define the Volcanic Explosivity Index (VEI).
- Describe the health effects of volcanoes.
- List the morbidity and mortality effects of volcanic ash.
- Explain the health hazards associated with a pyroclastic flow.
- Define the health risks associated with laze.
- Explain what can be done to prepare for a volcanic eruption.
- Describe what individuals who are outside should do during an eruption.
- Explain the actions that should be taken by individuals who are indoors following an eruption.
- Describe the health and safety risks associated with ash accumulation and ways to reduce risks.

Introduction—1991 Eruption of Mount Pinatubo, Luzon Island, Philippines

Raul grew up in the shadow of Mount Pinatubo in the town of Baguio. He had paid his dues and was anxious to retire from his 21-year career in the U.S. Air Force. His career path had taken him around the world but he was looking forward to wrapping up his career at Clark Air Base, Philippines, and settling nearby into a new home he was building for his wife and three kids. With his hometown so close, they would finally get to spend more time with family and end the regular moves and temporary duty assignments.

Nobody gave the risk of a volcanic eruption much thought. The volcano lay dormant for so many centuries that an eruption was simply not discussed. As Mount Pinatubo became active and early warning signs emerged that a large eruption could be approaching, the U.S. Air Force evacuated Clark Air Base. Small earthquakes began shaking the mountain throughout April and May. Several steam explosions emerged on the sides of the mountains. Even though Raul was doubtful that much more would happen, he sent his family to stay with his in-laws as he stayed behind with several hundred airmen to ride out the situation on the Air Base (Arana-Barradas, 2001).

On June 15, 1991, Mount Pinatubo experienced one of the largest and most dramatic volcanic eruptions in recent history (Figure 12–1). For the preceding week, magma slowly began to emerge and pool into a small lava dome and, finally, it exploded in a catastrophic eruption that ejected over a cubic mile (5 km) of ash and debris, the top thousand feet of the mountain, over 20 miles into the sky. At the same time, pyroclastic flows of

FIGURE 12–1 The June 12, 1991, eruption column from Mount Pinatubo taken from the east side of Clark Air Base. U.S. Geological Survey photograph taken on June 12, 1991, by Dave Harlow. Available at: http://vulcan.wr.usgs.gov/Imgs/Jpg/Pinatubo/Pinatubo91_eruption_plume_06-12-91.jpg.

hot ash and debris raced down the sides of the mountain and accumulated hundreds of feet deep in the surrounding valleys (Newhal et al., 2005). Had it not been for the initial clues and the evacuation of thousands, the death toll would have been in the tens of thousands. The official death toll in the days following the eruption was just over 300 (Spence et al., 1991). The majority of these fatalities were attributed to structural collapses that occurred under the tremendous weight of the accumulating ash on rooftops.

Raul's new home was not spared. As a major storm brought heavy rains to the area in the days following the eruption, the moisture met with the ash cloud causing it to rain mud. His home was blanketed in layers of ash and mud. He shoveled 20 dump truck loads from his home and began rebuilding (Arana-Barradas, 2001). The lush tropical vegetation across the region was buried and the terrain looked more like the surface of the moon. As the years have passed, the lush, green terrain returned but the economy never fully recovered. The Air Force closed Clark Air Base and returned the property to the Philippine Government. Although some uses have been found for major buildings left behind, the lives of area residents and the local economy will never be the same. Despite the setbacks, many local residents, like the family of Filipino-American and retired U.S. Air Force Technical Sergeant Raul Baon, have worked hard over the years to rebuild the region. Another eruption will happen someday. This is a fact of life that everyone there must live with every day.

Volcano-Related Definitions

Acid rain: Rain that becomes acidic by falling through volcanic emissions. It can sometimes have a pH less than 4 and can cause environmental damage.

Ash (volcanic): Refers to the particle size applied to the finest pyroclastic material, fragments less than 1/8 inches (2 mm) in diameter, ejected into the air by volcanic explosions.

Ash cloud: The fine volcanic material ejected into the air from a volcanic explosion or rising from a pyroclastic flow to form a cloud.

Ashfall (airfall): Volcanic ash and debris that falls from an eruption cloud.
Ballistic fragment: Rock fragments that become projectiles as they are ejected from a volcanic explosion.
Cinder cone volcano: Gas-filled lava is ejected into the air breaking into small cinders and accumulating around the vent.
Composite volcano: Also called a stratovolcano, usually has a crater at the summit that has a vent(s) that erupts lava, cinders, and ash. These pose a substantial risk to surrounding residents and can produce pyroclastic flows and lahars.
Lahar: A mass of volcanic debris mixed with water displaced from the volcano resulting in a mudslide of volcanic particles and fragments. It is also called a mudflow or debris flow.
Lateral blast: A horizontal explosion from or in the direction of the side of a volcano.
Lava: Molton rock or magma that reaches the Earth's surface.
Lava dome volcano: A dome of lava forms around the volcano vent. As it cools and hardens, it is cracked by internal pressure and fragments break free and accumulate around the dome.
Laze: A white plume of hydrochloric acid and vaporized seawater that occurs when lava enters the salt water causing chemical reactions that emit an acidic lava haze, or laze.
Magma: Molten rock containing dissolved gases and minerals; called lava when it reaches the Earth's surface.
N-95 respirator: An air purifying respirator that cleans 95% of the particles from the air before it is inhaled by the wearer.
Particulate matter: A suspension of fine solids or droplets in the air such as dust, ash, smoke, and fumes.
Pyroclast: The literal definition is "fire rock fragment." It is simply a piece of material formed by a volcanic explosion or ejected from a volcanic vent.
Pyroclastic flow: An extremely hot, fast moving mixture of fragmented volcanic material and gas traveling at high speeds down the slope of a volcano. It is formed during an eruption or following the collapse of a lava dome.
Shield volcano: Lava is slowly emitted from the volcano vent(s). Most of this volcano is accumulated lava flows that are nonexplosive.
Tephra: The ash fall following an eruption that includes sand-sized or finer particles of volcanic rock and other fragments.
Volcano: A vent or opening in a planet's surface through which volcanic material is erupted.

Volcanoes

Volcanoes are responsible for the formation of much of the Earth's surface. The origin of the word can be traced back to the Island of Vulcano, near Sicily. It was named after Vulcanus, the Roman god of fire and the forge (metal working). They are typically hills or mountains that surround a vent leading to reservoirs of molten rock below the Earth's surface. There are four primary types of volcanoes: Cinder Cone, Composite, Shield, and Lava Dome (U.S. Geological Survey, 1997a).

There are also levels of volcanic activity intensity that are categorized by a Volcanic Explosivity Index (VEI) (Table 12–1). This is a composite, semiquantitative estimate of a volcano's eruptive power. It includes the quantitative measurements of variables such as the eruption cloud height and amount of ejected tephra. It combines those quantitative values with qualitative eruption observations to place an eruption on a scale from 0 to 8. Each increase in value of one represents a 10-fold increase in eruption intensity (Newhall and Self, 1982).

Table 12–1 Estimated Volcanic Explosivity Index (VEI): History of Major Volcanoes

Pinatubo	Luzon, Philippines	3550 BC	5
Mount St. Helens	Washington, USA	2340 BC	5
Mount St. Helens	Washington, USA	2860 BC	6
Mount St. Helens	Washington, USA	1770 BC	5
Mount Fuji	Honshu, Japan	1350 BC	5
Pinatubo	Luzon, Philippines	1050 BC	6
Mount Fuji	Honshu, Japan	1030 BC	4+
Mount Fuji	Honshu, Japan	930 BC	5
Mount St. Helens	Washington, USA	530 BC	5
Mount Rainier	Washington, USA	250 BC	4
Mount Fuji	Honshu, Japan	800 AD	4
Pinatubo	Luzon, Philippines	1450	5
Mount St. Helens	Washington, USA	1480	5+
Mount St. Helens	Washington, USA	1482	5
Mount Fuji	Honshu, Japan	1707	5
Mount St. Helens	Washington, USA	1800	5
Mount St. Helens	Washington, USA	1980	5
Chichon, El	Mexico	1982	5
Pinatubo	Luzon, Philippines	1991	6

Source: Global Volcanism Program, Large Holocene Eruptions. Available at: http://www.volcano.si.edu/world/largeeruptions.cfm.

During a volcanic eruption, there are a variety of hazards produced in the air. Gases and particles are rapidly ejected. The gases are primarily comprised of water vapor, carbon dioxide (CO_2), and sulfur dioxide (SO_2). There are lesser volumes of other gases such as hydrogen sulfide (H_2S), hydrogen (H_2), carbon monoxide (CO), hydrogen chloride (HCl), hydrogen fluoride (HF), and helium (He) (U.S. Geological Survey, 2008). These gases and particulate debris can quickly reach heights of 10–20 miles and can literally be carried around the planet. Tephra particles immediately begin to fall out with large particles landing near the vent and smaller ones being carried downwind as a large cloud. Tephra is the ash that covers everything causing the area immediately downwind from an eruption to appear as a lunar surface. As tephra material is being ejected upwards, there are direct blasts and flows being projected laterally from the volcano as well. Large rocks and enormous volumes of materials can be hurled at a high rate of speed for many miles (Federal Emergency Management Agency, 2006). Pyroclastic flows may also form. These are clouds of hot particles and gases that spill down the slopes of a volcano (Figure 12–2). Depending on conditions, they can travel several hundred miles per hour and consume everything in their path (U.S. Geological Survey, 1997b). As the plume rises into the air, particles separate into layers of different charges generating lightning similar to what is seen during thunderstorms. There is new research emerging that there is also a new form of low energy lightning generated from arcing particles emitted from a volcanic vent (Reilly, 2007). This could pose particular hazards to passing aircraft.

FIGURE 12–2 Pyroclastic flow sweeps down the side of Mayon Volcano, Philippines during a September 15, 1984, eruption. Photo courtesy of U.S. Geological Survey by C. Newhall. Available at: http://volcanoes.usgs.gov/Imgs/Jpg/Mayon/32923351-018_large.jpg.

As a volcano fills the air with toxic gases and particles, the downwind areas are covered with ash. In addition, lahars are often formed. Lahars are flash floods or mudslides with heavy debris from a volcano (Figure 12–3). They can cause massive destruction for many miles and leave behind large volumes of sediment. The debris complicates the flooding by making it more destructive and more difficult to clean up during recovery.

Health Effects

There are a variety of health hazards produced by volcanic activity. Some are direct hazards such as the rocks, ash, gases, and heat from an eruption. The larger solid debris can cause ballistic injuries for miles, whereas the gases can cause respiratory damage. Pyroclastic flows and lava flows can cause thermal injuries, whereas the carbon dioxide and debris from an eruption can cause asphyxiation. Other hazards result from secondary events

FIGURE 12–3 Mount St. Helens, lahar damage. More than 200 homes and over 185 miles (300 km) of roads were destroyed by the 1980 lahars. USGS photograph taken on July 16, 1980, by Lyn Topinka. Available at: http://vulcan.wr.usgs.gov/Imgs/Jpg/MSH/Images/MSH80_red_house_in_mud_07-16-80.jpg.

triggered by a volcano, including earthquakes, flooding, landslides, lightning, and even tsunamis. The risks posed by an eruption are unique to each volcano and to each surrounding community. Some volcanoes will pose a larger acute threat from a massive eruption, while others may pose less acute risks but more chronic risks from gases and other emitted hazards. The type of volcano, geological composition of the area, distance from an eruption, weather patterns, topography, and preexisting health status of the exposed population are several important factors in the morbidity and mortality produced by an eruption.

The volcanic ash or tephra is comprised of very small rock and glass particles. They are hard, abrasive, and acidic (Figure 12–4). These materials can cause illness and injury through inhalational, mucosal, and direct dermal exposure. The severity of inhalational injuries depends upon the duration of exposure, the dose received, and the physical and chemical characteristics of the tephra. The smaller particles will travel deeper into the lungs and the acidity of the particles will have an immediate irritant effect. The presence of crystalline silica in the particles may pose additional risks. Although the silica content may be high, the respirable fraction of free silica determines the risk, not the total volume. Inhalation of the ash can cause respiratory distress and extended exposures can cause fibrogenic pulmonary effects (Nicol et al., 1985). These problems are more likely to occur in those with preexisting conditions like asthma, bronchitis, and emphysema. Particular attention should be given to limiting exposures of infants, the elderly, and those with chronic cardiovascular and respiratory conditions (Washington State Department of Health, 2008).

Gas and acidic particles can pose a variety of health risks. Emissions of sulfur dioxide (SO_2), hydrogen chloride (HCL), hydrogen fluoride (HF), hydrogen sulfide (H_2S), radon, and other gases can occur during eruptions or through daily degassing activities near a volcano. Some chemical risks are unique to specific volcanic scenarios. For example, laze is a hydrochloric acid mist that results from lava meeting sea water. When lava enters sea water, it vaporizes it so quickly that a series of chemical reactions occurs emitting a white plume that consists of hydrochloric acid and seawater. It is called lava

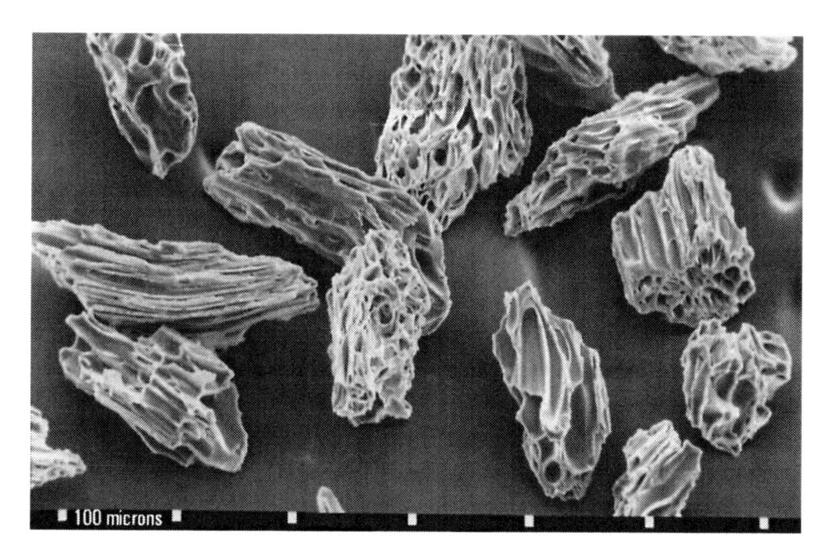

FIGURE 12–4 Volcanic ash glass shards were found in California in a layer of ash known as the Rockland ash bed. These are "pumiceous" from their gas-bubble holes or vesicles. Available at: http://volcanoes.usgs.gov/Imgs/Jpg/Tephra/SarnaSem_60-000_large.jpg.

haze, or laze (U.S. Geological Survey, Volcanic gases and their effects, http://volcanoes.usgs.gov/hazards/gas/index.php#reference).

Like the ash hazards, the gas, fume, and acid particle hazards will vary greatly between volcanic eruptions and impact vulnerable populations more severely than others. These gases are contained in molten rock under pressure. As the magma reaches the surface, it is like opening a shaken bottle of soda. As the pressure is released, the gases expand quickly causing an explosive display as the gases are released. This also creates the tephra particles mixed with acid aerosols that make the volcanic ash acidic.

If one cubic meter of magma containing 5% water is brought from below the surface where it is under extreme pressure to atmospheric pressure as it exits the volcano during eruption, it will expand to 670 m of water vapor, gases, and particulates. This is the primary force behind explosive volcanic eruptions (Sparks et al., 1997).

Following the initial eruption, the falling ash can interrupt a variety of critical infrastructures. The power supplies may be interrupted by ash blocking air intake systems and insulators at power facilities. This can have a serious impact on area residents depending upon home medical equipment and other critical equipment. It can also reduce the effectiveness of public communication during the crisis because many will be without telephones, radios, and televisions. Other infrastructures including the drinking water supply and wastewater treatment can be compromised. Drinking water from open reservoirs can experience heavy ash contamination that is difficult to manage with the existing infrastructure. There is also an excessive demand on the water supply during clean-up that can lead to shortages. The wastewater systems are also vulnerable. Both sewage and storm water systems can become clogged as ash accumulates as a heavy mud in the system. It can also compromise pumping systems and other machinery. The infrastructure damage resulting from volcanic ash can have a substantial impact on morbidity and mortality through the interruption of critical infrastructures such as these.

Top Ten Morbidity and Mortality Effects from Volcanic Ash

- Runny nose and nasal irritation.
- Dry coughing and throat irritation.
- Exacerbation of existing respiratory problems (e.g., severe bronchitic or asthmatic symptoms).
- Exposure to respirable free crystalline silica particles leading to silicosis (scarring of the lungs and impairment of function).
- Painful, itchy, or bloodshot eyes.
- Corneal abrasions or scratches.
- Acute conjuctivitis.
- Minor skin irritations.
- Crush injury and trauma from collapsing structures due to the weight of the ash on structures.
- Traumatic injuries from motor vehicle accidents resulting from poor driving conditions.

Adapted from USGS, Volcanic Ash: Effects & Mitigation Strategies, Health Section. Available at: http://volcanoes.usgs.gov/ash/health/index.php

Transportation may also be disrupted. The ash can obscure road signs and create slick roads. It also creates poor visibility. The ash is sometimes so thick that even brake lights and tail lights are not visible. This was the case following the 1980 eruption of Mount St. Helens in Washington. About 75 miles (120 km) away from the eruption, several centimeters of ash accumulated on the roads resulting in hundreds of motor vehicle accidents. A first-hand account of the situation was in a May 28, 1980, newspaper article that appeared in the *Seattle-Post Intelligencer:*

> *Brake lights could be faintly seen through the dust which rose 30 feet (9 m) above the freeway. Often, though, lights couldn't be seen, and all that could be heard was the tinkle of broken glass and the crunch of crumpled metal as cars from the rear headed into the ash clouds, and rammed vehicles already hidden in the clouds (Blong, 1984).*

Prevention

Like many other natural disaster scenarios, preparedness measures for volcanoes begin with a home preparedness kit and an understanding of the risk for volcanic eruptions where you live, work, and travel. Unlike many other natural disaster threats that are focused

FIGURE 12–5 David Johnston at Coldwater II, near Mount St. Helens at 1900 hours. Dave did not survive the next day's eruption. Coldwater II would eventually be renamed "Johnston Ridge" in honor of Dave. USGS photograph taken on May 17, 1980, by Harry Glicken. Available at: http://vulcan.wr.usgs.gov/Imgs/Jpg/MSH/Images/MSH80_david_johnston_at_camp_05-17-80.jpg.

tightly on a specific geographic region, volcanoes can reach hundreds of miles downwind and cause severe interruption in critical infrastructure. The risks are not restricted to active volcanoes. Even if you are near a dormant volcano, you should have a plan and be prepared to evacuate quickly. For those volcanoes that are active, people should be encouraged to avoid them. Unfortunately, curious spectators and even scientific researchers have been lost when they attempt to observe an eruption. If you are close enough to observe a volcanic eruption, you are in unnecessary danger (Figure 12–5). As technology improves the ability of volcanologists to predict an imminent volcanic eruption, the most important preventive measure is heeding warnings and evacuation announcements. These predictions and evacuations have saved thousands of lives during eruptions in recent history.

In preparation for the falling ash, those who live near or downwind from volcanoes should also add disposable breathing masks and goggles for each member of the family to their disaster kit. These are available through local hardware stores and are the same protective equipment used for protection from inhaling dust during sanding and other common home improvement activities. Also, since a volcano may cause earthquakes, flooding, and other secondary disasters, it is recommended that the preventive actions specific to these related hazards be followed as well.

Before a Volcano

- Be familiar with local and regional volcano threats, both active and dormant.
- Understand local plans and recommendations for preparedness.
- Establish an emergency kit. It should include an emergency supply of food, water, and other essentials, such as goggles and respiratory protection.
- Develop an emergency communication plan with family members.

- Stay away from active volcano sites.
- Plan evacuation routes away from areas prone to flooding or low-lying areas.
- Monitor television and radio, using a battery-powered radio as a backup.
- Evacuate immediately if instructed by authorities.
- Assist vulnerable neighbors that may not be able to prepare or evacuate.
- Understand how to shelter safely in place if evacuation is not an option.

Adapted from various sources, including the American Red Cross, Federal Emergency Management Agency, and Washington State Department of Health.

Immediate Actions

The initial actions taken during a volcanic eruption depend upon your proximity to the eruption and the availability of shelter. Ideally, there should be an evacuation order given and everyone should evacuate from the area. During evacuation, you should avoid areas downwind and avoid river valleys downstream from the volcano.

If you are caught in the area of a volcanic eruption, and you are outside, seek inside shelter immediately. Beware of landslides and mudflows. Avoid low-lying areas where mudflows are possible during the eruption. Once indoors, close all the windows, doors, fireplace dampers, etc. Close down your heating, ventilation, and air conditioning (HVAC) system and place damp, rolled towels along the base of the doors and anyplace else where a draft may introduce ash into the building. Remember to help neighbors that may need assistance, such as the elderly or disabled. It is also recommended that you bring in all animals and livestock under cover in closed shelters and put all machinery inside closed garages or barns. This includes parking vehicles under cover when possible.

Response and Recovery Challenges

Once an eruption has occurred, the proximity to the eruption will dictate the hazards. If you are outside the area of ash fall, avoid that area. Local emergency management officials will provide detailed instructions to area residents, so listen to your television or radio. If the power goes out, make sure you have a battery or crank operated radio as a backup. If it is necessary to go outside, during or immediately after the ash fall, you should cover your mouth and nose to filter out the larger particles and protect your respiratory tract. A properly fitted N-95 respirator is ideal. If those are not available, use a dust mask or other basic covering of your nose and mouth to limit particulate inhalation. This could be as simple as a damp wash cloth. Those who have preexisting respiratory conditions should avoid ash exposure as much as possible.

The ash fall from a volcano is heavy and accumulations on a roof can lead to collapse. The only time it should be cleared while ash is still falling is if it poses a risk for roof collapse. This is usually when the accumulation is approaching 4 inches. Clearing the ash from the roof should be done carefully because the ash will pose a slipping hazard and even minimal respiratory protection will limit physical abilities. All exposed

FIGURE 12–6 Dark channels created by lahars streak down the sides of Mount St. Helens during the May 18, 1980, eruption. The lahars were generated by the sudden melting of snow and ice from hot volcanic explosion and pyroclastic flows. USGS image available at: http://volcanoes.usgs.gov/Imgs/Jpg/MSH/30212265-041_large.jpg.

skin should be covered and goggles should be worn when clearing the roof of ash accumulation, and outer clothing should be removed before entering the building to limit the transport of ash into the structure.

Top Ten Volcano Ash Management Tips

- Those with respiratory conditions, such as asthma or bronchitis, should avoid exposure.
- Stay indoors and close windows, doors, dampers, etc.
- Replace or clean HVAC system filters often.
- Use an N-95 respirator or similar protection if outdoors.
- Wear goggles and avoid using contact lenses.

FIGURE 12–7 Volcanic ash falls to ground and creates darkness, Mount Pinatubo in the Philippines. Photograph by E. Wolfe on June 24, 1991. Available at: http://volcanoes.usgs.gov/Imgs/Jpg/Pinatubo/32923351-040_large.jpg.

- Cover exposed skin with long sleeves and long pants; remove outer clothing before entering the building.
- Do not go outside while ash is falling unless it is necessary to clear the roof to avoid collapse.
- Clear roof of ash more than 4 inches.
- Minimize travel to avoid vehicle damage; frequently change oil and air filters.
- Avoid contaminated food or water; have bottled water available; wash garden vegetables before eating.

Adapted from Washington State Department of Health, Volcano Preparedness Guidance (2008)

Summary

Volcanoes are one of the most powerful and destructive forces on the planet. They threaten those residing many miles around their immediate vicinity with massive destruction from pyroclastic flows, ballistic projection of materials, thermal injuries from volcanic emissions, and subsequent mud flows and ash fall. Fortunately, volcanologists have made substantial advancements in understanding and predicting eruptions. These advances have saved thousands of lives in recent eruptions and will save many more in the future. The best way to reduce volcano-related morbidity and mortality is simply educating the public of the importance of following the volcanologist's warnings and the emergency management official's recommendations. For those that do not or cannot evacuate, avoiding low-lying areas, reducing exposure to the ash, and making sure accumulations do not lead to structural failure, are the most important protective measures. Community preparedness should also focus on assisting vulnerable populations in preparing in advance and taking appropriate actions when eruptions threaten.

Websites

Alaska Department of Health and Social Services, Health Effects Associated with Volcanic Eruptions: www.epi.hss.state.ak.us/bulletins/docs/b2006_05.pdf.

American Red Cross, Volcano Preparedness: www.redcross.org/services/disaster/0,1082,0_593_,00.html.

Centers for Disease Control and Prevention, Emergency Preparedness and Response, Volcanoes: www.bt.cdc.gov/disasters/volcanoes/.

Dartmouth College, The Electronic Volcano, Volcano Study Resources: www.dartmouth.edu/~volcano/.

Durham University, Institute of Hazard & Risk Research, International Volcanic Health Hazard Network: www.dur.ac.uk/claire.horwell/ivhhn/index.html.

Federal Emergency Management Agency, Volcano Preparedness: www.fema.gov/hazard/volcano/index.shtm.

Oregon State, Volcano World, Teaching and Learning Resources: http://volcano.oregonstate.edu/.

Pan American Health Organization, Volcano -Preparedness Videos: www.paho.org/English/DD/PED/volcano.htm.

Smithsonian National Museum of Natural History, Global Volcanism Program: www.volcano.si.edu/world/.

U.S. Geological Survey, Volcano Ash: http://volcanoes.usgs.gov/ash/.

U.S. Geological Survey, Volcano Hazards Program: http://volcanoes.usgs.gov/.

University of Tokyo, Volcano Research Center, Volcanological Society of Japan, Internet Links: http://hakone.eri.u-tokyo.ac.jp/vrc/links/links.html.

Washington State Department of Health, Public Health Emergency Preparedness and Response, Volcanoes: www.doh.wa.gov/phepr/handbook/volcano.htm.

References

Arana-Barradas, L. A. (2001). *Out of the ashes: after Mount Pinatubo nearly buried it, air base bounced back.* Airman.

Blong, R. J. (1984). *Volcanic hazards: a sourcebook on the effects of eruptions*. Australia: Academic Press; p.424.

Federal Emergency Management Agency. (2006). Disaster information, Volcano. www.fema.gov/hazard/volcano/index.shtm.

Newhal, C., Hendley, J. W., & Stauffer, P. H. (2005). The Cataclysmic 1991 Eruption of Mount Pinatubo, Philippines, U.S. Geological Survey Fact Sheet 113–97. http://pubs.usgs.gov/fs/1997/fs113–97/.

Newhall, C. G., & Self, S. (1982). The Volcanic Explosivity Index (VEI)—an estimate of explosive magnitude for historical volcanism. *J Geophys Res* 87(C2):1231–1238.

Nicol, E. R., Lee, E. S., & Murrow, P. J. (1985). Respirable crystalline silica in ash erupted from Galunggung volcano, Indonesia, 1982. *Atmos Environ* 19(7):1027–1028.

Reilly, M. (2007) Erupting volcano crackles with low-energy lightning. *New Scientist*. February 23, 2007. www.newscientist.com/article/dn11250-erupting-volcano-crackles-with-lowenergy-lightning.html.

Sparks, R. S. J., Bursik, M. I., Carey, S. N., et al. (1997). *Volcanic plumes*. New York: Wiley; p.574.

Spence, R. J. S., Pomonis, A., Baxter, P. J., Coburn, A. W., White, M., & Dayrit, M. (1991). Building damage caused by the Mount Pinatubo eruption of June 15, 1991. http://pubs.usgs.gov/pinatubo/spence/index.html.

U.S. Geological Survey. (1997a). Principal types of volcanoes. http://pubs.usgs.gov/gip/volc/types.html.

U.S. Geological Survey. (1997b). Pyroclastic flows. http://pubs.usgs.gov/gip/msh//pyroclastic.html.

U.S. Geological Survey. (2008). Volcanic gases and their effects. http://volcanoes.usgs.gov/hazards/gas/.

Washington State Department of Health. (2008). Public health emergency preparedness and response, volcanoes. www.doh.wa.gov/phepr/handbook/volcano.htm.

13 Wildfires

Objectives of This Chapter

- Describe the difference between a wildfire and other fires.
- List the factors that influence the composition of wildfire smoke.
- Explain the effects of wildfires on human health.
- List the components of wildfire smoke.
- Describe the toxicity of wildfire smoke components.
- Recognize plants that are highly flammable and less flammable.
- List individual actions that can reduce the impact of wildfires.
- Describe the EPA Air Quality Index and how it is used to manage wildfire smoke hazards.
- List the populations at greatest risk of adverse health effects from wildfire smoke.
- Explain how wildfire ash should be cleaned up during recovery.

Introduction—2007 California Wildfires, San Diego

Considered by many as one of the most idyllic cities in the United States, San Diego is the second largest city in California and is famous for beautiful beaches and perfect weather. It is not a city that most people would associate with catastrophic disasters. In 2007, the Santa Anna winds stirred multiple fires throughout the region burning more than 300,000 acres, over 2500 buildings, and forcing the evacuation of over a half-million residents (Figure 13–1). By the time the 5000 plus firefighters battling the wildfires got them under control, 10 people had died and 122 were injured (KPBS, San Diego Public Radio, Wildfire Coverage, http://www.kpbs.org/news/fires). However, that was just part of the overall picture.

The impact of wildfires is far reaching, especially for vulnerable populations such as those with respiratory or cardiovascular conditions. Even among those who live many miles away from the fires, the health effects quickly became apparent as the blue skies filled with smoke and debris, taking on an ominous reddish hue. Many watching the smoke clouds roll in from a distance described it as having a menacing appearance and carrying an apocalyptic feeling. Health warnings filled the local airwaves recommending that those with allergies and respiratory conditions remain indoors. While the national media focused on the spectacular flames and heroic firefighters, a silent surge of smoke-related ailments swept across the state. It took a melodramatic collapse of singer Marie Osmond on a famous dance competition television show to bring more attention to the impact of the smoke and debris. She blamed her collapse on poor air quality and allergies triggered by the fires.

FIGURE 13–1 San Diego, California, October 25, 2007—Helicopters drop water and retardant on the Harris fire, near the Mexican border, to stop the wildfire from advancing. Photo by Andrea Booher/FEMA. Available at: http://www.fema.gov/photodata/original/33306.jpg.

Wildfire smoke exposures cause increased visits to area Emergency Departments and a surge in hospital admissions. The resulting illnesses include chest pain, asthma, bronchitis, and chronic obstructive pulmonary disease (CDC, 1999; Johnson et al., 2005; (Monitoring Health Effects of Wildfires Using the BioSense System—San Diego County, California, http://www.cdc.gov/mmwr/preview/mmwrhtml/mm5727a2.htm); Sutherland et al., 2005). Like most disasters, wildfires have a disproportionate impact on vulnerable populations and require targeted health communication efforts and a variety of local measures, such as school closures and evacuations.

Wildfire-Related Definitions

Air Quality Index (AQI): A measure of harmful substances present in the air. It is used by meteorologists and weather forecasters to report the purity of the air in a community on a specific day. A higher index indicates more pollution on a scale of 0–500. Once the AQI exceeds 100, it has exceeded national standards for key toxic substances. The higher it climbs, the greater the public health impact.

Fire mitigation: Actions taken to reduce the impact of a wildfire on human health and property.

Fuel: Both living and dead combustible materials that burn when provided heat and oxygen.

Ladder fuels: Vegetation providing fuel (like shrubs and branches) that carry the fire from the ground to the tops of trees, the same way a person would climb a ladder.

N-95 respirator: An air purifying respirator that cleans 95% of the particles from the air before it is inhaled by the wearer.

Particulate matter: A suspension of fine solids or droplets in the air such as dust, ash, smoke, and fumes.

Prescribed burn: A forest management tool where fire is carefully applied to forest fuels, for a specific purpose, under precise conditions, to achieve manageable objectives, such as to improve forage and habitat for wildlife and livestock or to reduce hazardous build-up of fire fuels.

Wildfire: Rapidly spreading, unwanted fires that rage through forests or wildland areas threatening human lives, property, and natural resources.

Wildland: A natural area without human development and habitation other than roads, power lines, and railroads.

Wildfires

Wildfires have always been part of the ecosystem and are often natural occurrences started by lightning strikes. They are also sometimes man-made through carelessness or intentional fire starting. These fires thrive best in areas that have vegetation, winds, and periods of dry weather. This includes heavily wooded areas, grasslands, and regions with thick scrub or underbrush. Although these fires can naturally occur and actually have long-term benefits for some ecosystems, human habitation near susceptible areas places populations at risk and in the path of dangerous flames and unhealthy smoke and debris. These effects include major public health, economic, and environmental damage.

Prescribed Burning Benefits

While wildfires are unwanted fires in undesirable places, there are benefits to prescribed burning as part of land management (Long, 2002). This includes

- Reducing the risk of future wildfires by establishing new growth that burns less intensely
- Promoting germination and flourishing of native, fire-adapted plants
- Enhancing quality of grazing lands
- Controlling pest problems
- Improving access for tree planning and other beneficial changes

The human impact of wildfires includes a range of health threats. There are hazards of heat injuries and illness, especially among those battling fires. Populations most vulnerable to heat, the effects, and strategies for dealing with it are described in the Heat Wave chapter of this text. There are certainly risks of burn injuries as well. Protecting the public from these injuries is the primary focus of first responders. However, the most challenging and far reaching public health aspect of wildfires is the smoke.

Wildfire smoke is a complex assortment of gases and particles. This includes carbon dioxide, carbon monoxide, nitrogen oxides, water vapor, particulates, and thousands of other compounds. The specific composition of wildfire smoke depends on a variety of meteorological and environmental factors that are unique to every situation. Although there are differences in smoke composition from one wildfire to another, local management of the associated smoke risks uses the same local air quality standards that are used in major cities across the United States to inform the public about air quality so decisions can be made concerning the risks of outdoor activities. The AQI is determined by the Environmental Protection Agency (EPA) based on measurements of five regulated pollutants (ground level ozone, particulate pollution, carbon monoxide, sulfur dioxide, and nitrogen dioxide) (U.S. Environmental Protection Agency, Office of Air and Radiation, 2000). The AQI takes the complex sum of risks associated with these primary pollutants and assigns a single number to communicate the level of risk to public health. If the AQI is 100 or less, the conditions are good for most people. The value of 100 corresponds to the national standards for these pollutants. As it climbs above 100, warnings are shared for vulnerable populations and at an even higher level, warnings apply to everyone.

Factors Influencing Wildfire Smoke Composition

- Types of vegetation being burned
- Types of structures being burned
- Topographical conditions
- Season
- Meteorological conditions (temperature and humidity)
- Smoke dilution by wind conditions
- Fire stage
- Fire temperature

FIGURE 13–2 A wildfire near a Montana ranch.

Source: Montana Multi-Agency Wildfire Rehabilitation and Recovery Program. Available at: http://dnrc.mt.gov/forestry/Assistance/FireRehab/Images/Ranchfire.gif.

Effects on the Human Populations

Wildfires can cause a variety of injuries and illnesses. Beyond the obvious risk of burns and heat injury, the most far-reaching and challenging health issue is smoke. The smoke produced by wildfires can produce effects ranging from airway and eye irritation to death, especially among individuals with conditions that make them more susceptible to inhalational exposures. Although there are potentially thousands of chemicals in the smoke from any given wildfire, the primary short-term inhalational challenges are particulate. Particles irritate the airway and cause symptoms even among healthy exposed individuals. These symptoms include coughing, wheezing, and difficulty breathing. Other smoke components have toxicity potential as allergens, carcinogens, mutagens, and teratogens. Some of these components are listed in Table 13–1. The majority of the exposed population will not experience long-term effects from wildfire smoke. However, there are some vulnerable populations who may experience severe short-term effects and some may have chronic symptoms as well.

There is a new research emerging from the 2007 California wildfires suggesting that the particles in smoke produced by wildfires are potentially more dangerous than the smoke that normally exists in an urban setting from traffic and other sources. The fire emissions are able to penetrate into indoor environments more readily than vehicle emissions. The wildfire smoke particles also have greater bioavailability or are more easily absorbed by human tissue (Esciencenews.com, 2008). These findings warrant more investigation and should be considered when making decisions on evacuation and sheltering.

Table 13–1 Wood Smoke Toxic Component Examples

Smoke Component	Toxicity
Inorganic gases	
Carbon monoxide (CO)	Asphyxiant
Ozone (O_3)	Irritant
Nitrogen dioxide (NO_2)	Irritant
Aromatic hydrocarbons	
Benzene	Carcinogen/mutagen
Styrene	Carcinogen/mutagen
Oxygenated organics	
Aldehydes	Carcinogen/mutagen/irritant
Organic alcohols	Teratogen/irritant
Phenols	Carcinogen/mutagen/teratogen/irritant
Quinones	Oxidation reduction/irritant/allergen/possible carcinogen
Chlorinated organics	
Methylene chloride	Possible carcinogen/neurological depressant
Methylene chloride dioxin	Possible carcinogen
Free radicals	Oxidation reduction/inflammatory/possible carcinogen
Particulate matter	Irritant/allergen/inflammatory/oxidative stress

Adapted from Naeher et al. (2007).

The EPA AQI is based on day to day urban pollution from traffic and industrial sources. If it is used as the absolute guidance for the public health response to wildfire smoke, it may understate the level of risk.

Prevention

Preparing for wildfires begins with an assessment of local risk and history of wildfires in your region. This information is available through the local emergency management office or by simply contacting a nearby fire department. Keep in mind that a lack of history does not necessarily mean the risk is zero. A long dry weather period can introduce the risk of wildfires to areas with little or no history. Home preparedness recommendations include basic fire safety such as maintaining fire alarms on each floor and near every bedroom. Ingress and egress plans are also essential. An ingress plan for first responders should consider how accessible the structure is to large fire trucks and how easily the home can be identified from the road. If the road is not wide enough or if the home cannot be identified from the road, it will hamper response and rescue efforts.

To reduce the wildfire risks to a home or other structure, it is recommended by the Federal Emergency Management Agency that a 30-foot safety zone be established around the home. If a residence is in an area known to have wildfire risks, the amount and type of vegetation in this zone should be carefully considered. Although all plants can burn, some are far less flammable than others, burn more slowly, have higher moisture content, and are generally safer to have close to a structure (See Tables 13–2 and 13–3). In this safety zone, vines and shrubs should not be growing close to or in contact with the structure. Tree branches in this zone should be trimmed up to at least 15 feet off the ground to reduce the risk of fire climbing up the trees like a ladder. Tree tops should be at least 15 feet apart to reduce the likelihood of fire spreading from one tree top to another. The lawn should be kept cut and watered and any accumulation of leaves or branches should be removed. These measures will help suppress the advancement of fire close to a home or other structure.

The safety zone around any structure should be adjusted according to local terrain. If a building sits on a hill, fire moves more aggressively uphill and the safety zone should be extended further than 30 feet downhill according to the slope. The removal of "flammable plants" should extend out to at least 100 feet from any structure and firewood should be stacked outside this distance as well. When constructing buildings in high-risk areas, nonflammable materials should be used as much as possible.

Table 13–2 Examples of Highly Flammable Plants

Trees	Shrubs	Groundcover and Other Plants
Bamboo	Acacia	Deer grass
Cedar	Arborvitae	Eulalia grass
Cypress	Gallberry	Pampas grass
Eucalyptus	Hopseed bush	Pine straw
Fir	Hollies	Rosemary
Pines	Junipers	Sage

Table 13–3 Examples of Less Flammable Plants

Trees	Shrubs	Groundcover and Other Plants
Ash	Azalea	Blue-eyed grass
Birch	Barberry	Daylily
Cherry	Camellia	Honeysuckle
Elm	Forsythia	Iris
Hickory	Gardenia	Lantana
Maple	Hydrangea	Periwinkle
Oak	Indian hawthorne	St. Augustine grass
Plum	Oleander	Stonecrop
Redbud	Rose	Trumpet creeper
Magnolia	Spirea	Yarrow
Tulip	Viburnum	Yellow jessamine
Willow	Weigela	Yellow-eyed grass

Plant flammability examples adapted from South Carolina Forestry Commission, Fire Smart Plant List for SC: Native and Landscape. Available at: http://www.state.sc.us/forest/scplants.pdf and the San Diego County Undesirable Plant List available at: http://www.co.san-diego.ca.us/dplu/docs/UndesirablePlants.pdf.

Ten Individual Preparedness Actions for Wildfires

1. Identify the wildfire risks where you live.
2. Make sure your home is accessible for a large fire truck and clearly marked for easy identification.
3. Avoid open burning.
4. If you build a fire, make sure it is away from trees or bushes and can be quickly extinguished.
5. Install smoke detectors on every level and near all sleeping rooms in your home.
6. Have an evacuation plan.
7. Create a 30-foot safety zone around your home clear of debris, tall grass, and flammable plants.
8. Create a secondary safety zone of 100 feet with less flammable plant varieties.
9. Clear combustible items at least 30 feet away from all structures.
10. Get trained in first aid and be prepared to help others during a crisis.

Immediate Actions

The initial actions taken during a wildfire are based on the current and anticipated fire and weather conditions. If there is a reasonable chance that a wildfire will threaten human lives and property, evacuation orders are issued. For those outside the area of risk

FIGURE 13–3 A female firefighter hauling hose on a Florida wildfire. Note the flame lengths. Available at: http://www.srh.noaa.gov/tlh/images/female_fire_fighter.jpg.

for fire hazards but close enough to experience smoke exposure, the EPA AQI may be used to communicate the level of risk from smoke exposure (See Table 13–4). The AQI provides a familiar and an understandable set of instructions for the public. As previously mentioned, this approach may not be as protective as preferred since emerging research suggests that the smoke from wildfires is potentially more toxic than the typical urban smoke and smog the AQI is based on. However, the benefit of using the AQI is that it facilitates better understanding of how local air quality can threaten your health. Using this index, there are six categories that associate the AQI values, the quality of air or level of concern that should be exercised, and an associated color.

If you are ever trapped in a stranded vehicle during a wildfire, it is usually better to stay in the vehicle than to try and outrun the advancing fire on foot. If the vehicle is operational, you should drive slowly with your headlights on so others can see you better through the smoke. Keep windows rolled up and close the air vents of the vehicle. If the vehicle must be stopped during a wildfire, park as far away as possible from tall trees and heavy brush. If you are ever surrounded and trapped in a vehicle, it is best to stay

Table 13–4 A Summary of the EPA Air Quality Index

Air Quality Index Value	Air Quality	Warning Colors
0–50	Good air quality; little or no risk	Green
51–100	Moderate; acceptable levels of pollutants will affect very few people	Yellow
101–150	Unhealthy for sensitive groups; general public not likely affected	Orange
151–200	Unhealthy for everyone; more serious for sensitive groups	Red
201–300	Very unhealthy for everyone; everyone may experience some effect	Purple
301–500	Hazardous to everyone; the entire population is affected	Maroon

Adapted from the EPA Guide to Air Quality and Your Health (Esciencenews.com, 2008).

inside the vehicle until the worst of the fires pass. Although the engine may stall and the car may rock from air currents generated by the fires, it is still safer to remain inside a vehicle. It is very unlikely that the gas tank will explode, even when the temperatures in the vehicle increase due to the passing fire.

If you are trapped inside a home or other structure, it is best to remain inside long enough for the worst of the wildfire to pass by. It will likely pass before the remaining structure burns down and the building can provide a temporary shelter as the fires pass. Once outside or if you are outside when fires approach, the best action is to find the closest area with the least amount of vegetation or fuel to burn. It may be a road, ditch, or open field. Take cover there and cover up with anything available while waiting for the flames to pass.

Neither evacuees nor professional firefighters should rely on respirators during wildfires given the current understanding of risks and benefits. Although many people purchase dust masks from local home supply stores to wear during wildfire smoke exposures, those are not adequately protective. In fact, there are no air-purifying respirators that offer adequate protection against wildfire smoke. Every situation needs to be evaluated to determine respiratory protection recommendations. The minimum protection is a properly fitted N-95 respirator (Figure 13–4). However, no air purifying respirator is capable of filtering out carbon monoxide and several other wildfire smoke contaminants. The current recommendations call for an air purifying respirator and a carbon monoxide alarm (Austin and Goyer, 2007) . Anyone relying upon an air purifying respirator for protection from wildfire smoke must consider the hazards that cannot be filtered. Carbon monoxide is one of several but it serves as an excellent surrogate for other smoke hazards.

Response and Recovery Challenges

Once a wildfire passes through an area, it is important to check the roofs of affected buildings for fires and embers. The attic should also be inspected for any sparks or fire hazards. If minor fire hazards are found, they should be neutralized by residents or local first responders. If local firefighters are not available, neighbors may assist in extinguishing fire hazards. A garden hose may be used to keep sparks and other hazards down. If the power is out, a garden hose may be connected to the outlet of a water heater. For

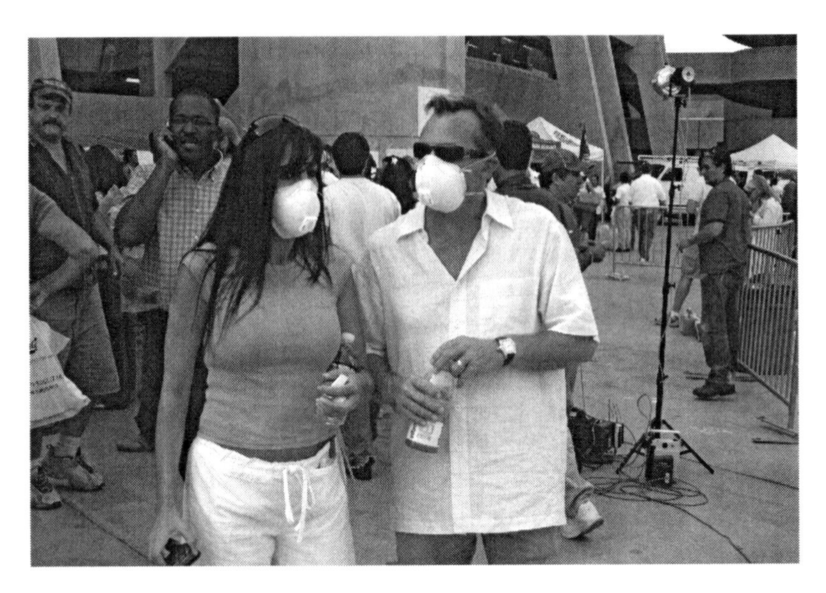

FIGURE 13–4 San Diego, California, October 24, 2007—Masks designed to reduce effects of smoke from Southern California wildfires were distributed as well as worn by Sonia and Aaron Brubaker of San Diego and other volunteers at the evacuation center at Qualcomm Stadium. Photo by Michael Raphael/FEMA. Available at: http://www.fema.gov/photodata/original/33270.jpg.

FIGURE 13–5 San Diego, California, October 26, 2007—Northern California fire crew works into the night clearing fire line and monitoring the back burn that was set to stop the Poomacha fire from advancing westward. Photo by Andrea Booher/FEMA. Available at: http://www.fema.gov/photodata/original/33373.jpg.

FIGURE 13–6 Ramona, California, November 9, 2007—Art Garcia views the remains of his home in Oak Tree trailer park following the wildfires of Southern California. Photo by Andrea Booher/FEMA. Available at: http://www.fema .gov/photodata/original/33644.jpg.

several hours after a wildfire, there should be a "fire watch" with regular monitoring for signs of smoke and fire throughout nearby structures.

Once the immediate risk of fire is controlled, and the smoke begins to clear, the primary residual hazard is the fire ash accumulation. It is very similar to fireplace ashes and is not considered to be a significant hazard to most people. However, it can pose a risk to susceptible groups including those with allergies and other sensitivities and can cause skin and respiratory reactions among some exposed populations. It is important to limit the exposure of children by cleaning play areas, toys, and pets. Any sensitive individuals working on clean-up operations should cover exposed skin, wearing long sleeves, pants, and gloves. Efforts should be made to limit airborne spread by using a high efficiency particulate air (HEPA) vacuum or wet cloth method for clean-up. Fitted N-95 respirators may also be used during dust clean-up.

Summary

Wildfires can threaten any area of human habitation near wildlands. The expansion of man-made structures into these areas will continue and more buildings and people will be at risk in the future. As these regions are developed, it is vital that mitigation and preparedness measures be incorporated into neighborhood planning. This includes the selection of less flammable landscaping plant varieties and construction materials. Safe zones must be established around buildings to reduce the risk of human occupants and property.

Even long established urban, suburban, and rural communities should have enhanced wildfire preparedness planning. Although the flames may not reach into some areas, the smoke may impact the health of large downwind populations. This is a particular risk for vulnerable populations such as the very young, elderly, or those with existing respiratory or cardiovascular illnesses. Community preparedness must focus on assisting these vulnerable populations in preparing in advance and in taking appropriate actions when fire threats are imminent.

Community awareness and education measures can reduce wildfire associated morbidity and mortality by communicating the basic mitigation steps to protect property and the appropriate individual actions to take when a wildfire threat is looming. If residents in high-risk areas take the recommended actions to reduce risk and then follow the instructions of local officials during an incident, the risk of illness and injury from wildfires will be considerably reduced.

Websites

American Red Cross, Wildfires: Are you prepared? www2.redcross.org/static/file_cont258_lang0_123.pdf.

American Red Cross, Wildfire Preparedness: www.oc-redcross.org/article.aspx?&a=627.

Firewise Communities, Firewise Community Brochure: www.firewise.org/resources/files/fw_brochure.pdf.

National Wildfire Coordinating Group, Glossary of Wildfire Terminology: www.nwcg.gov/pms/pubs/glossary/pms205.pdf.

U.S. Fire Administration, Wildfire: Are you prepared? www.usfa.dhs.gov/downloads/pdf/publications/fa-287–508.pdf.

U.S. Forest Service, Fire Consortia for the Advanced Modeling of Meteorology and Smoke: www.fs.fed.us/fcamms/.

U.S. Forest Service, Fire policy and reports, programs and priorities: www.fs.fed.us/fire/management/index.html.

State of California, Department of Public Health. Wildfire smoke: a guide for public health officials: www.arb.ca.gov/smp/progdev/pubeduc/wfgv8.pdf.

State of California, Safe clean-up of fire ash: http://bepreparedcalifornia.ca.gov/NR/rdonlyres/13A9371E-2B6C-48DC-8F62-FA5CAA340670/0/SafeCleanupofFireAshfactsheet10_251500.pdf.

References

Austin, C. C., & Goyer, N. (2007). Wildfire 2007, Respiratory protection for wildland firefighters—much ado about nothing or time to revisit accepted thinking? www.fire.uni-freiburg.de/sevilla2007/contributions/doc/SESIONES_TEMATICAS/ST6/Austin_Goyer_CANADA.pdf.

CDC. (1999). Surveillance of morbidity during wildfires—central Florida, 1998. *MMWR* 48:78–79.

Esciencenews.com. (2008). Where there's wildfire smoke, there's toxicity. http://esciencenews.com/articles/2008/11/20/where.theres.wildfire.smoke.theres.toxicity.

Johnson, J. M., Hicks, L., McClean, C., & Ginsberg, M. (2005). Leveraging syndromic surveillance during the San Diego wildfires, 2003. *Morb Mortal Wkly Rep* 54(Suppl):190.

Long, A. (2002). Benefits of prescribed burning. FOR 70, School of Forest Resources and Conservation, Cooperative Extension Service, Institute of Food and Agricultural Sciences, University of Florida. http://edis.ifas.ufl.edu/pdffiles/FR/FR06100.pdf.

Naeher, L., Brauer, M., Lipsett, M., et al. (2007). Woodsmoke health effects: a review. *Inhal Toxicol* 19:67–106.

Sutherland, E. R., Make, B. J., Vedal, S., et al. (2005). Wildfire smoke and respiratory symptoms in patients with chronic obstructive pulmonary disease. *J Allergy Clin Immunol* 115:420–422.

U.S. Environmental Protection Agency, Office of Air and Radiation. (2000). *Air quality index: a guide to air quality and your health*. Research Triangle Park, NC: Environmental Protection Agency; Publication no. EPA-454/R-00-005. http://airnow.gov/index.cfm?action=aqibroch.index.

14 Winter Storms

Objectives of This Chapter

- Recognize the types of winter storm warnings and watches.
- List the elements needed to cause a winter storm.
- Describe the primary and secondary hazards created by winter storms.
- Define the four mechanisms responsible for heat depletion from the body.
- List the physical risk factors for cold injury.
- Recognize the severity levels of frostbite injury.
- Describe freezing and nonfreezing cold injuries.
- Explain how to prepare a home and a vehicle for severe winter weather.
- List the actions to take when stranded in a vehicle during a winter storm.
- Describe the secondary causes of winter weather morbidity and mortality.

Introduction—Midwest U.S. Ice Storm, 2007

An interesting phenomenon occurs in the aftermath of most disasters. It may be human nature to move forward and forget but some catastrophes stay with those who experience them first hand. With massive hurricanes, earthquakes, volcanoes, and other violent displays of nature it seems expected that people will remember and discuss them for months and years to come. An ice storm seems that it would be far less memorable. However, the Midwest ice storm of 2007 was one event that seems to be staying on the minds of those who experienced it.

In January 2007, a series of winter storms swept across Oklahoma and Missouri. It started on Friday the 12th and continued throughout the weekend. The Midwest area is famous for tornadoes and other weather phenomenon resulting from the convergence of weather systems and this system was no exception. A warm, saturated cloud rose above a layer of cold air. As the rain fell through freezing temperatures below, the droplets became "supercooled." In other words, the droplets remained in a liquid state but were below freezing temperatures on impact and quickly turned to ice. This encased everything with ice, including streets, power and utility lines, and trees (See Figure 14–1).

As ice accumulated, power lines began to drop. Many were knocked down by falling branches as the trees buckled and split under the tremendous weight of the ice. In Springfield, Missouri, alone about 75,000 residents were without power for up to 2 weeks during freezing temperatures. In this one community, more than 300 utility poles had to be replaced. National Weather Service meteorologist John Gagen called it a "50-year event" and said that the accumulated weight of the ice on local utility poles and lines was more than 2 tons per pole; that is equivalent to the weight of a Chevy Suburban added to every utility pole in town (Johnson, 2008). Although a significant ice storm is expected in the area about every 20 years, this one was considered as an exceptional event that may occur once every 50 years.

The ice coating on the streets made it difficult to hold a vehicle on the road for those who ventured out. Although few wanted to get out, it was necessary for those who were not prepared. Trips to pick up essential items were difficult, if you could find a store open. As with most disasters, it was particularly difficult for vulnerable populations and those with preexisting conditions. Individuals in rural areas that depend upon home health visits were isolated. Some in critical need of dialysis, medical equipment, medications, and other essentials experienced potentially life-threatening delays. Special needs shelters were established to care for many individuals that needed special assistance. Fortunately, no fatalities were attributed to the interruption in medical services and support but it was certainly considered a wake-up call for the individuals and the communities affected.

In Oklahoma, the ice storms cut off power to over 100,000 homes and were responsible for nearly 3000 trips to local hospitals. At least 32 deaths were attributed to the storms, including 19 motor vehicle accidents, 8 cases of hypothermia, 2 smoke inhalation, and 3 falls (Oklahoma Department of Emergency Management, 2007). The situation was similar in Missouri where thousands were without power and at least 15 people died from storm-related conditions. Like Oklahoma, many were due to traffic fatalities. Seven died in traffic accidents and an equal number in Missouri died from carbon monoxide poisoning (Missouri State Emergency Management Agency, 2007). There was also one tragic case of 68-year-old Mary Maud East of Clinton, Missouri, who died from injuries she suffered when her barn collapsed under a heavy accumulation of ice and snow.

Winter storms can be deceptive. Although this scenario of an ice layer across two states does not appear as dramatic as many other natural disasters, it is just as deadly. The missing

FIGURE 14–1 The trees surrounding the National Weather Service office in Springfield, Missouri, were damaged or fallen due to the January 2007 one and one half inches of ice accumulation.

Source: National Weather Service, Available at: http://www.crh.noaa.gov/sgf/?n=icestormjan07photos.

branches and downed trees have been cleared. The evidence of the storm was apparent for a season but the utilities are repaired and the trees are growing back. Residents in hard hit areas still talk about the storm but less frequently than in the past. It cannot be known for sure if this waning interest is because of their confidence in preparedness measures taken since the storms or if it is just the beginning of the next round of apathy.

Winter Storm Definitions

Blizzard warning: Sustained winds or gusts of 35 miles per hour (56 km/hour) or greater with significant falling or blowing snow limiting visibility to less than a quarter mile (400 m); expected to continue for 3 or more hours.

Blizzard watch: Sustained winds or gusts of 35 miles per hour (56 km/hour) or more with significant falling or blowing snow limiting visibility to less than a quarter mile (400 m); expected to continue for 3 or more hours is possible within the next 36 hours.

Blowing snow advisory: Widespread blowing of snow; winds 25–35 mph (40–56 km/hour), with occasional visibility restrictions to less than a quarter mile (400 m).

Chilblains (also pernio): An inflammation of the skin with painful, itchy patches caused by an abnormal reaction to cold temperatures.

Conduction: The transfer of heat through a substance due to a heat difference between the two (e.g., loss of heat from sleeping on cold ground).

Convection: The transfer of heat in a gas or liquid (e.g., loss of body heat from high winds).

Evaporation: The transfer of heat when liquid is converted into a vapor (e.g., loss of body heat from conversion of moisture on the body to vapor).

Freezing rain: Rain that freezes when it hits the ground; may result in a coating of ice on roads, walkways, trees, and power lines.

Freezing rain advisory (also "freezing drizzle"): Freezing rain is expected to accumulate from trace amounts up to a quarter inch (1–6 mm).

Frostbite: Damage to skin and underlying tissue from freezing. Ice crystals form inside the cells causing them to rupture and destroy the cells.

Frostnip: Freezing of the top layers of skin. Usually seen on fingers, toes, cheeks, and earlobes. It is reversible if treated. If not warmed, it can progress to frostbite.

Frost/freeze warning: Widespread frost during growing season; temperatures at or below 32°F (0°C) expected. A "hard freeze" has temperatures below 28°F (–3°C).

Heavy snow warning: Heavy snowfall amounts are imminent. The definition of "heavy" varies by region.

Hypothermia: Abnormally low body temperature.

Ice storm warning ("freezing rain warning" in Canada): Heavy ice accumulation is occurring or imminent. The definition of "heavy" varies by region but usually ranges from ¼ to ½ inch (6–12 mm) or more of freezing rain.

Lake effect snow warning: Heavy lake effect snow with amounts of 6 inches (15 cm) or more in 12 hours or less or 8 inches (20 cm) in 24 hours or less is occurring or imminent.

Radiation: The transfer of heat due to surrounding objects having lower surface temperatures than the body. It is not due to air motion (convection) or surface contact (conduction).

Sleet: Rain that turns to ice before reaching the ground; may also cause moisture on roads to freeze and become slippery.

Sleet warning: Sleet accumulations of 2 inches (5 cm) or more are occurring or imminent in the next 12 hours.

Snow advisory: Moderate snowfall amounts are occurring or imminent. The definition of "moderate" varies by region.

Trenchfoot: A nonfreezing cold weather injury resembling frostbite. It is a condition caused by prolonged exposure to cold wet conditions and was a particular problem for soldiers involved in trench warfare during World War I.

Wind chill: An apparent temperature that describes the combination of low temperatures and the effect of wind on exposed skin.

Wind chill warning: Extreme wind chill temperatures are occurring or imminent. The temperatures vary by region.

Winter storm warning: A winter storm is imminent or occurring in your area. Hazardous conditions may include snow, freezing rain/sleet, and strong winds.

Winter storm watch: A winter storm is possible in your area within the next 36 hours. Hazardous conditions may include snow, freezing rain/sleet, and strong winds. Listen for possible announcements of winter storm warning.

Winter weather advisory: Hazardous winter conditions are likely or imminent; used for a combination of two or more conditions including snow, freezing rain/drizzle, sleet, and blowing snow.

Winter Storms

No other potential natural disaster threat has been romanticized and celebrated like winter weather. As songs like "White Christmas" and "Winter Wonderland" fill the airwaves, it is easy to see why many people are lulled into forgetting the risks associated with winter storms. These storms include a variety of hazards including low temperatures, high winds, freezing rain, sleet, and heavy or blowing snow. The storms form in a variety of ways but their formation has three common components including cold air, moisture, and lift. Obviously for snow and ice to form, the temperature must be below freezing and moisture must be present. In addition, there must be lifting of moisture so it rises to form clouds and precipitation. This usually occurs when a mass of warm air collides with cold air and is forced above it. It can also occur from air flowing up a mountain side or off a lake.

Extreme cold is defined by each region. Though it may be normal for Fairbanks, Alaska to have sustained cold that freezes a local river solid enough for local residents to drive their cars on it, it is also considered extreme cold in Orlando if the weather reaches freezing and damages citrus crops. Although the perceptions and definitions change by region, the one common factor is vulnerable populations. The very young and old are affected most by low temperatures regardless of local norms and extremes. Regional infrastructures are established around their norms. Though some areas can sustain weeks or months of extremely low temperatures with no apparent impact on local utilities and other essential services, other areas begin seeing pipes burst with a single night of low temperatures because they are not insulated and homes are not well heated.

High winds are also common with winter storms. If they accompany snow, it can result in reduced visibility, deep snow drifts, and hazardous wind chill factors. During ice storms, the winds can knock down ice laden trees and power lines. Near coastal areas the winds can cause flooding, property damage, and endanger the lives of local residents. In mountainous areas, winds can descend down the side of a mountain causing gusts of over 100 miles per hour.

Heavy accumulations of ice or snow that accompany some winter storms can lead to fatal motor vehicle and other accidents. Large regions can be paralyzed by downed utilities, disrupted power and communication, and isolation of vulnerable individuals from critical services. Even a small accumulation of ice can cause dangerous conditions for motorists and pedestrians. These deceptively dangerous ice accumulations will typically result in a patient surge of slip and fall injuries at local emergency departments.

FIGURE 14–2 A blizzard in Negaunee, Michigan, reducing visibility.
Source: National Weather Service. Available at: http://www.wrh.noaa.gov/images/pqr/blizzard.jpg.

Larger accumulations of ice or snow can collapse buildings and cut off local residents from home, healthcare, fire, police, ambulance, and other vital services. Livestock can be lost as they are cut off from food and water. Avalanches can result from heavy snow in mountainous regions.

The cost associated with winter storms is often extremely high. The snow removal efforts alone can be staggering. By late January 2005, Massachusetts had exhausted their $37.6 million budget for snow removal from the state's highways and government properties. The Governor requested an additional $28 million when over 42 inches (106 cm) of snow fell in a single month, breaking a 113-year-old record for Massachusetts snowfall in a single month (Johnson and Levenson, 2005).

Effects on the Human Body

Although humans have some ability to acclimate to a variety of environments, we cannot adapt well to extreme cold. The environment can quickly draw heat from the body leading to localized or systemic injuries. As the body produces metabolic heat, it moves from the muscle tissue to the skin. The ambient air temperature, wind speed, moisture, sunlight, clothing, and other factors influence how much heat is retained by the body and how much is lost to the environment.

The exchange of body heat takes place through the actions of four mechanisms, including convection, conduction, evaporation, and radiation. Convection is the transmission of heat from the body to the surrounding air or water. This includes the loss of body heat due to high winds. Conduction is the loss of heat from contact between two surfaces. For example, heat passes from an individual's feet to the ground if they have poorly insulated shoes. Evaporation is the loss of heat from water turning to vapor. This includes the loss of body heat from sweating and respiration. Finally, radiation heat loss is when surrounding surfaces have lower temperatures and the body experiences a loss of heat that is not from direct contact or wind (U.S. Army, 2005). These processes work together to draw heat from the body and can result in three possible cold injury scenarios. They include localized injuries, systemic injuries, or a combination of both.

Risk Factors for Cold Injury

Physical Factors

- Elderly (age > 65).
- Infants (age < 1).
- Physically disabled.
- Mentally impaired.
- Concurrent infectious disease.
- Fatigue.
- Smoking.
- Malnourished.
- Engaged in winter sports, outdoor exercise, or work.
- Users of alcohol or illegal drugs.
- Medical conditions that affect the body's ability to produce heat such as spinal cord injury, diabetes, stroke, hypothyroidism, Parkinson's Disease, arthritis.

Social Factors

- Low income.
- Homeless.
- Socially isolated.
- Urban residence.
- Poor access to healthcare or warming shelters.

Weather Factors

- Low ambient temperatures.
- Strong winds.
- Wet conditions.

Localized cold injuries range in severity from frostnip to frostbite. Frostnip is a minor peripheral injury that occurs when the top layers of skin begin to freeze. It is typically seen on fingers, toes, cheeks, and earlobes. Although it is a minor injury where tissue remains pliable and blood circulation continues to the affected area, it can still be painful and can progress to frostbite if the area is not warmed. Frostbite occurs as blood flow to the affected area is restricted or stopped. The variables that determine the severity of the injury are temperature, exposure time, wind speed, and moisture. Skin can be exposed to temperatures below 32°F (0°C) for several hours without injury if there is little moisture and winds are low. However, if the wind speed increases or if moisture is present and evaporating, it can diffuse the layer of warm air that normally surrounds the body and lead to localized cold injuries (See Table 14–1). If frostbite is not treated quickly, the damage will be permanent. The lack of oxygen from poor blood circulation leads to nerve damage. The skin becomes discolored for lighter skinned cold injury victims ranging from bluish to purple and eventually to black. The severity of the injuries is determined by the length of time the skin remains frozen and by the individual risk factors of those exposed. Health conditions like diabetes or habits like smoking or alcohol use that can restrict blood flow can make some exposed populations more susceptible to serious localized cold injuries than others.

Frostbite Severity

1st degree: Epidermal or surface involvement; causes some redness, swelling, and sensitivity for a few weeks.
2nd degree: Full thickness of the skin freezes; swelling and blisters for weeks progressing to dark eschars.
3rd degree: Freezing goes deeper than the skin; hemorrhagic blisters form, bluish-grey skin, painful rewarming; gangrenous eschars.
4th degree: Muscle, bone, and tendons involved.

Adapted from Reamy BV. Frostbite: review and current concepts. *J Am Board Fam Pract* 1998;11:34–40.

Table 14–1 National Weather Service Wind Chill Chart

Wind (mph)	Temperature (°F)																	
Calm	40	35	30	25	20	15	10	5	0	-5	-10	-15	-20	-25	-30	-35	-40	-45
5	36	31	25	19	13	7	1	-5	-11	-16	-22	-28	-34	-40	-46	-52	-57	-63
10	34	27	21	15	9	3	-4	-10	-16	-22	-28	-35	-41	-47	-53	-59	-66	-72
15	32	25	19	13	6	0	-7	-13	-19	-26	-32	-39	-45	-51	-58	-64	-71	-77
20	30	24	17	11	4	-2	-9	-15	-22	-29	-35	-42	-48	-55	-61	-68	-74	-81
25	29	23	16	9	3	-4	-11	-17	-24	-31	-37	-44	-51	-58	-64	-71	-78	-84
30	28	22	15	8	1	-5	-12	-19	-26	-33	-39	-46	-53	-60	-67	-73	-80	-87
35	28	21	14	7	0	-7	-14	-21	-27	-34	-41	-48	-55	-62	-69	-76	-82	-89
40	27	20	13	6	-1	-8	-15	-22	-29	-36	-43	-50	-57	64	-71	-78	-84	-91
45	26	19	12	5	-2	-9	-16	-23	-30	-37	-44	-51	-58	65	-72	-79	-86	-93
50	26	19	12	4	-3	-10	-17	-24	-31	-38	-45	-52	-60	67	-74	-81	-88	-95
55	25	18	11	4	-3	-11	-18	-25	-32	-39	-46	-54	-61	68	-75	-82	-89	-97
60	25	17	10	3	-4	-11	-19	-26	-33	-40	-48	-55	-62	69	-76	-84	-91	-98

FROSTBITE OCCURS IN 15 MINUTES OR LESS

Other localized injuries result from a combination of cold temperatures and moisture. These nonfreezing cold injuries include chilblains and trenchfoot. Chilblains, or pernio, are painful, itchy patches of inflamed skin. They are caused by an abnormal reaction to cold temperatures (Simon et al., 2005). They can also occur if cold skin is warmed too quickly. Trenchfoot is a more serious nonfreezing injury that was a particular problem during World War I among troops fighting in the trenches. Their mission required them to spend long hours standing in cold water leading to a condition that resembled frostbite.

> *"Your feet swell to two or three times their normal size and go completely dead. You could stick a bayonet into them and not feel a thing. If you are fortunate enough not to lose your feet and the swelling begins to go down. It is then that the intolerable, indescribable agony begins. I have heard men cry and even scream with the pain and many had to have their feet and legs amputated."*
>
> World War I Veteran Sergeant Harry Roberts
> (Kelly and Whittock, 1995)

The most important direct cause of winter weather morbidity and mortality is hypothermia or low body temperature. This is a systemic cold injury that can quickly become life threatening. Hypothermia is considered mild when the core body temperature is greater than 90°F (32°C) and less than 95°F (35°C), moderate when the temperature is from 82°F (28°C) to 90°F (32°C), and severe when the temperature is less than 82°F (28°C) (Danzl and Pozos, 1994). One of the initial signs of hypothermia is the impairment of mental function. It is often compared to a person appearing "drunk" with slurred speech, tiredness, and disorientation. The body will also eventually lose the ability to shiver. Shivering is a protective reflex and is lost with moderate hypothermia. Eventually, the hypothermia victim can no longer move at all. However, even many victims that appear deceased with low body temperatures, dilated pupils, and no detectable breathing or pulse, have sometimes been revived through CPR and slow warming of the body. This is most frequently observed with individuals who are experiencing hypothermia from exposure to cold water versus ambient temperatures. According to the U.S. Search and Rescue Task Force, cold water reduces body temperature 32 times faster than cold air (See Table 14–2) (United States Search and Rescue Task Force, Cold water survival, http://www.ussartf.org/cold_water_ survival.htm).

The majority of winter weather morbidity and mortality is the result of secondary causes such as motor vehicle accidents, heart attacks from exertion, slips and falls, and carbon monoxide poisoning. The leading cause of winter storm deaths is motor vehicle accidents. They account for about 70% of winter storm deaths (U.S. Department of Commerce, 1991). Heart attacks from shoveling snow are also common as people who rarely exercise and have a variety of risk factors overdo it as they try to clear heavy snow. Falls are most severe among the elderly and can result in traumatic head injuries or fractures. Carbon monoxide (CO) is a colorless, odorless gas that is in exhaust from vehicles, stoves, generators, and other sources of combustion. As people turn to alternative sources of heat when there are power outages during winter storms, they sometimes use equipment such as generators without properly venting exhaust leading to CO poisoning. CO poisoning also occurs as people stranded on the roads continue to run their vehicles for warmth without clearing the area around the exhaust pipe causing exhaust to back up inside the vehicle. These secondary winter storm deaths are preventable through basic awareness for the populations at risk.

Table 14–2 Time to Unconsciousness and Death in Cold Water

Water Temperature	Time to Exhaustion or Unconsciousness	Time to Death
<32°F (<0°C) Visible ice in freshwater Anchorage, AK, Coast in winter	<15 minutes	<15–45 minutes
32–40°F (0°–4.5°C) Boston Harbor in winter Great Lakes in winter	15–30 minutes	30–90 minutes
41–50°F (5°–10°C) North Pacific Coast in winter North Atlantic Coast in spring	30–60 minutes	1–3 hours
51–60°F (10.5°–15.6°C) Southern California Coast in winter Carolina Coasts in winter	1–2 hours	1–6 hours
61–70°F (16.1–21.1°C) Southern California Coast in summer Great Lakes in Summer	2–7 hours	2–40 hours
71–80°F (21.7–26.7°C) Hawaiian Coast in summer Puerto Rican Coast in winter	3–12 hours	Indefinite
>80°F (>27°C) Gulf of Mexico in summer South Atlantic Coast (GA/FL) in summer	Indefinite	Indefinite

Adapted from United States Search and Rescue Task Force, Cold Water Survival. http://www.ussartf.org/cold_water_survival.htm.

Winter Storm Deaths (U.S. Department of Commerce, 1991)

- Cold injuries.
  - 75% are male.
  - 50% are older than 60 years.
  - 20% occur at home.
- Snow- and Ice-Related Deaths
  - 70% are motor vehicle accidents.
  - 25% are people who get caught out in the storm.

Prevention

The preventive steps necessary to reduce winter storm morbidity and mortality begin with and build upon the basic preparedness measures described in Chapter 1. Preparedness kits for homes should focus on things needed if heat, power, and other utilities are lost for several days. If residing in a rural area, it is a good idea to be prepared for the possibility of a couple weeks of isolation. When winter weather is anticipated, there should be an adequate supply of prescription medication kept on hand at all times. For those who have medical conditions that require home care, dialysis, and other vital services, a discussion needs to occur with providers on what should be done if winter weather results in isolation of those needing care. The home preparedness kit should focus on items needed to keep warm, fed, hydrated, and safe and a similar kit should be placed in each vehicle. The supplies in each vehicle should be established with a worst case scenario in mind. Those residing in rural areas that could potentially be isolated for an extended period of time should have more preparedness supplies in their vehicle than those who seldom drive outside densely populated urban and suburban areas.

Winter Weather Home Preparedness

- Winterize your home to retain heat by insulating walls and attics, caulking and weather-stripping doors and windows, and installing storm windows or covering windows with plastic.
- Also winterize barns, sheds, or any other structures that may provide shelter for your family, neighbors, livestock, or equipment.
- Clean out rain gutters, repair roof leaks, and cut away tree branches that could fall on a house or other structure during a storm.
- Insulate pipes with insulation or newspapers and plastic and allow faucets to drip a little during cold weather to avoid freezing.
- Keep fire extinguishers on hand and make sure everyone in your house knows how to use them.
- Know where to shut off water valves in case a pipe bursts.
- Know ahead of time what you should do to help elderly or disabled friends or neighbors.
- Establish a winter storm supply kit that includes:
  - First-aid supplies and essential prescription medications.
  - Battery-powered NOAA Weather Radio, portable radio to receive emergency information, flashlights, and extra batteries.
  - Food supply (dried fruit, canned food with a can opener, or other food that does not require cooking).
  - Bottled water (at least 1 gallon per person per day for at least 3 days).
  - Warm clothing, including hat, gloves, and boots.
  - Extra blankets.
  - Essential baby items if there is a child in the home.

- Necessary pet supplies for at least 3 days.
- Emergency heating source, such as a fireplace, wood stove, space heater, etc.
- Learn to use properly to prevent a fire and ensure proper ventilation.
- Fire extinguisher and smoke detector. (Test units regularly to ensure they are working).

Winter Weather Vehicle Preparedness

- Winterize your vehicle before the winter season begins. Check or have a mechanic check the following items on your car:
  - Antifreeze levels: sufficient to avoid freezing.
  - Battery and ignition system: in top condition with clean battery terminals.
  - Brakes: check for wear and fluid levels.
  - Exhaust system: check for leaks and crimped pipes; repair or replace as necessary.
  - Fuel and air filters: replace and keep water out of fuel lines using additives and maintaining a full tank of gas.
  - Heater and defroster: ensure they work properly.
  - Lights and flashing hazard lights: check for serviceability.
  - Oil: check for level and weight. Lighter oils may be better at lower temperatures.
  - Thermostat: ensure it works properly.
  - Windshield wiper equipment: repair any problems and maintain washer fluid level.
  - Install good winter tires: Make sure the tires have adequate tread. All-weather radials are usually adequate for most winter conditions. However, some jurisdictions require that to drive on their roads, vehicles must be equipped with chains or snow tires with studs.
  - Maintain at least a half tank of gas during the winter season.
- Fully charge and carry a cell phone.
- Keep your gas tank near full to avoid ice in the fuel lines and to avoid running out of gas if stranded.
- Try to avoid traveling alone, especially in rural areas.
- If you must travel, let someone know your timetable and routes.
- Carry a winter storm supply kit that includes:
  - An extra change of warm clothing.
  - Blankets/sleeping bags.
  - Flashlight with extra batteries.
  - Windshield scraper, brush, and shovel.
  - A bag of sand or cat litter to pour under tires for traction if they are stuck.
  - A brightly colored, preferably red, piece of cloth.

- Tow rope or cable, small tool kit, and booster cables.
- A small supply of extra water and nonperishable food.
- First-aid kit

Adapted from multiple sources including the American Red Cross, National Oceanic and Atmospheric Administration, Federal Emergency Management Agency, and National Highway Traffic Safety Administration.

It is important to dress appropriately for winter weather. Layers of loose clothing should be worn and if sweating begins, layers should be removed to stop perspiration and avoid a subsequent chill. The trapped air in loosely fitting clothes provides insulation. A major key to reducing the risk of cold injuries is keeping dry and that includes accumulation of moisture produced by your body. There is a popular saying among hikers that "cotton kills and nylon thrills." The problem with cotton is when it gets wet, is does not wick moisture away from the body. Instead, it holds the moisture against the skin and draws body heat. It also becomes heavy and abrasive when it gets wet.

Clothing should be also worn loosely and in layers so that tight clothing does not restrict blood flow to the extremities or restrict movement. It is also important to wear tightly woven, waterproof fabrics on the outside to keep external moisture from getting close to the skin. Protecting the extremities is essential. Waterproof, insulated boots should be worn to keep the feet warm and dry. Gloves should be worn to protect the hands and fingers. Mittens that fit well around the wrists are better at keeping the hands

FIGURE 14–3 Red Cross workers search for victims buried in cars following snowfall during the "Blizzard of 77" outside Buffalo, NY.
Source: National Weather Service. Available at: http://www.photolib.noaa.gov/htmls/wea00952.htm.

warm than gloves. A hat should be worn and the ears and face should be covered. If temperatures are extremely cold, you should also avoid touching any metal surfaces with bare skin. The small amount of moisture on the skin can immediately freeze to cold metal surfaces and result in serious injury.

Immediate Actions

When threatening winter weather approaches, it is important to stay informed by listening to radio, television, or a NOAA weather radio. Populations at risk should eat and hydrate well but avoid excessive use of alcohol or caffeine. Public messages should be reinforced addressing three scenarios. These include, what to do if caught outside, if stranded in a vehicle, and if isolated at home without power and other vital utilities and services.

If you are caught outside during a winter storm, the priority is staying dry and warm so seeking shelter is essential. If shelter cannot be found, you should limit exposure to the wind by preparing a makeshift shelter with anything available, even if that means using snow accumulation to block the wind. If possible, you should build a fire for warmth and to draw attention.

If you must travel in your vehicle during winter weather, try to drive only during the day, on main roads, and avoid traveling alone. Be especially cautious on overpasses and bridges because they are often the first to freeze in winter weather. If you are stranded in a vehicle, your best bet is usually to stay there. You should pull off the road when it is safe and turn on hazard lights. You also may want to raise the hood and tie a brightly colored piece of fabric (preferable red) to the vehicle antenna or hang it out the window. As long as you stay on main roads, your vehicle is likely to be spotted

FIGURE 14–4 A severe ice storm can make roads impassable and knock out utilities resulting in the isolation of many for days or weeks.

Source: National Weather Service. Available at: http://www.srh.noaa.gov/lch/prep/icestorm.jpg.

sooner by rescuers so it is usually safer to stay in the vehicle than to set out on foot. It is very easy to get lost in wind driven snow, especially as hypothermia compromises your ability to think clearly. If a building is visible, be careful when making the decision to proceed on foot to take shelter there. Distances can be easy to misjudge when there is heavy snowfall and the effort required to get there through the snow may be far more than what is initially thought. You should run the vehicle about 10 minutes of each hour to keep warm and at night, leave the inside dome light on while the engine is running to increase the chances of being seen. Before starting the vehicle, make sure the snow is not accumulating around the exhaust pipe and leave one window on the downwind side of the vehicle open slightly for fresh air and to lessen the risk of CO poisoning. If there is more than one person in the vehicle, it is recommended that each person should take a turn resting while the other watches for rescuers. Keep moving around using small exercises moving your arms and legs to maintain blood circulation to extremities.

If you are home during a winter storm, it is best to stay inside. To conserve heat, you may want to close off unused rooms of the home and place rolled up towels along the cracks at the base of the doors. If power is out and you use a generator or alternative heat sources, such as a fireplace or space heater be sure they are vented properly to avoid CO poisoning. Stay well nourished and hydrated. Also avoid excessive use of alcohol or caffeine. If you go outside to clear snow, make sure you stretch before going out and take it slowly. Those who do not have a regular exercise program are particularly susceptible to a heart attack from overexertion. You should keep dry and if it is extremely cold, cover your mouth to protect your respiratory system. You should also be aware of the signs and symptoms of cold injury and quickly get dry and warm if those signs are observed. If the symptoms persist, seek medical care. If the sun is reflecting off the snow, there are increased risks for solar keratitis or snow blindness, and sunburn. Exposure should be limited and sun block and eye protection should be used.

Summary

Winter storms can be deceptively hazardous. Small accumulations of ice can make roads impassable and bring down utilities. Low temperatures and cold winds can quickly cause injury on what may otherwise appear to be a pleasant day. Again, the impact is most severe among vulnerable populations such as the very young, elderly, and those with existing health issues. As with heat waves, winter weather morbidity and mortality may never be fully characterized because it can escalate existing conditions resulting in fatal outcomes that are not always attributed to the weather. What may be the biggest challenge of severe winter weather is the loss of essential services when roads may be impassable resulting in the isolation of vulnerable populations. The basic elements of a prepared community are achievable with the proper level of community awareness and the motivation of residents to look out for each other. A widespread loss of services during a major winter storm will quickly overwhelm available response resources and make neighbor to neighbor collaboration essential to reduce overall morbidity and mortality.

Websites

American Red Cross, Talking About Disaster: Guide for Standard Messages, Winter Storm: www.redcross.org/static/file_cont265_lang0_127.pdf.

Canadian Centre for Occupational Safety and Health, Cold Weather Worker Safety Guide: www.ccohs.ca/products/publications/pdf/coldweathersampleguide.pdf.

Centers for Disease Control and Prevention, Extreme Cold: A Prevention Guide to Promote Your Personal Health and Safety: www.bt.cdc.gov/disasters/winter/pdf/cold_guide.pdf.

National Institute for Occupational Safety and Health, NIOSH Safety and Health Topic: Cold Stress: www.cdc.gov/niosh/topics/coldstress/.

Occupational Safety and Health Administration, OSHA Offers Tips to Protect Workers in Cold Environments: www.osha.gov/html/cold_protection_2004.html.

The Ohio State University, Agricultural Tailgate Safety Training, Cold Weather Exposure Module: www.cdc.gov/nasd/docs/d001601-d001700/d001677/d001677.pdf.

Princeton University, Outdoor Action Guide to Hypothermia and Cold Weather Injuries: www.princeton.edu/~oa/safety/hypocold.shtml.

U.S. Army, Technical Bulletin TB MED 508, Prevention and Management of Cold-Weather Injuries, April 2005: www.usariem.army.mil/download/tbmed508.pdf.

U.S. Army, Unit Leaders' and Instructors' Risk Management Steps for Preventing Cold Casualties: chppm-www.apgea.army.mil/documents/coldinjury/ColdRiskManual9-22-08-3jr.pdf.

U.S. Department of Commerce, National Oceanic and Atmospheric Administration, National Weather Service. Brochure: "Winter Storms Deceptive Killers: A Guide to Survival". November 1991: www.nws.noaa.gov/om/brochures/wntrstm.htm.

U.S. Department of Health and Human Services, Winter Ice Storm Advice and Resources: www.hhs.gov/ice/.

U.S. Search and Rescue Task Force. Hypothermia and Cold Weather Injuries: www.ussartf.org/hypothermia_cold_weather_injuries.htm.

References

Danzl, D. F., & Pozos, R. S. (1994). Accidental hypothermia. *N Engl J Med* 331:1756–1760.

Johnson, G., & Levenson, M. (2005). 1-month snowfall a 113-year high. *The Boston Globe*. January 27, 2005. www.boston.com/news/weather/articles/2005/01/27/1_month_snowfall_a_113_year_high/.

Johnson, W. (2008). It was terrible. *Springfield News-Leader*. January 12, 2008. www.news-leader.com/article/20080112/NEWS01/801120381.

Kelly, N., & Whittock, M. J. (1995). *The twentieth century world (Heinemann history study units)*. Portsmouth, NH: Heinemann Educational Books; p.17. http://books.google.com/ books?id=cEkFL7LuiVUC.

Missouri State Emergency Management Agency. (2007). Situation Report, Winter Storm, January 24, 2007. http://sema.dps.mo.gov/SitReps/SITREP1242007.pdf.

Oklahoma Department of Emergency Management. (2007). Situation Update 35, State/Federal Ice Storm Response Continues. January 23, 2007. www.ok.gov/OEM/Emergencies_&_Disasters/Winter_Weather_Event_ 20070112_Master/Winter_Weather_Event_20070112-39.html.

Simon, T. D., Soep, J. B., & Hollister, J. R. (2005). Pernio in pediatrics. *Pediatrics* 116(3):e472–e475.

U.S. Department of Commerce. (1991). National Oceanic and Atmospheric Administration, National Weather Service. Winter storms deceptive killers: a guide to survival. www.nws.noaa.gov/om/brochures/wntrstm.htm.

U.S. Army. (2005). Technical Bulletin TB MED 508, Prevention and Management of Cold-Weather Injuries. www.usariem.army.mil/download/tbmed508.pdf.

Index